ALGORITHMS AND ARCHITECTURES FOR REAL-TIME CONTROL 2000
(AARTC'2000)

A Proceedings volume from the 6th IFAC Workshop,
Palma de Mallorca, Spain, 15 – 17 May 2000

Edited by

V. HERNÁNDEZ
Departamento de Sistemas Informáticos y Computación,
Universidad Politécnica de Valencia, Spain,

and

G.W. IRWIN
Department of Electrical and Electronic Engineering,
The Queen's University of Belfast, UK

Published for the

INTERNATIONAL FEDERATION OF AUTOMATIC CONTROL

by

PERGAMON
An Imprint of Elsevier Science

UK   Elsevier Science Ltd, The Boulevard, Langford Lane, Kidlington, Oxford, OX5 1GB, UK

USA   Elsevier Science Inc., 660 White Plains Road, Tarrytown, New York 10591-5153, USA

JAPAN   Elsevier Science Japan, Tsunashima Building Annex, 3-20-12 Yushima, Bunkyo-ku, Tokyo 113, Japan

Copyright © 2000 IFAC

First edition 2000

Library of Congress Cataloging in Publication Data

A catalogue record for this book is available from the Library of Congress

British Library Cataloguing in Publication Data

A catalogue record for this book is available from the British Library

ISBN 0-08-043685 4

Transferred to Digital Printing 2007

IFAC WORKSHOP ON ALGORITHMS AND ARCHITECTURES FOR REAL-TIME CONTROL 2000

Sponsored by
International Federation of Automatic Control (IFAC)
Technical Committee on Algorithms and Architectures for Real-Time Control

Organized by
Departamento de Sistemas Informáticos y Computación, Universidad Politécnica de Valencia.
Departament de Ciències Matemàtiques i Informàtica, Universitat de les Illes Balears.

Workshop Chairman
Hernández, V. (SP)

International Programme Committee (IPC)
Irwin, G.W.   (Chairman) (UK)

Albertos, P. (SP)
Alonso, A. (SP)
Atherton, D.P. (UK)
Bass, J. (UK)
Boullart, L. (B)
Bulsari, A. (FIN)
Carelli, R. (ARG)
Crespo, A. (SP)
Dodds, G. (UK)
Fleming P. (UK)
Garcia Nocetti, F. (MEX)
Goodall, R.M. (UK)
Halang, W. (D)
Hernandez, V. (SP)
Holding, D.J. (UK)
Ionescu, T.C. (ROM)
Jones, D.I. (UK)
Kopacek, P. (A)
Kwon, W.H. (ROK)
Lauwereins, R. (B)
Lewis, F. (USA)
Levy, D. (AUS)

Magnani, G. (I)
Man, K. (HK)
Marcos, M. (SP)
Masten, M.K. (USA)
Motus, L. (EST)
Pashkevich, A.P. (BY)
Pereira, C.E. (BR)
Puente, J. (SP)
Puigjaner, R. (SP)
Ruano, A.E. (P)
Sa da Costa, J. (P)
Sanz, R. (SP)
Shin, K.G. (USA)
Skubich, J. (F)
Solano, J. (MEX)
Spong, M.W. (USA)
Tokhi, O. (UK)
Tornero, J. (SP)
Tyrrell, A. (UK)
Verbruggen, H.B. (NL)
Zomaya, A.Y. (AUS)

National Organizing Committee (NOC)
Llamosí, A.   (Chairman)

Alonso, J.M.
Alvarruiz, F.
Arias, E.
Blanquer, I.
de Alfonso, C.

Guerrero, D.
Román, J.E.
Ruiz, P.A.
Vidal, A.M.

PREFACE

The 6th IFAC Workshop on Algorithms and Architectures for Real-Time Control (AARTC'2000) was held at Palma de Mallorca, Spain. Previous workshops in this area have been held at Bangor-UK (1991), Seoul-Korea (1992), Ostend-Belgium (1995), Vilamoura-Portugal (1997) and Cancun-Mexico (1998). The objective, as in previous editions, was to show the state of the art and to present new developments and research results in software and hardware for real-time control, as well as to bring together researchers, developers and practitioners, both from the academic and the industrial world.

The AARTC'2000 Technical Program consisted of 11 presented sessions, covering the major areas of software, hardware and applications for real-time control. In particular, sessions addressed robotics, embedded systems, modeling and control, fuzzy logic methods, industrial process control and manufacturing systems, neural networks, parallel and distributed processing, processor architectures for control, software design tools and methodologies, and SCADA and multi-layer control. A total of 38 papers were selected from high-quality full draft papers and late breaking paper contributions (consisting of extended abstracts). Participants from 15 countries attended the AARTC'2000 workshop.

The technical program also included two plenary talks given by leading experts in the field. Roger Goodall (Department of Electronic & Electrical Engineering, Loughborough University, UK) presented "Perspectives on processing for real-time control", and Ricardo Sanz (Universidad Politécnica de Madrid, Spain), focused on "CORBA for Control Systems". Another highlight in the program was the final session on industrial presentations which was held in common with the Workshop on Real-Time Programming (WRTP'2000). In this session, Abel Jiménez (Industria de Turbo Propulsores S. A., Spain) presented the "Thrust Vectoring System Control Concept", Ulrich Schmid (Technische Universität Wien, Austria) made a presentation with the title "Applied Research: A Scientist's Perspective", and Harold W. Lawson (Lawson Konsult AB, Sweden) addressed "Systems Engineering of a Successful Train Control System".

We are grateful to the members of the International Program Committee, who did a remarkable job in reviewing the papers submitted to the workshop within the established schedule. Special thanks also to the chairman of the Workshop on Real-Time Programming, Alfons Crespo, and the chairmen of the IPC of the same workshop, Wolfang Halang and Juan Antonio de la Puente, for the co-organization of the industrial session. We are also indebted to Juan Antonio de la Puente, for the development of the SW that made possible the organization of the workshop via Internet.

Finally, thanks to the contributing authors of the AARTC'2000 Workshop.

Vicente Hernández   George W. Irwin
Workshop Chairman.   International Program Committee Chair

CONTENTS

PLENARY PAPER II

INDUSTRIAL PROCESS CONTROL AND MANUFACTURING SYSTEMS

NEURAL NETWORKS

ROBOTICS II

PARALLEL AND DISTRIBUTED PROCESSING

PROCESSOR ARCHITECTURES FOR CONTROL

SW DESIGN TOOL AND METHODOLOGIES

SCADA AND MULTI-LAYER CONTROL

INDUSTRIAL PRESENTATION

PERSPECTIVES ON PROCESSING FOR REAL-TIME CONTROL

Roger Goodall

*Department of Electronic & Electrical Engineering,
Loughborough University, Loughborough,
Leicestershire, LE11 3TU, UK*

Abstract

An overview is provided of the issues relating to processor implementation of real-time control. The paper concentrates upon the practical rather than theoretical aspects and has a focus upon the requirements for high-performance embedded-control applications. Controller formulations are reviewed to clarify algorithmic requirements, after which issues relating to quantisation both in time and amplitude are identified, before moving on to a summary of processing options which can be used. Finally, future possibilities and trends are discussed. *Copyright © 2000 IFAC*

1. INTRODUCTION

The title of this paper has been chosen carefully, because the use of the word "perspectives" implies a personal and perhaps not completely comprehensive presentation. The author's expertise in the area of real-time control is practical rather than theoretical, with an approach which is pragmatic rather than rigorous. Nevertheless the paper's content will be of general interest to control engineers, because some of the more practical issues of implementing high performance complex controllers are often not appreciated. Indeed many control engineers want their control law to be implemented as straightforwardly as possible, and generally they have neither the will nor the skill to achieve efficient, cost-effective solutions.

The author's background is in applications of advanced control, principally for high-performance electro-mechanical systems, examples being Maglev vehicle controllers, railway vehicle active suspension controllers, active magnetic bearings and more recently some work relating to aircraft flight control. The focus of the paper is therefore towards embedded control operating at high sample frequencies, but also applied to systems which are dynamically complex, often made more difficult by effects such as vibration modes close to the required bandwidth. The other characteristic is that they are substantially linear, or at least sufficiently linear for a linear time-invariant (LTI) controller to work very effectively. The main message of the paper therefore applies to systems with these four characteristics: embedded, high sample rate, high complexity, but linear.

The paper will first follow the trends in controller formulations before moving on to discuss digital processing for control from the viewpoint of requirements and algorithms. These will then be mapped onto where the state-of-the-art is at present with processing solutions to meet the requirements. Finally the paper will look forward, both in terms of controller requirements and forthcoming implementation options. Throughout a strong link is made between the algorithm and the architecture, ie the form of the equations to be computed and the manner of implementation, because the synergy between the two is vital for efficient implementation.

2. CONTROLLER FORMULATIONS

Over the last 50 years or so control law formulations have progressively developed: classical control, modern (optimal) control using model-based approaches, then on to adaptive control, and with concepts such as learning control and intelligent control starting to emerge. The first two categories are the most commonly used in embedded applications, and it's therefore useful to identify in a qualitative sense their characteristics with respect to implementation.

2.1 Classical control law formulations

Classical controllers are generally designed in the s-domain and then digitised via emulation, for example using Tustin's method which is purely algebraic. There are ways of designing directly in the discrete domain, but for demanding high-performance controllers the sample frequency needs to be high (this requirement is clarified in a later section), and once this condition is satisfied the emulation approach gives very similar results to the direct digital design methods.

Although many examples of classically-designed controllers are relatively simple, they can become complex when there are multiple feedback signals and controls. In this case they are normally created in a block-structured form such that the individual transfer functions which contribute to the whole are relatively simple, and this certainly makes implementation more straightforward, combined with which it is usually possible to have a range of standard elements such as PID compensators which can be brought together for a particular application.

2.2 Modern model-based control law formulations

Since the 1960s control engineers have exploited the idea of control law design based explicitly upon a mathematical description of the system to be controlled: optimal controllers with estimators, observers or Kalman filters; H-infinity and other robust control approaches; multivariable approaches in general. The characteristics of these techniques are that the controllers themselves generally have a similar or higher complexity to the system being controlled, and for the dynamically-complex systems of the kind mentioned in the Introduction the controllers will therefore also be complex, in addition to which they are structurally complicated and so simplification in the multi-input multi-output case is not straightforward. The controllers are most naturally produced in state-space form, and while as a whole this provides for a standardisation of implementation, it's not the same as having the standard smaller compensator elements which exist for classical control formulation.

More complex formulations arise with adaptive control and other more sophisticated approaches, but in practice things like adaptation are often best as a slower, background process. Therefore, if it is possible to handle the LTI computations such as arise from classical or modern design efficiently, it's a natural extension to vary the controller parameters to provide adaptation, or to interfere with its internal variables to introduce non-linearity.

3. PROCESSING ALGORITHMS

No matter what its origin, the control law must be converted into an algorithm, ie the set of real-time equations which are calculated each time the inputs are sampled. There are two key issues in respect of this process: quantisation considerations, and the form of the equations. Although in practice these two issues are linked, here they are discussed with independently.

3.1 Quantisation

Discrete-time implementation requires quantisation in two respects: quantisation in time, essentially the sampling process, and quantisation in amplitude. The former is concerned with choosing the sample frequency, the latter with determining wordlengths for the coefficients (which prescribe the actual functionality of the control law to be implemented), and the internal variables (values which inevitably must be stored from one sample period to the next to provide the dynamic properties of the controller).

Sample frequency. A common rule of thumb is to sample at ten times the loop bandwidth, and other statements suggest that it's not possible to sample at rates of more than perhaps 30 times the bandwidth, otherwise numerical problems arise, principally relating to finite wordlength effects (ie amplitude quantisation). However a rigorous control systems approach is firstly to ask "What sample frequency is needed for the required performance?", and if that turns out to be, for example, 100 times the bandwidth, then the next question is "How can this be achieved?" There is really no problem with achieving a higher sample frequency, but it is necessary to have a proper understanding of the recursive numerical processes, and design the computation accordingly. Work by the author involved implementation of a general-purpose filter which could have a cut-off frequency as low as 1Hz using a sample frequency of 20kHz (Goodall and Donaghue, 1993). There was a particular reason for this requirement, and to make it work it was necessary to think carefully about the numerical aspects, but it was tested experimentally and worked effectively. Some of the implementation constraints which have been stated in the literature are therefore not correct, but it is a matter of understanding properly what is happening in the computations.

It is relatively straightforward to develop criteria for determining sample frequency, and these are introduced diagrammatically because it helps to visualise what is going on, but the same result emerges from a mathematical description of the effect of sampling.

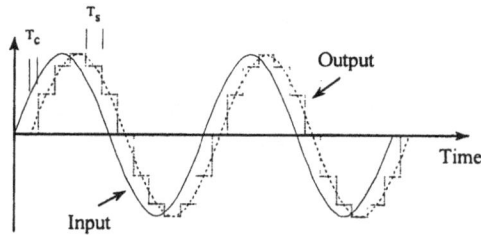

Fig 1. Effect of sampling and computation

The sampling process, illustrated in Fig. 1, can be analysed by deriving the phase lag introduced in response to a sinusoidal input, which can be considered to have a period T which corresponds to the loop bandwidth so that "over-sampling" factors mentioned in the previous paragraph can be compared. In practice there will also be a computation time, in which case the phase lag is increased compared with sampling above. It can be seen from the diagram that the fundamental component of the sampled signal is delayed by half the sample period T_s plus the full computation time T_c, and Eqn 1 below gives the corresponding phase lag.

$$\text{Phase lag} = 360\,(T_S / 2 + T_C) / T \quad (\text{deg}) \qquad (1)$$

This phase lag will degrade the stability margin compared with a continuous controller. Bearing in mind that the aim usually is to provide 40-45° phase margin, a reasonable criterion is to ensure that this degradation is no more than 5°. The computation may in fact take the whole sample period, but a more typical value is to assume $T_C = T_S/2$ and using this results in a requirement that $T_S \leq T/72$, in other words that the sampling frequency should be 72 times the bandwidth, and in practice a factor of 100 is an effective target for high performance controllers. Less than this and the stability will increasingly be compromised by the use of digital control.

It should be mentioned that a number of digital controller design techniques can be used to compensate for these delays, essentially by introducing forward extrapolation into the digital algorithm, and thereby enabling a lower sample frequency to be employed. For "well-behaved" systems this is sometimes possible, but in most practical electro-mechanical systems there are unmodelled dynamics at high frequencies (eg structural resonances, drive shaft flexibility, etc), and the adverse high frequency characteristics of such prediction processes as the sample frequency is approached (ie rapidly-changing phase and increased gain) mean that in practice this is not appropriate.

Coefficient wordlength. The coefficients are quantised in magnitude, and it is necessary to ensure that they are sufficiently accurate when this process occurs, ie that a suitable coefficient wordlength is chosen. It has been recognised for some time that normal forms of recursive digital filters have serious sensitivity problems (Liu, 1971), and various methods for quantifying this sensitivity have been proposed. A practical approach is to think in terms of tolerances, in other words how the fractional accuracy of the basic characteristics of the controller such as gains, time constants, cut-off frequencies, etc are affected by fractional changes in the discrete coefficients, and this leads to the use of a normalised sensitivity matrix illustrated in Eq 2.

$$\begin{pmatrix} \delta G / G \\ \delta \tau / \tau \\ \delta \omega / \omega \\ \text{etc} \end{pmatrix} = \begin{pmatrix} 200 & & & 300 \\ & & 500 & \end{pmatrix} \begin{pmatrix} \delta a_0 / a_0 \\ \delta a_1 / a_1 \\ \delta b_1 / b_1 \\ \delta b_2 / b_2 \end{pmatrix} \qquad (2)$$

Using the kinds of sample frequencies mentioned in the previous sub-section it's not uncommon that the discrete coefficients must be expressed hundreds of times more accurately (ie in a fractional or percentage sense) than the actual accuracy needed for the characteristics of the controller. These values shown are not for a specific example, they are simply typical values to emphasis the problems. (See Goodall (1992) for more detail of this method for analysing coefficient sensitivity.)

Quantisation of controller variables. Whereas the coefficient sensitivity problem is one which can be determined at the design stage, the quantisation of the controller's internal variables occurs as part of the real-time computation processes. This effect is often referred to as quantisation "noise", but in a practical control system there's usually enough noise already, and in practice it's unlikely to be acceptable if the digital computation process is adding significantly to this.

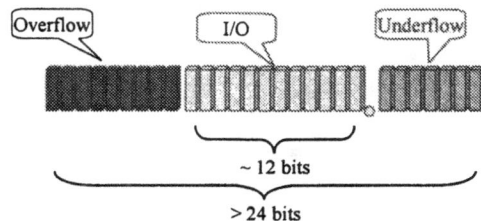

Fig. 2 Internal variable numerical requirements

The issue really are concerned with making sure there are enough overflow and underflow bits in the variable wordlength, compared with the wordlength chosen for the input and output variables - Fig. 2 is a generalised diagram. The internal variables can often become significantly larger than the inputs, not because of high gains but as a consequence of the recursive computations, and if there's not enough overflow space then large inputs will not be properly processed. In addition, it's also necessary to allow for underflow, otherwise small variations in the inputs will not propagate through the processing - the input changes but nothing happens on the output, which therefore compromises the resolution chosen for the A to D conversion process.

With a realistic sampling rate for control as discussed earlier, and using 12 bit input/output wordlength, a guideline is that the internal variables need to have at least 24 bits. It's worth noting in this respect that a so-called "32 bit floating point" processor isn't really 32 bit precision - the IEEE standard, which generally corresponds to "float" variables specified in C, only has a 23 bit mantissa (one bit is allowed for the sign and 8 bits for the exponent). The floating-point nature of the variable mitigates some of the effects, but there are undoubtedly a number of circumstances where this is not accurate enough.

3.2   Numerical algorithms

In terms of the actual real-time equations, the discussion focuses mainly upon the discrete operator which is to be used. However it's useful to consider generalised discrete state-space structures, which not only reflect the choice of operator, but also extend the possibilities beyond this simple choice. These two issues are discussed in the sub-sections which follow.

Choice of discrete operators. Most people are familiar and comfortable with the z operator, because z transform techniques provide the basis for much of our digital control analysis. For real-time control we must implement the inverse operation z^{-1} because we can't go forward in time, and Eq. 3 is the calculation which corresponds to the diagram in Fig. 3 - essentially z^{-1} becomes a shift operation within the algorithm.

Fig. 3 The shift operator z

$$x_2(\underline{n+1}\,T_s) = x_1(nT_s) \tag{3}$$

Fig. 4 The difference operator δ

$$\delta = z-1 \quad \text{or} \quad \delta = (z-1)/T_s \tag{4}$$

$$x_2(\underline{n+1}\,T_s) = x_1(nT_s) + x_2(nT_s) \tag{5}$$

However the use of z creates serious coefficient sensitivity problems and demanding requirements for the internal variables, and it's been recognised since the 1980s that alternative formulations using the δ operator are computationally superior. The algebraic relationship between δ and z is very simple - two possibilities are given in Eq 4. The first one is simpler and more appropriate from the implementation view point; the second, which includes a division by the

sample period T_s, is what Middleton and Goodwin (1990) propose. At high sample rates the delta operator converges on the Laplace operator s, and this property provides an important unification between continuous and discrete time. Whichever is used, the equivalent operation is "Shift and Add", rather than the simple shift of the z operator. Using the δ operator results in a total transformation of the coefficient sensitivity problem, and significant reduction in the internal variable wordlength requirement.

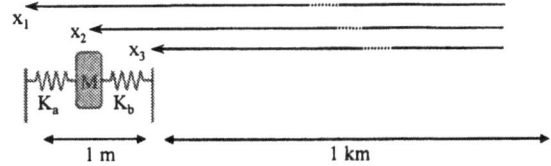

Fig. 5 Mass/spring analogy

$$F = a_1x_1 + a_2x_2 + a_3x_3$$
$$a_1 = K_a \quad a_2 = (-K_a + K_b) \quad a_3 = (-K_b) \tag{6}$$

It's interesting to use an analogy to emphasise the distinction between z and δ. Imagine that, around a km away, there is an assembly consisting of a mass connected by two springs, perhaps a metre long in total. It is necessary to know the force being exerted by those two springs on the mass. So, by some means the distances to the two fixing points and to the mass are measured, then the equations based upon those three measurements are formulated as shown in Eq 6. It's easy to show that the coefficients applied to the measurements, which depend upon the spring constants, will be extremely sensitive; a small independent variation in one creates large errors in the calculation, and if these are implemented digitally high accuracy requires long wordlengths. Of course it's obvious that mathematically, physically and experimentally this method is inappropriate - the better approach is simply to measure the deflections across the two springs, in which case the spring constants are used directly as parameters in the equation. Now the first method corresponds to using the shift operation z, whereas the direct measurement of spring deflections corresponds to the δ operation. However probably 99.9% of digital controllers in the world use the first, the equivalent of which would be rejected instinctively and immediately for the situation described. What is needed when implementing recursive digital filters is the small differences between successive samples of the variables, and using the delta operator does this inherently. It is numerically superior, just as easy to use, and it's difficult to understand why people persist in using z for implementation!

Discrete state-space formulation. Although the control law to be implemented may emerge from the design process in a variety of forms, it's always possible to reformulate it as a discrete state-space matrix equation - see Fig 6 and Eq 7. This enables the state variables to be updated as function of their past values and the new values of the inputs, and also to calculate new values for the outputs. This form of course is well known, but here the measurements are the inputs and the controls are the outputs, because the equations are for the controller, not the physical system.

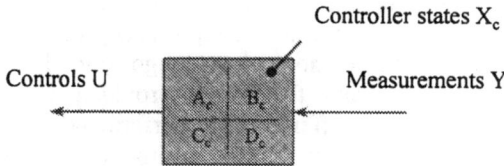

Fig. 6 Discrete state space form

$$X_c(n+1\,T) = A_c X(nT) + B_c Y(nT)$$
$$U(nT) = C_c X(nT) + D_c Y(nT) \tag{7}$$

$$X' = TX$$
$$A'_c = TA_c T^{-1} \quad B'_c = TB_c \quad C'_c = C_c T^{-1} \tag{8}$$

Eq 8 shows that it is possible to define a transformation matrix T to give a new set of states X', with correspondingly modified values for the various matrices. The important thing to appreciate is that the choice of controller states, ie the internal variables, is a design freedom; with very high numerical accuracy any value of T will give the same overall result from the measurements to the outputs (subject to some mathematical constraints on T). The formulation can therefore be chosen to minimise the numerical processing requirements. The numerical problem can be shown to be principally in the state update equation matrix because therein lie the recursive computations, and below are a few possibilities, illustrated using a fourth order system.

$$
\begin{bmatrix} x_1 \\ x_2 \\ x_3 \\ x_4 \end{bmatrix}_{n+1} =
\begin{bmatrix}
33.2023 & 2.4852 & 19.8231 & 0.0003 \\
364.0115 & 24.4260 & 219.1156 & -23.9984 \\
-84.5713 & -6.1371 & -50.6463 & 1.9994 \\
47.8690 & 3.1519 & 28.8231 & -2.9997
\end{bmatrix}
\begin{bmatrix} x_1 \\ x_2 \\ x_3 \\ x_4 \end{bmatrix}_n
$$
$$
+ \begin{bmatrix} 1 \\ 5 \\ -2 \\ 1 \end{bmatrix} y_n \tag{9}
$$

Equation 9 shows an A matrix with no "structure"; it is fully populated with values, and it becomes necessary to make twenty multiplications and additions to carry out the state update, in other words this is a matrix multiplication and addition. This kind of structure emerges from the work undertaken by Gevers and Li (1993), in which they seek to identify what are called "minimal realisations" giving optimal parameterizations. However, although the precision required for the computations may be low, because so

many are required it is unlikely to be the most efficient solution overall, despite the mathematical optimality.

$$
\begin{bmatrix} x_1 \\ x_2 \\ x_3 \\ x_4 \end{bmatrix}_{n+1} =
\begin{bmatrix}
3.9824 & -5.9474 & 3.9475 & -0.9826 \\
1 & 0 & 0 & 0 \\
0 & 1 & 0 & 0 \\
0 & 0 & 1 & 0
\end{bmatrix}
\begin{bmatrix} x_1 \\ x_2 \\ x_3 \\ x_4 \end{bmatrix}_n
+ \begin{bmatrix} 1 \\ 0 \\ 0 \\ 0 \end{bmatrix} y_n
$$
$$\tag{10}$$

Eq 10 gives the controller canonic form using z, in which it can clearly be seen that the expressions for x_2 to x_4 are simple shift operations. This structural form has substantially simplified the computations because of all the ones and zeros, which can be called "trivial" coefficients because multiplication is not needed. However the use of the z operator means that this will certainly have difficult coefficient sensitivity and internal wordlength problems as mentioned earlier. The coefficients will probably have fractional sensitivity factors of 1000 or more - to achieve say 1% accuracy overall will require them to be expressed to much better than .001% accuracy.

$$
\begin{bmatrix} x_1 \\ x_2 \\ x_3 \\ x_4 \end{bmatrix}_{n+1} =
\begin{bmatrix}
3.9824 & -0.0002 & -0.0000 & -0.0000 \\
1 & 1 & 0 & 0 \\
0 & 1 & 1 & 0 \\
0 & 0 & 1 & 1
\end{bmatrix}
\begin{bmatrix} x_1 \\ x_2 \\ x_3 \\ x_4 \end{bmatrix}_n
+ \begin{bmatrix} 1 \\ 0 \\ 0 \\ 0 \end{bmatrix} y_n
$$
$$\tag{11}$$

A third option is shown in Eq 11 - it has some similarities with Eq 10 in that it preserves the large proportion of the "trivial" coefficients, but it can be seen that the new value of x_2 is now its old value plus the old value of x_1, and similarly the new value of x_3 is its old value plus x_2, and so on. In other words this is now implementing the δ operator, and in this case it's easy to show that the coefficients are not sensitive at all. If 1% overall accuracy is required, all that is necessary is a coefficient wordlength which also provides 1% accuracy. (The coefficients on row 1 are in fact small numbers, not zeros.)

Many other structures are possible: some can be used to ease the difficulties with the z operator forms, others can build further upon the advantages provided by the δ operator, but the important point is to understand that the basic structure of the discrete state-space form can be manipulated to provide large advantages for implementation.

4. PROCESSOR OPTIONS

Having understood the control law requirements and the corresponding algorithms, it is possible to consider the processor options which are available. The possibilities which exist are described historically in increasing levels of sophistication, and both the devices and software issues are discussed.

4.1 Processing devices

Analogue processors were used in the early days of control, based upon the analogue operational amplifier. The operational amplifier devices which are available today are cheap and very effective, offering infinite-resolution, delay-free, parallel processing for a few cents per processor. Of course there are significant limitations of these devices, principally their lack of programmability and the difficulty of making them adaptive. Nevertheless they are mentioned here because most modern control engineers forget that they are an option; in some circumstances they may be the correct choice, particularly for the more simple fast-acting inner loops which are frequently used in embedded control applications. The message is "Don't forget the analogue processor" - sometimes it is the right solution. One of the problems is that analogue design skills are rare nowadays and so people instinctively turn to the digital solution.

Next came the microprocessor, initially 4 bit versions, then 8 bit and now 16 bits and higher. A standard device could be programmed through software to meet a number of applications. However it was necessary to add memory for programme and data, timers for handling the real-time issues, digital and analogue input and output etc, and so the hardware solution could be quite complex. Accordingly microcontrollers were developed. These included the memory but also had a range of peripherals on the chip, and provided a much more appropriate solution for control than the microprocessor. However it was relatively easy to run out of processing power as more was demanded of the control algorithms. One solution was the transputer, which was designed to provide fast and easy inter-processor communication such that efficient parallel processing could be achieved. There was a lot of interest in these in the late 1980s, but in many senses they never fully realised their manufacturer's expectations. The more significant development has proved to be the Digital Signal Processor, with an architecture aimed at computationally-intensive applications by incorporating multiplication hardware on the chip, and by a bus structure which separated instructions from data. Many of the DSP families include versions which have the peripherals provided by a microcontroller to provide a minimal hardware solution. Initially these were fixed point, but now we have powerful floating-point devices as well, although the unit price of these is relatively high. It's interesting that Texas Instruments' device which is targetted for embedded real-time control is a fixed point device. Even though floating-point DSPs are commonly used during development, in production the majority of cost-critical embedded applications use fixed-point microcontrollers.

4.2 Software design

It's useful to review the programming aspects, because although software development tools have come a long way since the early days of microprocessors, there are still issues of importance. There are two principal ways of programming: firstly the use of the chosen processor's Assembly language, highly efficient but more difficult to develop and not transportable to a different processor; secondly the use of a high level language, converted into object code by a compiler - relatively easy to develop and fairly transportable to another processor, but not as computationally efficient.

The author undertook some computation time comparisons between low and high level programming for microprocessors some years ago (Goodall and Goodall, 1993), and this work provided a useful indication of the difference in performance. If the equations were written in C using floating-point variables, computation took around 20 times as long compared with code written in assembly language. Using integer variables it was possible to reduce this factor so the compiled code took perhaps 3 or 4 times as long, but this approach required a good understanding of the numerical requirements and variable scaling, so it's nothing like as easy as with floating point variables

It is important to be careful of compilers, because the user is in the hands of the persons who wrote the compiler. For example, as mentioned earlier a 32-bit floating point DSP strictly only provides 23 bit precision, and the use of "doubles" instead of "floats", ie 64 bit variables instead of 32, might be required. Unfortunately these higher precision variables are sometimes not supported by DSP compilers, and the author is aware of one instance where "doubles" are implemented as "floats" without any warning!

A third software option is the use of products such as Matlab's real-time workshop which generates code, or links to a DSP card directly, often straight from a Simulink description of the controller. This of course is one level higher again, and even more trust is being placed in the manufacturer of the product, but many workers have found that this provides a very easy method of implementation. However the overall efficiency may be quite low, and so although this may be very suitable for a development environment, it's probably not suitable for a high-volume, embedded control application.

5. TRENDS FOR THE FUTURE

The previous sections have outlined the main issues and given an interpretation of where real-time processing for control is today. But what of the future? How are the requirements likely to develop? What will be the implications, for the algorithms? What sort of processing architectures will be available? This final section will give a personal view which tries to answer these questions, at least in part.

5.1 Advanced requirements

In addition to the computationally-intensive part of the processing, there are other functions which usually have to be provided as part of an embedded real-time controller but which haven't been mentioned. It's important to identify these requirements as well as looking to the future, and Fig 7 provides a vision for the full range of functions which will be needed for future embedded control applications.

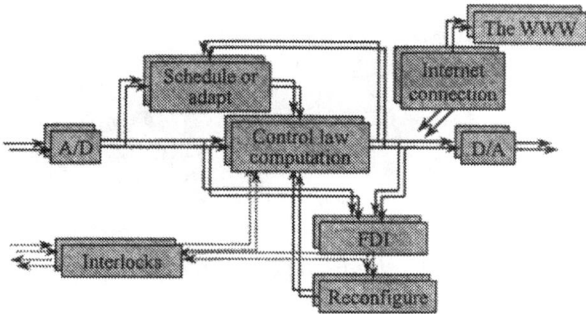

Fig. 7 Future embedded controller functionality

Probably the most important aspect is what is often called the interlocking, ie all the logical functions which have to be carried out to monitor logic inputs such as limit switches, to provide enable signals for power amplifiers, to switch on small indicator lights, to provide simple diagnostics, and so on. It is mainly a stand-alone task, largely independent of the control law processing, although it may be that some interaction with the computational function is needed, for example to limit the action of integrators until the control system is enabled. This is not of course the kind of function which a device such as a DSP is particularly efficient at handling, but it usually doesn't have the same degree of time criticality as the control law computation so it can usually be provided as a background task.

Adaptation, particularly in the form of gain scheduling, is also something which is not uncommon, and that is relatively straightforward to introduce by adjusting the coefficients in the controller. Other forms of adaptation also include the effects of the outputs, or perhaps state or parameter estimates from the system. If the adaptation process is relatively slow (ie compared with the closed loop system bandwidth), then the LTI computational framework still applies; faster adaptation probably requires a re-think of the way the equations are formulated, although in this case provability of stability margins etc is more difficult.

Another function which is increasingly being incorporated is fault detection and isolation. It's shown in Fig. 7 as simply taking the controller inputs and outputs, but obviously more complex strategies are possible. It's also possible to have reconfiguration of the control law to respond to the identified fault. In the case of FDI alone the output will interact with the logical operations being carried out, perhaps to initiate a safe shutdown; if reconfiguration is included then there will be change to the coefficients (and perhaps the structure) of the control law computations, and both these links are shown on the diagram.

Those mentioned so far are principally internal functions, but external connections are also important. Many microcontrollers include a standard bi-directional serial communication peripheral, to facilitate easy connection to a computer, for example; a few include a more complex interface to provide direction connection to an external bus such as CAN or Fieldbus. However communications seem to be becoming oriented around the internet, and so it is probable that internet connectivity will become an increasingly important feature.

As control extends its influence, safety and reliability become increasingly important. Fault detection is of course an essential part of this, but almost inevitably duplication or triplication of the processing channel is needed to ensure safe and/or reliable operation in the presence of faults, and so there will probably be an interest in the future of providing a certain amount of the parallel redundancy within the processing device, and Fig. 7 has a superimposed copy of the various functions to indicate this.

5.2 Future algorithms

The advanced structural forms of algorithm described above, especially those based upon the δ operator, are about as efficient as can be expected for might be called the conventional approaches for control system processing, and more information regarding implementation using the δ operator is described by Goodall et al (2000). There are however two ideas which might yield further efficiencies, and the first of these is what is called "one-bit processing".

One-bit processing: Conventionally a continuous signal is digitised into a parallel binary number, multiplied with the coefficient to produce a parallel binary output, as shown in Fig 8 which has only a single multiplication to illustrate the point.

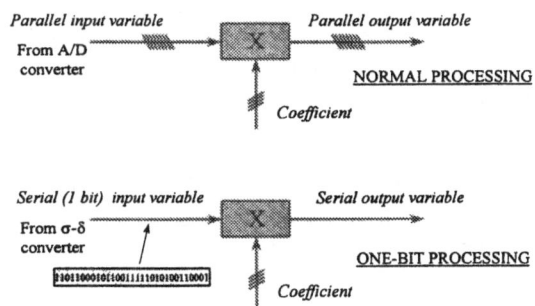

Fig. 8 One-bit signal processing

However, A/D converters can be simplified for integration onto silicon using an approach called sigma-delta modulation (Candy, 1985), in which the continuous signal is digitised in a variable bit stream, a single input albeit varying at high frequency. Again this must be multiplied by the coefficient, but consider this process: the input is either one or zero, so the multiplication becomes an addition if it's 1, no action if it's a zero. In other words the process is not a multiplication at all, and since the multiplier array is the dominant feature on a processor chip there are potentially substantial benefits to be gained. Conventional (ie non-control) signal processing people have been investigating this (Johns and Lewis, 1993), but the disadvantage for such applications is that everything has to be done much more frequently because of the nature of the incoming bitstream. However for control it is already necessary to sample fast to minimise sampling delays (as described in Sect 3.1), so it's potentially a very interesting possibility for control processing, although as yet the benefits are unproven.

Variable sample rate processing: The normal approach is to use a constant sample rate because it makes everything more straightforward. Although there have been studies about the consequences of the inevitable variations away from precise sampling (Albertos and Crespo, 1999), there's a new technique which uses carefully designed variable sampling (Bilinskis and Cain, 1996). It is known as "alias-free signal processing", and the implications of the phrase "alias-free" are that it is possible to extract information at frequencies well in excess of the average sample rate. The corollary for control, perhaps, is that the effect of sampling delays can be reduced such that the average sample frequency doesn't then need to be as high. Of course it's necessary to do extra calculations compared with fixed rate processing, so whether the total computation requirement is reduced is not clear, but it's another interesting possibility for investigation.

5.3 Future architectures

Control engineers have mainly had to use general-purpose processor architectures which were usually designed with other applications in mind. The DSP is a classic example of this, because its design was very much driven by the needs of signal processing for communications applications, and although there are control-specific versions which have appropriate peripherals on the chip, the basic processing architecture is certainly not optimised for control. One possibility therefore is the author's own work on a control-specific processor with a targetted architecture, known as a CSP rather than a DSP. It is however unlikely that a manufacturer will produce a CSP, the main reason being that nowadays the trends are more in the direction of exploiting the increased accessibility to programmable silicon.

Programmable silicon: There are now large Field Programmable Gate Arrays (FPGAs) and Programmable ASIC devices which offer hundreds of thousands of gates, and it's therefore possible to use these to provide complex processing functions. Combined with this is the move towards so-called "Intellectual Property cores", ie specific designs existing in a hardware description language such as

VHDL which can be transferred between different silicon solutions, just as compiler source code can be used on different processors.

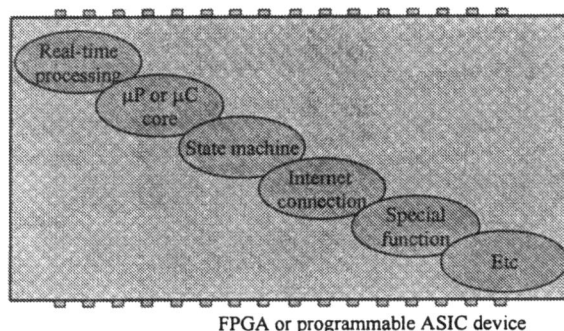

Fig. 9 Customised control processors

Future devices might be highly customised: as seen in Fig. 9 there is a processing core to handle the computationally-intensive aspects, a more simple general-purpose processor core to provide some of the general housekeeping (timing, analogue I/O, etc), perhaps some of the gates might be configured into a state machine to handle the Boolean interlocking function, and there might even be something which provides internet connectivity. A number of other functions are possible, but all provided on a single chip.

Co-design: It is necessary to use co-design of hardware and software to achieve the best possible solution for a particular application. At one time operational amplifiers were used as processing elements in analogue processors, and their cheapness meant that a customised circuit was created for each application. Then general-purpose digital processors of one kind or another became the norm, which created a commonality of hardware, and the customisation came in the software. New generations of integrated circuit technology enable a return to fully customised hardware, this time digital rather than analogue, to meet the particular controller needs. In practice a blend of hardware and software is more appropriate, emphasising the idea of effective co-design to achieve an optimised solution.

The subject is complex and not one in which the author has much expertise, but it is certainly possible to see an inevitable move away from the general-purpose processor solutions which have served us well for two decades, towards a much higher degree of device customisation for a particular requirement. The key to it is the availability of design tools which enable this to be achieved easily. Overall this will inevitably provide more cost-effective implementation of advanced control approaches which at present may be difficult to implement, and this must be a positive influence for Control Engineering in general

6. CONCLUSION

The paper has provided an overview of the issues, technologies and trends relating to processing for real-time control. As stated in the introduction, the focus has been mainly concerned with implementation solutions for embedded applications, and it is clear that the trends may differ in, for example, process control. Nevertheless it's useful to draw out the main points from the paper:-

- Powerful processors are available, but efficient implementation isn't straightforward.

- To achieve cost-effective solutions it is necessary to think about the algorithm and the architecture, together.

- It's likely that the general-purpose processor will progressively be replaced by more customised silicon solutions.

It's important to recognise that, despite the view which is prevalent among academic researchers, from a practical industrial viewpoint achieving efficient and effective implementations of high performance controllers is not solved. There are therefore many research areas for exploration, both relating to the algorithms for processing and the computing architectures.

REFERENCES

Albertos P, and A. Crespo, (1999), Real-time control of non-uniformly sampled systems, *Control Engineering Practice*, 7 (4), pp 445-458.

Bilinskis, I. and G. Cain, (1996), Digital alias-free signal processing tolerance to data and sensor faults, *IEE Colloquium on Intelligent Sensors*, pp 6/1-6/6

Candy, J. C., (1985), A use of double integration in sigma delta modulation, *IEEE Trans Comms,* **Com-33**, (3), pp 249-258.

Gevers, and Li, (1993), Parameterization in control estimation and filtering problems, *London: Springer Verlag.*

Goodall, R. M., (1992), A practical method for determining coefficient wordlength in recursive digital filters, *IEEE Trans on Signal Processing*, **40**, (4), pp 981-985.

Goodall, R. M. and B. Donaghue, (1993) Very high sample rate digital filters using the operator, *Proc IEE*, Pt G, **140**, (3), pp 199-206.

Goodall, R. M. and M. G. Goodall, (1993), Computational procedures for recursive digital filters, *Microprocessors and Microsystems*, **17**, (10), pp 587-596.

Goodall, R. M., S. R. Jones, R. Cumplido-Parra, F. Mitchell and S. Bateman, (2000), A control system processor architecture for complex LTI controllers, *6th IFAC Workshop on Algorithms and Architectures for Real-Time Control*, Palma De Mallorca, Spain, pp 167-172.

Johns, D. A. and D. M. Lewis, (1993), Design and analysis of delta-sigma based IIR filters, *IEEE Trans Circuits and Systems II*, **40**, pp 233-240.

Liu, B, (1971) Effect of finite wordlength on the accuracy of digital filters - a review, *IEEE Trans Circuit Theory*, **CT-18**, (6), pp 670-677.

Middleton, R. H. and G. C. Goodwin, (1990), Digital control and estimation - a unified approach, *Prentice Hall.*

REFERENCE ARCHITECTURE FOR ROBOT TELEOPERATION: DEVELOPMENT DETAILS AND PRACTICAL USE

Bárbara Álvarez*, Andrés Iborra*, Alejandro Alonso[†], Juan Antonio de la Puente[†]

*Universidad Politécnica de Cartagena, Dpto.Tecnología Electrónica,Paseo Alfonso XIII, 50, 30203-Cartagena, Spain

[†]Universidad Politécnica de Madrid, Depto. Ingeniería de Sistemas Telemáticos,Ciudad Universitaria s.n., E-28040 Madrid, Spain

e-mail: : balvarez@plc.um.es, aiborra@plc.um.es, aalonso@dit.upm.es, jpuente@dit.upm.es

Abstract: *The need to avoid a redundant effort in software development has been recognized since long time ago. Currently, some works are focused on the generation of products designed to be reused. The steps of the domain-engineering process have been applied and certain architectural patterns have been selected. These common software models have facilitated the development of reusable components. The architecture has been applied successfully to the development of different teleoperation platforms used in maintenance activities of nuclear power plants. In particular, this paper present how the reference architecture has been implemented for different products of ROSA, to the TRON (Teleoperated and Robotized System for Maintenance Operation in Nuclear Power Plants Vessels) system and to the IRV system. Copyright © 2000 IFAC*

Keywords: *Software architecture, robots control, real-time systems, teleoperation, software engineering*

1.INTRODUCTION

Reuse becomes an important factor in current developments due to the market competitiveness and time-to-market requirements. Although the interest of reusing software arises with the origins of the programming, this issue has not been taken to the practice with success for a long time (R. Prieto-Diaz, 1991) (W. Tracz, 1995). One of the reasons is the difficulty of combining existent software components. In addition, although reusing at this level is a good practice, it is not enough. One way of rising the reuse degree is to base this approach on software architecture, which comprises the software components, their visible properties, and the relationships among them (L. Bass et al., 1998).

The question that arises is how to develop a software architecture that can be effectively reused. The common approach is to concentrate on a domain area. It seems feasible to reuse a software architecture in a set of system with some common features. Then, a suitable practice is to develop a reference architecture that takes into account the special properties of a type of system in a domain area. A reference architecture is defined as a division of functionality together with data-flow between pieces and a mapping onto software components (L. Bass et al., 1998).

This approach was used by the authors and, as a result, (A. Alonso et al., 1997) a Domain Specific Software Architecture (DSSA) for robot teleoperation systems was developed. Since the publication of this article, the reference architecture has been used with great success in the development of a number of teleoperation systems for different robots and tools. The goal of this article is to revisit the approach taken from a higher abstraction and more mature point of view and to describe the cases where it has been used.

Section 2 shows the process followed to identify the components, relations and requirements of teleoperation systems. In section 3, the way certain architectural styles were selected for the interaction among subsystems is described. In sections 4, 5 and 6 several teleoperation robot systems developed with the proposed reference architecture are described. Section 7 includes some conclusions.

2.DOMAIN ENGINEERING

The process of domain engineering covers all the activities to build a common software core for a family of systems (J.V. Withey, 1994). This process were applied to obtain a reference architecture for robot teleoperation systems (A. Alonso et al., 1997). The following activities were performed (Figure 1 shows this process):

1) Domain analysis: in order to identify the set of components that are common to teleoperation applications. The result was a domain model.
2) Domain design: in order to obtain a generic design, based on the previous result and on the study of patterns or common models.
3) Implementation of the design: based on reusable components that can be used in different products.

There are several methods to perform a domain analysis but most them do not offer techniques to capture and to represent the information related to a family of systems.

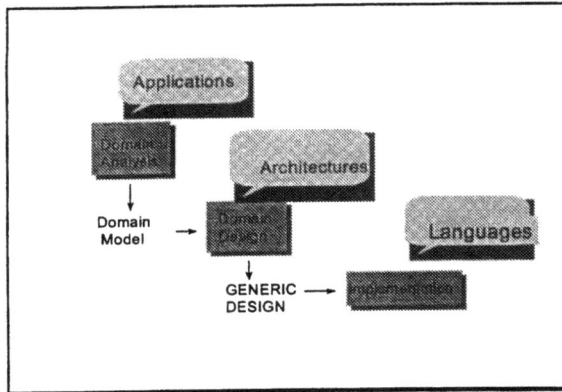

Figure 1: Domain-engineering process

FODA *(Featured-Oriented Domain Analysis)* (K.C. Kang et al., 1990) offers these techniques and supports reusability, not only at the pure functional level, but at the architectural level as well. The FODA method proposes the use of certain models for performing a domain analysis. In particular, it proposes the following activities: (1) a context analysis, that allows to define the environment of the domain and (2) a functional analysis, in order to identify the similarities and differences of the existent systems. As a result of the application of this method to the teleoperation systems, a generic specification of requirements has been obtained. In (A. Alonso et al., 1997) functional and non-functional requirements for a general teleoperation system are described, as well as the basic functional requirements or services that the system should provide to the end users, and the timing requirements. In these systems, time requirements should be met in order to ensure that the information that the operator receives is valid and reflects the current state of the robot.

The following step in the domain-engineering process is domain design. There are many works that describe how to perform a domain analysis, however they do not explain how to use the results of such analysis to obtain a generic design. The approach described in Peterson (A.S. Peterson et al., 1994) was used. It intends to group in subsystems the elements that work together for performing a certain task.

The next step in the definition of the reference model for teleoperation systems was to split the functionality among a number of components and to identify the data flow among the pieces. Figure 2 shows how the teleoperation reference model is mapped onto software components. In short, the functionality of these components or subsystems is the following:

- **Graphical representation**. This subsystem is in charge of showing to the operator the current state of the robot and the environment where it is operating. It provides operations for initializing the representation and for updating the status of the robot, according to the information received from the remote control unit.
- **Collisions detection**. This subsystem provides the required functions for checking whether a given movement command is safe, in the sense

that the robot does not collide with the operating environment or with itself.
- **User Interface**. It is in charge of interacting with the operator. It allows him to issue commands to the robot and to know their execution status.
- **Communications**. This subsystem embodies the communication protocol with the remote control unit. It provides means to send commands and to receive status information. In this module, a mechanism for fault treatment should be included for guaranteeing safety, in case of loss of communication with the control unit.
- **Controller.** It is the core of the system and it is in charge of executing the commands from the user, by using the other components and sending the appropriate orders to the robot. It also receives the robot status from the remote control unit, for checking its consistency and showing it to the operator.

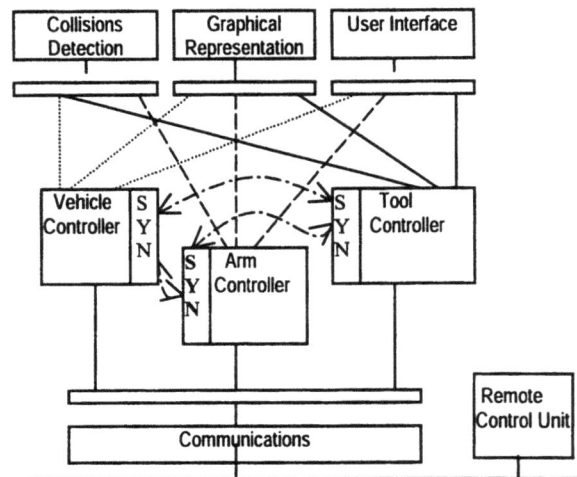

Figure 2: High level architecture description.

In this domain, it is common to have a robot with different controllers. This is the case of a moving arm, where it is necessary to consider the control of an articulated arm, whose base can be fixed on a vehicle and whose end joint holds a tool for performing specific operation. In this case, the teleoperation system has to deal with three controllers and the proper synchronization mechanisms should be implemented.

3. REFINEMENT OF THE REFERENCE MODEL

In order to get the reference architecture it is necessary to specify how the different components are going to interact. For this purpose, the approach followed is based on characterizing the interaction between the components and selecting an appropriate architectural style, which defines a particular pattern for the run-time control and data transfer mode. The use of architectural styles allows to use well-defined, well-understood and successfully tested interaction mechanisms. They help to develop systems with quality properties, such as maintainability and portability.

Two architectural styles were used for defining the interaction between the components: client-server (A. Berson, 1992) and communicating processes, according to the classification in (L. Bass et al., 1998). The client-server style is appropriate when there is a component that provides a service and another that requests it. It allows to change the server or the client, as long as the defined services are requested and provided following the designed interface. This is the case with the interactions between the *Graphical Representation* and the *Collisions Detection* with the *Controller*. The first component provides services for drawing the status of the robot. The second allows to check whether or not there are collisions in a possible robot movement. In these two cases, the interfaces were implemented based on messages, in order to allow distribution. The interaction is asynchronous, because the controller cannot be blocked for a long time waiting for a service to be completed.

A general communicating processes style was used for the interaction between the *User Interface* and the *Communication* components with the *Controller*. This is so because there is not necessarily a cause-effect relationship that guides their interaction Any of the components can take the initiative to send some data. For example, consider the information that the *Communication* component sends to the *Controller*. It can be periodic status data or aperiodic alarms, caused by some unexpected robot behavior. On the other hand, the controller may sent commands to the robot in any moment. The same holds for the relationship between user interface and the controller(s). A controller sends to the user interface sporadic messages, when robot status has changed or some failure occur in the system. The user sends commands to the controller when a certain robot behavior is required.

In the internal design of the components, two styles were used: layered and object oriented. The layered style is useful to achieve portability and modificability. It is based on structuring the components as a set of layers and it was used on the design of the *Communications* component. In case of having to port it to a different hardware or to use a different protocol, only some of the layers need to be changed.

The internal design and implementation of the rest of the subsystems is based on object-oriented and abstract data types styles. These paradigms emphasize the bundling of data and the knowledge of how to manipulate and access to that data. The data encapsulation promotes reuse and modifiability.

The robot control should not be interrupted by any reason. Therefore, certain mechanisms have been introduced for decoupling the control task from the rest of the subsystems and for receiving the information from the same ones. In this way, if a failure occurs in other subsystem, the controller can continue operating. These mechanisms are modules in charge of communicating with the rest and translating data. The communication of these modules with the interfaces is by means of procedure calls. When a message is sent to a controller, these decoupling modules store it in a buffer, so that the controller is not interrupted.

Each controller can be seen as a control loop which check continuously the received messages buffer.

The general controller designed performs the following operations: (1) receive commands from the operator, (2) check if the operations are feasible, (3) simulate the movements before executing them, (4) send commands to the remote control unit, and (5) update the state of the system. This design has shown to be general enough to be used for the implementation of controllers for different robots. In case of having several controllers for a robot, as mentioned in the previous section, step 2 includes the synchronization between them.

These types of systems use to have time requirements that must be met. The *Rate Monotonic Analysis (RMA)* (T. Klein et al.,1993) theory have allowed to reason with confidence about timing correctness at the tasking level and to analyze whether tasks deadlines can be met. In this way, it is offered a reference architecture and a framework for analyzing its timing response, which is described in (B. Álvarez et al., 1998).

The rest of the article presents the development of different teleoperation systems using this reference architecture.

4.THE ROSA SYSTEM

The ROSA (*Remotely Operated Service Arm*) system provides a remote user interface for controlling a jointed arm of six axis (figure 3). Different tools are used for inspection and repairing of the tubes in the steam generators.

Figure 3: The ROSA arm.

The development platform is a Hewlett Packard 9000 model 725 workstation, with the HP-UX operating system. A local area network connects the teleoperation platform with the robot control unit. The remote controller is based on an Heurikon HK68/V30XE processor, a VME bus, an amplifier, and several control cards attached to it.

Figure 4 shows a high level description of the architecture for the ROSA system. This scheme includes two controllers: robot and tool. A generic controller was developed in Ada (Ada, 1987), since its modular characteristics and generic facilities make the development of reusable code easier.

The implementation of the components for the

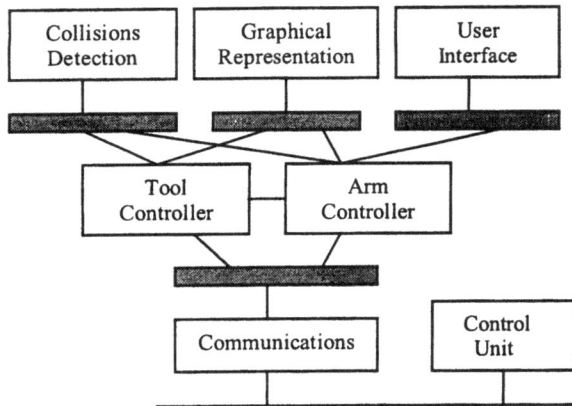

Figure 4: High-level architecture for ROSA.

graphics and the collision detection were based on a commercial tool called Robcad. It allows to define the robot and its operation environment in 3-D. It automates the operations for collisions detection and its graphical output is excellent. These components were developed in the programming language C++, since the libraries offered by this commercial tool were developed in this language.

User interface has been developed in the programming language ANSI-C. The resources offered by X Windows have been used by means of User Interface Language (UIL) and Motif.

The rest of the architecture has been implemented in Ada. In particular, the communications subsystem was based on TCP and UDP sockets. A package of asynchronous real-time Ada drivers for interconnected systems exchange (PARADISE) was used. This package offers an interface to the communication routines of the operating system Unix. Generic modules were developed for the interfaces among subsystems. They offer services for the communication among processes.

For the planned steam generator maintenance operations, it was necessary to develop a number of mechanical tools to perform specific operations in the tubesheet of steam generators, such as: (1) to detect wrong tubes. An inspection process based on eddy currents is used, (2) to cancel wrong tubes, (3) to recover them, (4) to drill plugs and (5) to place the nozzle-dams.

Tools controllers have been developed for dealing with this tools. They are different instances of the generic controller described in the previous section. The required parameters of these software packages are related with elements used to characterize the mechanisms (jointed arm and tools). These elements are specific commands and state machines. This allows to define: (1) robot status, (2) its evolution after a command execution and (3) the commands that are feasible for each state.

A controller was developed for an electro-disintegration machine (figure 5). This tool was designed for the disintegration of strange elements which could be located in the tubesheet (plugs, drills, etc.). The tool

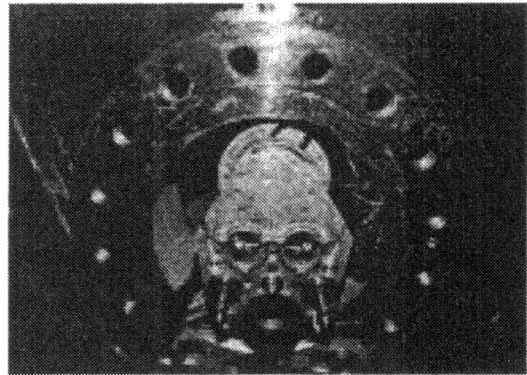

Figure 5: Electro-disintegration machine.

consists of the following parts: a cone for anchoring to the robot's end, a fixed platform on which the devices of vision are installed, lineal potentiometers for its alignment with the tubesheet and some camlocks for its anchorage to this one. The electro-disintegration head is located in the surface of a slider table. Commands were provided to fit the electro-disintegration head and activate it, to fit intensity and voltage parameters, and to load and unload the tool.

Two tool controllers are used to screw up and to control a gripper which is used for the placing of the nozzledams. Figure 6 shows a gripper. This tool consists of four pneumatic pistons. These activate four pincers that are used for placing the nozzledams in the primary circuits.

Figure 6: Gripper.

The fasteners and nuts that hold the nozzledams are installed by means of a screwdriver (figure 7). In the case of the gripper, commands are provided for controlling the pincers and anchoring the tool to the robot. To control the screwdriver, some commands are provided to adjust the head, to screw and to unscrew.

A commercial machine was adapted for welding plugs in the tubesheet of the steam generator. In this case, the subjection cone of the tool to the robot and the support for the plugs were designed. Figure 8 shows this tool. Some commands are provided to activate the head, to gauge the tool and to load and unload the tool.

For each tool, a common model was used for developing the user interface and the generic communication module. In this case, the generic parameters

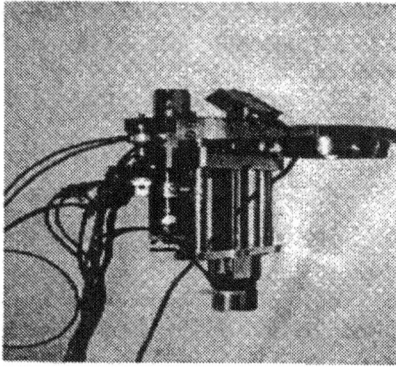

Figure 7: Screwdriver.

were instanced for connecting with the tool processes of the remote control unit.

Figure 8: Welding machine

5.THE IRV SYSTEM

The IRV (*Inspection Retrieving Vehicle*) system is a teleoperated vehicle with sensors, lights, cameras and interchangeable end-effectors. The vehicle can operate ten meters underwater, inside pipes of 400 mm and larger. This system is used for retrieving foreign objects from inside the primary circuit nozzles of the nuclear power plants. Figure 9 shows the vehicle

Figure 9: Vehicle of the system IRV.

The hardware platform for this teleoperation system is the same than in the ROSA system. Most of this system was directly reused. In particular, it was necessary to implement three controllers: vehicle, arm

and tool. These controllers communicate among them in order to synchronize the operations of the controlled elements. In order to implement the controllers, the generic packages developed in the ROSA system were instanced for the IRV system. The generic parameters are related with the operations of the devices (commands and state machines).

The *Graphic Representation* consists of a system of artificial vision. Its function was to identify the edges of the objects and the most significant elements of the environment and to superpose a wire frame representation of these with the real images provided by cameras (A.Iborra et al., 1993). The input information for this module is sent, due to efficiency reasons, through an independent channel directly from cameras. This subsystem is implemented on a different hardware platform (a personal computer). The rest of the teleoperation system runs on a HP 9000/725.

In the current implementation there are no collisions detection subsystem, and the graphical representation subsystem does not include a model of the environment and devices. They are not necessary for the current operations, but they can be added using commercial tools. The user interface subsystem is based on Motif. The vision system and the communications modules have been developed in the programming language C. It is important to note that the change of the *Graphical Representation* component did not require to modify the rest of the system.

6.THE TRON SYSTEM

TRON (*Teleoperated and Robotized System for Maintenance Operation in Nuclear Power Plants Vessels*) is a robotized system used for retrieving objects of the nuclear plants reactor vessels. Figure 10 shows this system. Due to human errors during the operations of recharge of fuel, objects may fall into the vessel. This system can be introduced trough holes, which are called bottom internals, to inspect them and to recover the objects without having to disassemble the nucleus.

The whole system comprises a jointed pole, the end-effectors and a navigation system based on artificial vision techniques, that aids the operator to move through a really complex environment (dark and full of obstacles). The pole consists of four joints. The end-effector and the inspection cameras are attached to the end link. The reduced dimensions of the inlet (1.5 inches) prevent the use of more complex mechanisms.

In this system, a real time operating system was not been required. The motion speed of the pole is very low. Therefore, a potential collision with the environment can be detected and the motors can be disable before it happens. For this reason, the controllers and user interface components were implemented on a Pentium PC, running Windows. The programming language was C++. New interface modules among subsystems were developed without modifying the communication mechanisms among them.

Figure 10: TRON system.

In this case, the design of the architecture was reused. The implementation language and the execution platform were different than in the previous cases, hence there was no code reuse. The generic controller class was implemented at the top of the hierarchy class. The pole and end-effector controllers have been derived from such mechanism controller class. The object-oriented programming paradigms allow to design software to be adapted or extended if new functionality has to be added. As in the previous systems, the pole and the end-effector have been described in terms of their basic commands, their state machine and their structural and dynamic models.

Graphical representation and collisions detection servers run on a HP 9000/725 workstation and the utilities provided by Robcad have been used. In this case, the collisions detection module is very simple due to not require inverse kinematics. Communications links between processes running on the PC (user interface and controllers) and the processes running on the workstation (graphical representation and collisions detection) are done with TCP/IP sockets.

7.CONCLUSIONS

The application of domain-engineering process and software architecting techniques were the basis for the development of a reference architecture for robot teleoperation systems. Several architectural styles have been selected for describing the interaction rules among components. Each style depends on which qualities are required.

The suitability of this architecture has been validated by its use in the development of different products with the same basis and its application to other robots. The use of the Ada language facilitated the development of generic components that could be reused for the different products of a same system. Furthermore, the interaction with other subsystems written in C has not been a problem.

In the three systems, the software has been tested successfully with the real robot and a 1:1 mock-up of the nuclear plant part where it will work. The devel-

opment of these tools and their control software results in the acquisition of a proprietary technology for maintenance works in nuclear plants. Due to the reference architecture qualities, the cost of this development was affordable.

8.ACKNOWLEDGMENTS

This work has been partially supported by the Spanish Government Programmes for Research in Electrical Power (project PIE-041049), and for Technological Actuation in Industry (PAUTA projects 753/95 y 53/96). TRON is supported inside EUREKA–MAINE program (EU1565).

9.REFERENCES

R. Prieto-Diaz (1991), *"Reuse in the U.S.A.".* In Proceedings of the 13th Annual International Conference on Software Engineering. IEEE Computer Society Press.

W. Tracz (1995) *"Confessions of a Used Program Salesman: Institutionalizing Software Reuse".* Addison Wesley Publishing Company.

Len Bass, Paul Clements, Rick Kazman (1998), *Software Architecture in Pratice*, Addison-Wesley.

A. Alonso, B. Álvarez, J .A. Pastor, J.A. de la Puente and A.Iborra (1997). *"Software Architecture for a Robot Teleoperation System".* 4th IFAC Workshop on Algorithms and Architectures for Real-Time Control.

James V. Withey (1994). *Implementing Model-Based Software Engineering in Your Organization : An approach to Domain Engineering.* Technical report, Software Engineering Institute (CMU).

K.C. Kang and S. Cohen (1990), *Feature-Oriented Domain Analysis (FODA) feasibility study.* Technical report, SEI (CMU).

A. Spencer Peterson, Jay L. Stanley Jr. (1994), *Mapping a Domain Model and Architecture to a Generic Design.* Technical report, (CMU).

Alex Berson (1992). *"Client/Server Architecture".* McGraw-Hill.

T. Klein, M. Ralya and M. Gonzalez Harbour, (1993). *A Practitioner's Handbook for Rate Monotonic Analysis.* Kluwer Academics Publishers.

B. Álvarez, A. Alonso, and J.A. de la Puente (1998). *"Timing Analysis of a generic robot teleoperation software architecture",* Control Engineering Practice, vol.6, no.6, pp.409-416.

The *Ada Reference Manual* (1987), ISO 8652:1987.

A.Iborra, M.A. Lázaro, S. Dominguez, P. Campoy, M.Alvarez, R.Aracil (1993). *"An automatic system for the real time integration of live action and synthetic 3-D computer images"* 4th Eurographics Animation and Simulation Workshop.

INDUSTRIAL ROBOT MULTIRATE CONTROL WITH THE VXWORKS REAL-TIME OPERATING SYSTEM

J.Salt, A.Valera, A.Cuenca, A.Ibañez

Departamento de Ingeniería de Sistemas y Automática
Universidad Politécnica de Valencia
Camino de Vera 14, 46022 Valencia (Spain)
e-mail: {julian, giuprog, acuenca, tibanez}@isa.upv.es

Abstract: This paper deals with the control of an industrial robot by means of a real-time operating system: VxWorks. One of the main goals of this paper is to explain the use of one cross-development tool: Tornado-VxWorks, which helps in the control implementation. Moreover, it will be designed a multi-rate controller and implemented with the cross-development tool in order to control an industrial robot. *Copyright © 2000 IFAC*

Keywords: Robot Control, Computer Aided Control System Design, Digital Computer Applications, Real-Time Computer Systems, Multirate.

1. INTRODUCTION

This paper shows several important concepts of real-time operating systems and multi-rate control.

On one hand, they are presented several tasks schemes, where these tasks are interconnected to make available the control. In these schemes, it can be observed tasks with different sampling periods. In this way a multi-rate control can be implemented.

On the other hand, it is also used a multi-rate technique to design multi-rate controllers for any polynomial reference (Kranc G.,1957), (Salt, J., *et al.*, 1994), (Albertos, *et al.*, 1996).

In this work, the industrial cartesian robot with three axis X-Y-Z of Figure 1 has been used. There is access to the robot positions by means of linear encoders and it can generate control actions, which activate on the robot axis by means of actuators.

Because of the process followed in this work is similar for the three axis, it will be shown later only the controller designed for Y axis and its implementation in a real-time operating system.

2. CROSS-DEVELOPMENT TOOL

The cross-development tool consists of two differentiated sides. Each one of them runs on one computer:

- Tornado (WindRiver Systems, Inc., 1999a): it is also known as host. It can be used to edit, compile, link, debug and store real-time code.
- VxWorks (WindRiver Systems, Inc., 1999b, c): it is also known as target, and it can be used to execute real-time code, since actually, on the target runs the real-time operating system. Moreover, it must incorporate the HW, which is needed to interact with the industrial cartesian robot (mainly encoder and counter card and data acquisition card). Concretely, it is used PCL-833 (Advantech Co., Ltd., 1994) card to obtain digital positions of the axis, and PCL-726 (Advantech Co., Ltd., 1993) card to convert D/A

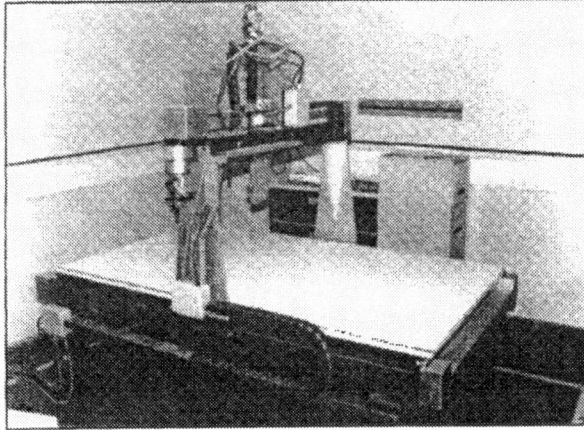

Figure 1: Cartesian robot.

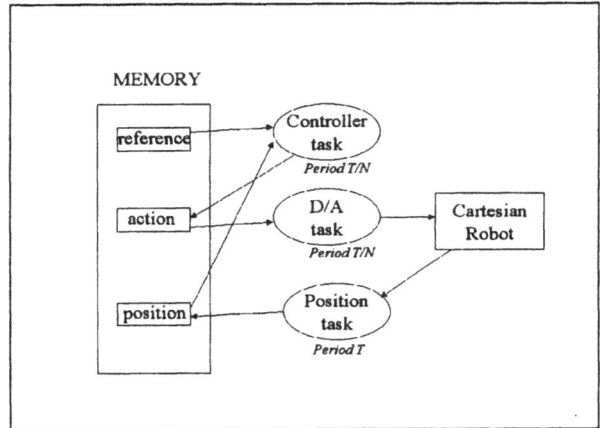

Fig 2: Basic tasks scheme.

the adequate control actions in order to move the different elements of the robot. Both sides are connected by means of a twisted pair cable, allowing the host to send object modules to the target and execute them in this latter.

3. REAL-TIME TASKS STRUCTURE

As it is known, one of the cases where multi-rate control is used is when it is wanted to implement a control system, in which there is only access at low frequency to the system measurement due to the measurement processing (visual feedback, network delays, etc). To compensate this low frequency access, they are generated control actions at high frequency in order to control correctly the system. It is clear that there are several frequencies working at the same control scheme.

In the studied case, there is an industrial environment, in which it is accessed only at low frequency to the axis position output. Therefore, control actions must be generated at high frequency in order to control the industrial robot. For these reasons, the implementation of a multi-rate control is proposed.

Since VxWorks is a real-time operating system, it can execute several tasks at the same time with different sampling periods. Consequently, it is able to create a multi-task multi-rate application, which will implement the multi-rate control.

Not only it is necessary having several tasks, but also it is needed these tasks are synchronised, in this case by means of timers. Concretely, it is worked with POSIX (Portable Operating System Interface) timers, because they are able to make portable code. It is associated one timer to each task and later fixed the sampling period of each timer. In this way, when the timer expires, the task will be executed. After the execution, the timer is initialised again so that the task can be reexecuted at next period.

Because of it is required to implement a control, in which N control actions are generated (where N is an integer number, which is known as multiplicity), and one axis position output is taken, then clearly tasks with different sampling period are needed. Concretely, if T is the sampling period of the position output, it is required a *controller task* working at T/N seconds and, of course, another *position task* working at T seconds.

If it is watched closely, the tasks scheme indicated in Figure 2 is needed.

Figure 2 can be explained as follows:

- *Controller task*: this task reads the reference and the position from a memory array. In this way, it can be generated the position error. Later, the position error is expanded in order to calculate N control actions, which are saved into another array. Each one of N actions is obtained every T/N seconds and received by the *D/A task*.
  Expanding a signal means transform a low frequency signal into a high frequency signal. Performing this transformation, the error signal (working at T seconds) is converted into another error signal (working at T/N seconds), which has the error value in the first sampling and null error value in the N-1 following samplings. Therefore, it is adequated the error signal so that the controller task (which works at T/N seconds) can generate correctly N control actions.
- *D/A task*: this task reads each one of control actions from a memory array, and applies them on the cartesian robot at the adequate time (every T/N seconds).
- *Position task:* this task obtains a position output every T seconds and saves it in the array.

The control scheme is specified in figure 3.

As it has been explained before, in figure 3 it can be seen the error signal has been expanded. This is represented by means of $(E^T)^{T/N}$.

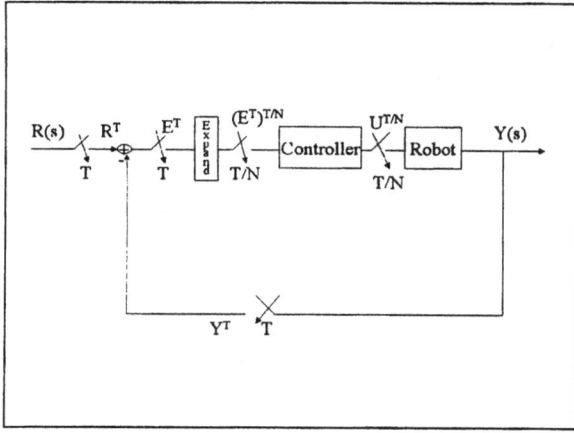

Fig 3: Basic control scheme.

Another way to carry out this application it is splitting up the controller in two sides: a slow side (working at T period) and a fast side (working at T/N period). Now slow control actions are expanded in order to transmit them from the slow side to the fast side.

In figure 4 it is shown the tasks structure. In this structure, it is replaced the *controller task* of figure 2 by a *slow controller* and by a *fast controller*.

In the figure 5 it is shown the control scheme.

4. MULTI-RATE CONTROLLER DESIGN

Using the method followed in (Kranc G.,1957), it can be designed multi-rate controllers easily, for any kind of references.

Concretely, it is wanted to design a multi-rate controller for the Y axis taking a ramp reference. This design will fix $N=2$ (multiplicity) and $T=0.01$ *sec* (slow sampling period). Finally it is taken $\lambda=0$ (inherent delay lag), $m=1$ (number of the first interval T/N after the delay lag λ) and $m_1=1$ (number of the first interval T after the delay lag λ).

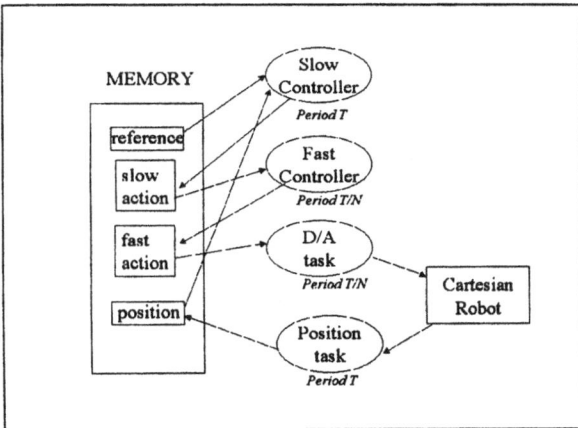

Fig 4: Slow/Fast tasks scheme.

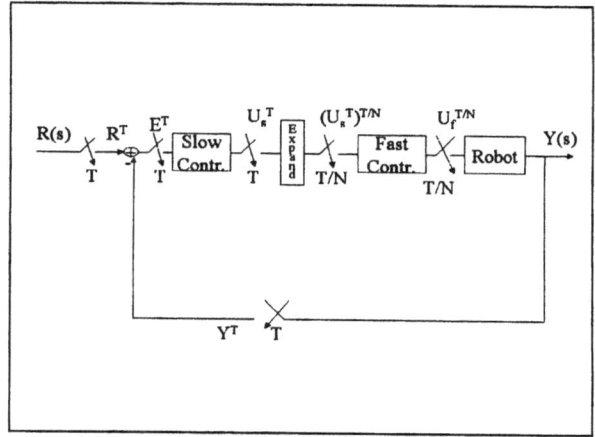

Fig 5: Slow/Fast control scheme.

The transfer function of Y axis is

$$G_p(s) = \frac{2240.85}{s} \quad (1)$$

Carrying out the Kranc's method, it is obtained the following multi-rate controller (2)

$$D(z_2) = \frac{1.5 - z_2^{-1}}{11.2 - 11.2 z_2^{-1}} = \frac{U(z_2)}{E(z_2)}$$

whose control law (at $T/N=0.005$ *sec*) is (3)

$$u(k_2) = \frac{11.2 u(k_2 - 1) + 1.5 e(k_2) - e(k_2 - 1)}{11.2}$$

This controller has been implemented with the tasks and control schemes presented in figures 2 and 3. If it is wanted to use the tasks and control schemes of the figures 4 and 5, it must split the controller up in a fast side and another slow side. To carry out the latter, it can be used the following expression (Kranc G.,1957)

$$D(z_N) = \frac{1}{1 - K(z)} \frac{K(z_N)}{G(z_N)} \quad (4)$$

where the slow side of the controller is the first member of the product and the fast side of the controller is the last member.

Because the following equations can be expressed (5), (6), (7)

$$K(z) = (2 - z^{-1}) z^{-1}$$
$$K(z_2) = (1 + z_2^{-1})^2 (1.5 - z_2^{-1}) z_2^{-1}$$
$$G(z_2) = \frac{11.2 z_2^{-1}}{1 - z_2^{-1}}$$

it can obtain (8), (9)

19

$$G_{SC}(z) = \frac{1}{1 - 2z^{-1} + z^{-2}} = \frac{U(z)}{E(z)}$$

$$G_{FC}(z_2) = \frac{1.5 + 0.5z_2^{-1} - 2.5z_2^{-2} - 0.5z_2^{-3} + z_2^{-4}}{11.2} = \frac{U(z_2)}{E(z_2)}$$

whose control laws are (10), (11)

$$u(k) = e(k) + 2u(k-1) - u(k-2)$$

$$u(k_2) = \frac{1.5e(k_2) + 0.5e(k_2-1) - 2.5e(k_2-2) - 0.5e(k_2-3) + e(k_2-4)}{11.2}$$

To demonstrate that both schemes obtain the same results, it has been implemented two different applications by means of Tornado-VxWorks. In the first application, it is implemented the basic tasks and control scheme (figure 2 and 3) with the control law (3). In the second application, it is implemented the slow/fast tasks and control scheme (figure 4 and 5) with the control laws (10) and (11).

The results are shown in figures 6, 7 and 8, being identical for both applications.

It can be observed that the position output follows the reference accurately enough. Only minimum errors are generated (the biggest one is of 5 counts of encoder that is approximately 0.067 mm) in the first iterations. After that, errors are of ± 1 count (approximately 0.013 mm). The ideal case (which is been validated by simulation procedure) is when it is not had error after first iterations. This ideal case is not reached due to the technologic error that appears when it is used linear encoders without decimal precision.

5. CONCLUSION

This paper has shown two ways to perform a multi-rate controller for an industrial cartesian robot. To carry out these two ways, it has been used a cross-development tool, Tornado-VxWorks. Tornado allows to develop real-time code and VxWorks execute it. It has been made two applications demonstrating that both are equivalent.

Also it has been detected the technologic error due to linear encoders. This technologic error does not allow the position output to follow accurately the reference. Nevertheless, the final error reference-position is minimum.

ACKNOWLEDGMENT

This work was supported by the Plan Nacional de I+D, Comisión Interministerial de Ciencia y Tecnología (CICYT), program TAP98-0252-C02-02.

Dr. Ismael Ripoll Ripoll (Departamento de Informática de Sistemas y Computadores, D.I.S.C.A., Universidad Politécnica de Valencia), because his help has been fundamental and very important to learn to work with real-time tasks.

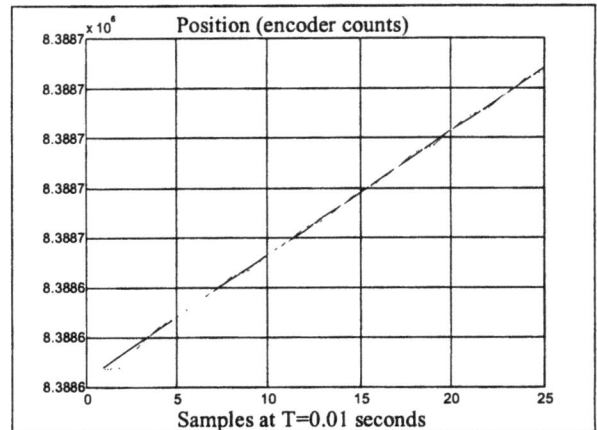

Fig 6: Position (cyan) versus reference (black).

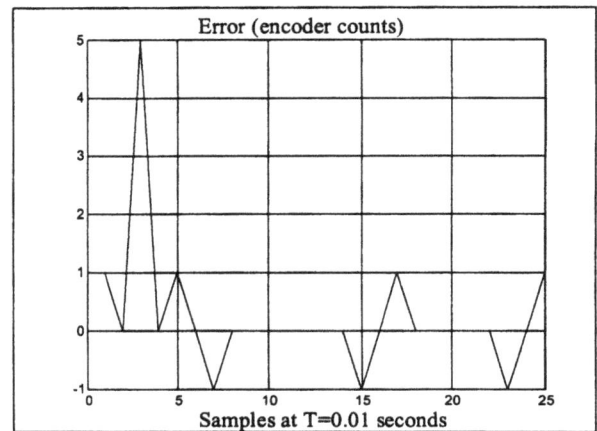

Fig 7: Reference-position error.

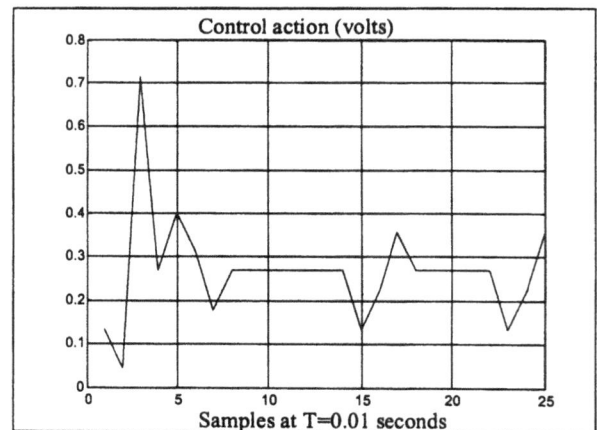

Fig 8: Robot control actions.

REFERENCES

Advantech Co., Ltd. (1993)
PCL-726 User's Manual. Edition 1. Taiwan.

Advantech Co., Ltd. (1994)
PCL-833 User's Manual. Edition 1. Taiwan.

Albertos, P., Salt, J., Tornero, J. (1996)
Dual Rate Adaptive Control. Automatica. **vol.AC-32, n. 7**, pp. 1027-1030.

Kranc G. (1957)
Compensating Error-Sampled System by Multirate Controller. Trans. AIEE, **vol.6, M.6, part II**, pp.149-155.

Salt, J., Albertos, P., Tornero, J., Ledesma, B. (1994)
Digital Controller Improvement by Multirate Control. Proc: The third IEEE Conference on Control Applications. Glasgow, U.K. ISBN 0 7803 1872 2. 1994.

WindRiver Systems, Inc. (1999a)
Tornado User's Guide, 2.0. Edition 1. USA.

WindRiver Systems, Inc. (1999b)
VxWorks Programmer's Guide, 5.4. Edition 1. USA.

WindRiver Systems, Inc. (1999c)
VxWorks Reference Manual, 5.4. Edition 1. USA.

REDUCING MEMORY ACCESS TIME IN THE REAL-TIME IMPLEMENTATION OF SIGNAL PROCESSING AND CONTROL ALGORITHMS

U. Kabir*, M. A. Hossain* and M. O. Tokhi**

* *Department of Computer Science, University of Dhaka, Dhaka-1000, Bangladesh.*
** *Department of Automatic Control and Systems Engineering,*
The University of Sheffield, UK

Abstract: This paper presents a method of enhancing the execution time of a processor in implementing signal processing and control algorithms, through a process of minimisation of memory access time of the processor. The method is based on the design and corresponding software coding of an algorithm, and thus can be adopted for a variety of real-time signal processing and control applications. The finite difference simulation algorithm of a flexible beam is considered to demonstrate the effectiveness of the proposed method. The algorithm is implemented on a general-purpose computing platform, and a comparative performance evaluation of the processor with the proposed method is presented and discussed. *Copyright © 2000 IFAC*

Keywords: Finite difference method, high performance computing, memory management, real-time signal processing and control, simulation of flexible beams.

1. INTRODUCTION

The high performance demand in real-time signal processing and control applications has motivated the development of advanced special-purpose and general-purpose hardware architectures. However, the developments within the software domain have not been at the same pace and/or level as within the hardware domain. Thus, although advanced computing hardware with significant levels of capability is available in the market, these capabilities are not fully utilised and exploited at the software level.

From the hardware point of view, current performance varies according to whether the processor possesses a pipeline facility, is microcode/hardwired operated, has an internal cache or internal RAM, has a built-in math coprocessor, floating point unit etc. In contrast, program behavior is difficult to predict due to its heavy dependence on application and run-time conditions. There are also many other factors affecting program behavior, including algorithm design, data structure, language efficiency, programmer skill, and compiler

technology (Ching and Wu, 1989; Hossain, 1995; Hwang, 1993). Moreover, in multi-processor architectures, partitioning, mapping and scheduling of an algorithm among processors, and inter-processor communication are important factors.

Performance of a processor is also related to program optimization facility of the compiler, which is machine dependent. The goal of program optimization is, in general, to maximize the speed of code execution. This involves several factors such as minimization of code-length, minimisation of memory access and optimising program code by removing redundancies (Hwang, 1993; Tokhi *et al.*, 1997).

It is essential for enhanced performance of a processor that a characteristic matching between the computing requirements of an algorithm and computing capabilities of the processor is made. Moreover, the memory management facility of the processor plays an important role in its overall performance in implementing an algorithm. This further includes the memory access time required during execution of a program code. Some special-purpose DSP devices, for example the Texas Instruments TMS320 devices, incorporate instructions, at the assembly language level, that allow executing commonly occurring operations in digital filtering applications, such as multiply, add and shift together. Such facilities attempt to minimise the memory access time and hence enhance the performance of the processor. In this paper, the issue of reducing the memory access time is addressed in a generic manner. The proposed methodology is exemplified and demonstrated with simulation of a flexible beam. Real-time simulation is important from a control perspective, as it provides a suitable platform for test and evaluation controller designs. The paper is organised as follows:

Section 2 describes the finite difference simulation algorithm of a flexible beam. Section 3 describes the formulation of the algorithm using a previously reported (conventional) strategy and the proposed method. Experimental results are presented in section 4, and the paper is concluded in section 5.

2. SIMULATION OF A FLEXIBLE BEAM

Consider a cantilever beam system with a force $U(x,t)$ applied at a distance x from its fixed (clamped) end at time t. This will result in a deflection $y(x,t)$ of the beam from its stationery position at the point where the force has been applied. In this manner, the governing dynamic equation of the beam is given by (Virk and Kourmoulis, 1988)

$$\mu^2 \frac{\partial^4 y(x,t)}{\partial x^4} + \frac{\partial^2 y(x,t)}{\partial t^2} = \frac{1}{m} U(x,t) \qquad (1)$$

where, μ is a beam constant and m is the mass of the beam. Discretising the beam in time and length using central finite difference (FD) methods, a discrete approximation to equation (1) can be obtained as (Virk and Kourmoulis, 1988)

$$Y_{k+1} = -Y_{k-1} - \lambda^2 S Y_k + \frac{(\Delta t)^2}{m} U(x,t) \qquad (2)$$

Where, $\lambda^2 = \left[(\Delta t)^2 / (\Delta x)^4 \right] \mu^2$ with Δt and Δx representing the step sizes in time and along the beam respectively, S is a pentadiagonal matrix (the so called stiffness matrix of the beam), Y_i $(i = k+1, k, k-1)$ is an $(n-1) \times 1$ matrix representing the deflection of end of sections 1 to n of the beam at time step i (beam divided into $n-1$ sections). Note that in evaluating the deflection of any point at the free end, the deflection of previous two points will be required. Equation (2) is the required relation for the simulation algorithm that can be implemented on a digital processor easily.

3. ALGORITHM DESIGN

In this section a formulation of the beam simulation algorithm with reduced memory access time during execution is presented. The simulation algorithm previously has been formulated and implemented as follows (Tokhi et al., 1997)

Algorithm-1 (conventional algorithm):

Step-1:

y[1][2] ← −y[1][0]−lumsq*(a*y[1][1]−
                    4*y[2][1]+y[3][1]);

y[2][2] ← −y[2][0] − lumsq*(− 4*y[1][1]+
                b*y[2][1] −4*y[3][1]+y[4][1]);

．．．．．．．．．．．．．．．．．．．．．．．

Step-2:

y[1][0] ← y[1][1]; y[1][1] ← y[1][2];

y[2][0] ← y[2][1]; y[2][1] ← y[2][2];

．．．．．．．．．．．．．．．．．．．．．．．

Go to Step-1 for N iterations.

Note in Step-2 of the above algorithm that at the end of every iteration, in shifting the deflection of the beam along the time scale, the value of y[1][1] is shifted to y[1][0] and the value of y[1][2] is shifted to y[1][1]. This requires memory shifting twice, and thus four times memory access. In the first memory access, the value is read from y[1][1] and in the second memory access the value is shifted to y[1][0], in the third memory access, the value is read from y[1][2] and finally, in the fourth memory access the value is shifted to y[1][1]. Thus, if in this process the memory access time is T sec. For a single iteration, in n segments of the beam, the total time for memory access will be $4nT$ sec. Thus, for N iterations, the total memory access time will be $4nNT$ sec. As an example, if $T = 60$ nsec, $n = 19$ and $N = 10000$ then the total memory access time will be $4 \times 19 \times 1000 \times 60 \times 10^{-9} = 0.0456$ sec. This will be a comparably substantial percentage of the total execution time of the algorithm. Thus, it is necessary to eliminate this extra memory access time. For this purpose, a modified formulation of execution of the algorithm is proposed. In the modified algorithm, instead of shifting the value from one memory location to another location, the memory cells are rearranged to calculate the result. Thus, there is no extra time required for memory access (shifting). Elimination of this shifting could offer the following advantages:

1. It would reduce the memory allocation, which in turn, leads to an increase in other activities of the

CPU and therefore, enhance the performance of the CPU.

2. Reduction of memory allocation might enhance the performance in implementing algorithms in real-time.

The proposed approach in implementing the algorithm is given as follows:

Algorithm-2 (modified algorithm):

y[1][2] ← −y[1][0] − lumsq*(a*y[1][1] −
                4*y[2][1]+y[3][1]);

y[2][2] ← −y[2][0] − lumsq*(− 4*y[1][1]+
            b*y[2][1] −4*y[3][1]+y[4][1]);

. . . .

y[1][0] ← −y[1][1] − lumsq*(a*y[1][2] −
                4*y[2][2]+y[3][2]);

y[2][0] ← −y[2][1] − lumsq*(− 4*y[1][2]+
            b*y[2][2] −4*y[3][2]+y[4][2]);

. . . .

y[1][1] ← −y[1][2] − lumsq*(a*y[1][0] −
                4*y[2][0]+y[3][0]);

y[2][1] ← −y[2][2] − lumsq*(− 4*y[1][0]+
            b*y[2][0] −4*y[3][0]+y[4][0]);

. . . .

Thus, instead of using a single code block and data shifting portion, as in Algorithm-1, to calculate the deflection, three code blocks, as in Algorithm-2, are used within the modified approach.

4. EXPERIMENTATION

To assess the impact of the modified approach in the performance of a processor implementing the algorithm, both the conventional and the modified algorithms were coded and implemented on a Pentium 133 processor. The execution times of the processor were obtained for over several number of iterations with the beam divided into 19 FD sections in each case. Fig. 1 shows the execution time evolution of the processor in implementing the algorithms over a range of iterations. It is noted that the amount of time wasted increases with increasing total number of iterations.

Table 1 shows the amount of time in seconds and in percentage of total execution time wasted due to memory access and shifting with algorithm-1. It is noted that in excess of 26% of the total execution time of the algorithm was spent on accessing the memory when implementing algorithm-1.

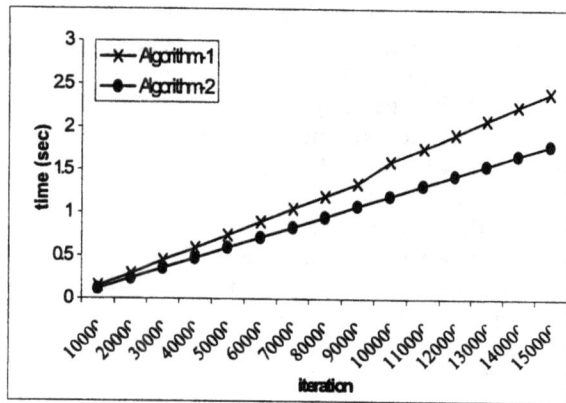

Fig.1: Execution time of the processor in implementing the convientional and modified algorithms.

Table 1: Time wasted due to memory access and shifting with algorithm-1.

Number of iterations	Memory access time with Algorithm-1 (sec)	Wasted time with Algorithm-1 (% of total)
10000	0.029	19.46
30000	0.09	20.22
60000	0.179	20.02
90000	0.26	25.42
120000	0.489	26.11
150000	0.607	25.29

Fig. 2 shows a graphical representation of the corresponding time wasted due to memory access in implementing algorithm-1 on the processor. It is noted that a significant amount of time is wasted due to memory access and shifting in implementing algorithm-1. This amount of time, however, is saved with the modified approach. It is also noted that at iteration 110000, the amount of wasted time suddenly becomes very high. This sudden change in wastage time may be due to the available cache. In case of Algorithm-1, the built-in cache (level 1) or level 2 size was not sufficient to provide the allocation of additional data at the 100000 to 110000 iteration increment.

5. CONCLUSION

An investigation into the performance enhancement of a processor, on the basis of reducing time of memory access and shifting, has been presented within the framework of real-time implementation of signal processing and control algorithms. An approach allowing elimination of the memory access and shifting time has been proposed and demonstrated in the implementation of a finite difference simulation algorithm of a flexible beam. It has been shown that substantial time saving and hence performance enhancement of a processor is achieved with this approach. The approach can have a significant impact in the real-time implementation of signal processing and control algorithms with either uni-processor or multi-processor architectures.

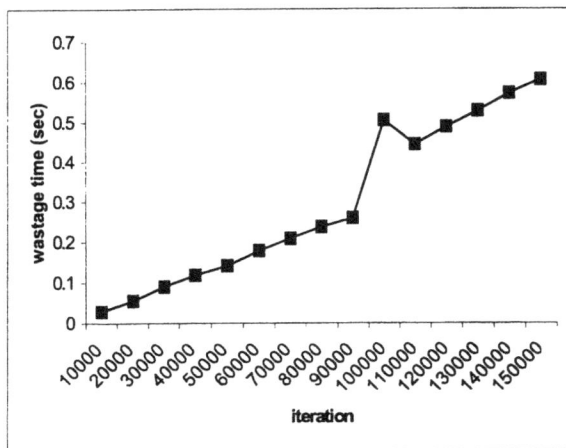

Fig. 2: Wasted time in implementing algorithm-1 on the processor.

REFERENCES

Ching, P. C. and Wu, S. W. (1989). Real-time digital signal processing system using a parallel architecture. *Microprocessors and Microsystems*, **13**, pp. 653-658.

Hossain, M. A. (1995). *Digital signal processing and parallel processing for real-time adaptive noise and vibration control*, Ph.D. thesis, Department of Automatic Control and System Engineering, The University of Sheffield, UK.

Hwang, K. (1993). *Advanced computer architecture – Parallelism, scalability, programmability*, McGraw-Hill, California.

Tokhi, M. O., Hossain, M. A., Baxter, M. J. and Fleming, P. J. (1997). Performance evaluation issues in real-time parallel signal processing and control. *Journal of Parallel and Distributed Computing*, **42**, pp. 67-74.

Virk., G. S. and Kourmoulis, P. K. (1988). On the simulation of systems governed by partial differential equations, *Proceedings of IEE Control-88 Conference*, pp. 318-321.

FORCE AND POSITION CONTROL OF ROBOTIC MANIPULATORS: AN EXPERIMENTAL APPROACH

Luis Filipe Baptista *,[1] **José M.G. Sá da Costa** **

* *Escola Náutica Infante D. Henrique*
Department of Marine Engineering
Av. Eng. Bonneville Franco, Paço de Arcos, Portugal
Phone: +351-21-4460010, E-mail: luisbaptista@enautica.pt
** *Technical University of Lisbon, Instituto Superior Técnico*
Department of Mechanical Engineering/GCAR
Av. Rovisco Pais, 1049-001 Lisboa, Portugal
Phone: +351-21-8418190, E-mail: sadacosta@dem.ist.utl.pt

Abstract: In this paper robot force control experiments in non-rigid environments are presented. The force control algorithms implemented in real-time are an impedance based control scheme with force tracking capability and a hybrid force/position control algorithm. A two-degree-of-freedom planar manipulator and a non-rigid contact surface were constructed for purposes of control evaluation. A low cost PC-based real-time software architecture was developed to fully control the robotic setup. The obtained results for a given force/position task reveal the successful implementation of the force control algorithms in real-time, especially for the impedance controller with force tracking. *Copyright ©2000 IFAC*

Keywords: Force control, real-time computer systems, impedance control, robot control, manipulators.

1. INTRODUCTION

In industry, many robotic tasks are generally related to manipulation or painting, requiring only position control of the arm, but some other tasks like assembly, pushing and polishing require the interaction between the manipulator's end-effector and the environment. This fact leads to the necessity of controlling the interaction force between the robot and the environment and consequently to develop more sophisticated force control algorithms (Zeng and Hemami 1997). Although a lot of different control schemes have been proposed in the literature the major force control approaches can be classified either as hybrid control (Raibert and Craig 1981) or impedance control (Hogan 1985). The hybrid control separates the robotic force task into two subspaces, a force con-

trolled subspace and a position controlled subspace. Two independent controllers are then designed for each subspace. In contrast, the impedance control does not attempt to control force explicitly but rather to control the relationship between force and position of the end-effector in contact with the environment (Singh and Popa 1995). Furthermore, when the environment has a non-rigid behavior it is possible to compute a virtual trajectory to apply the desired force profile on the environment. This approach is attractive since enables an effective application of force control algorithms in real-time with a low computational burden (Baptista 2000).

The development of new materials, the improvement of mechanical manipulator design and faster microprocessors have emerged the necessity of applying more sophisticated control algorithms in the new generation of industrial robots (Siciliano 1998). The implementation and validation of advanced control algo-

---

[1] This work was partly supported by PRODEP Program 5.2, Nº3/94 and PRAXIS XXI under contract 3/3.1/GEC/ 2707/95.

rithms require a flexible structure in terms of hardware and software. This equipment is of crucial importance to study and analyze the controller's performance. Unfortunately, one of the major difficulties in testing new force/position control algorithms relies in the lack of available commercial open robot controllers. In fact, industrial robots are equipped with digital controllers having fixed control laws, generally of PID type with no possibility of modifying the control algorithms to improve their performance. Generally, robot controllers are programmed with specific languages like ACL, VAL or VAL+ with fixed programmed commands like MOVE, SPEED, OPEN or CLOSE, having internally defined path planners, trajectory interpolators and filters among other functions (Baptista *et al.* 1998). Moreover, in general those controllers only deal with position and velocity control, which is insufficient when it is necessary to obtain an accurate force/position tracking performance.

In this paper real-time experiments with a proposed impedance control algorithm with force tracking capability and a hybrid force/position control algorithm are presented. In the first controller, a reference trajectory algorithm generate the virtual positions in the constrained direction to the impedance controller in order to apply the desired contact force on a non-rigid environment with uncertainties. The impedance control results are compared to the hybrid control results for a given force/position task. The real-time experiments were carried out with a two-degree-of-freedom (2-DOF) planar manipulator and a non-rigid mechanical environment built in the Robotics Laboratory. Also, a low cost open PC-based software architecture based on a commercial servo-axis interface board was developed to control the robotic setup in an effective and reliable way.

This paper is organized into 5 sections. Section 2 presents the experimental setup and the real-time software architecture used in the experiments. Section 3 present the impedance control approach with force tracking and the hybrid control algorithm implemented in real-time. Section 4 describe the obtained experimental results with both control algorithms for the given task. Finally, some conclusions are drawn in Section 5.

2. EXPERIMENTAL SETUP

The experimental setup shown in Fig. 1 consists of a 2-DOF planar manipulator, a non-rigid contact surface, a 6-axis force/torque sensor and the control hardware/software. The planar robot has two revolute joints driven by HARMONIC DRIVE actuators (HDSH-14). Each actuator has an electric d.c. motor, a harmonic drive, an encoder and a tachogenerator. The robot links parameters are given in Table 1 where l_{c_1}, l_{c_2} are the distances to the mass centers of the links 1 and 2, respectively, I_{zz} are the inertia moments related to the Z-axis and I_{m_1}, I_{m_2} are the actuator's inertia

Table 1 Planar 2-DOF robot parameters. (Note: link length l_2 includes the force sensor length and the end-effector device length.)

link i	l [m]	m [Kg]	l_c [m]	I_{zz} [Kg.m^2]	I_m [Kg.m^2]
1	0.320	5.00	0.163	0.120	0.081
2	0.330	1.20	0.112	0.015	0.021

moments related to the output shaft. In Fig. 1 the schematic diagram of the robotic setup is presented.

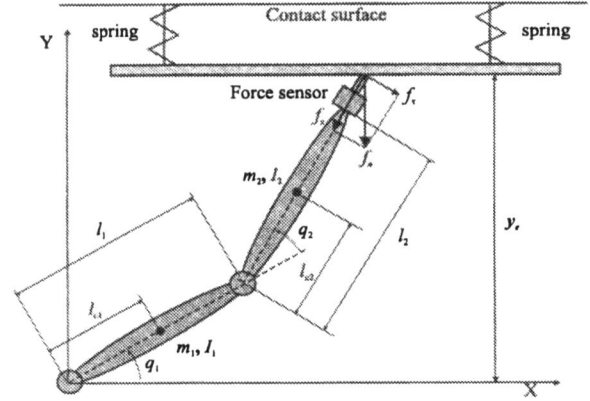

Fig. 1. Top view of the robotic setup schematic diagram. (Note: The force sensor arrow components f_x, f_z indicate the negative directions of the contact force measurements).

The contact surface used in the force control experiments is based on a steel plate with two springs in parallel guided by shafts with ball bearings. In general, a linear spring model is mostly used to represent the contact between the end-effector and the environment. Considering that this particular surface can be modeled as linear spring, the normal reaction force f_n on the end-effector is given by:

$$f_n = k_e \delta y \qquad (1)$$

where $\delta y = y - y_e$ is the displacement of the contact surface and k_e is the environment stiffness in the constrained direction. The term y_e is the distance between the contact surface and the base cartesian frame, as shown in Fig. 1. The estimated stiffness of the environment can be obtained experimentally by measuring the linear surface displacement when a force profile is applied by the end-effector. The collected pair of data $f_n - \delta y$ can be used to estimate the global environment stiffness value $\hat{k}_e$=1636 N/m, using the least-squares method. The obtained results are presented in Fig. 2, which exhibits a large surface deformation when a 10 N smooth step force profile is applied. Also, the environment has a nonlinear behavior of hysteresis type. The estimated environment stiffness coefficient is an important parameter that will be used in the implementation of the impedance control algorithm with force tracking (see Section 3 for details).

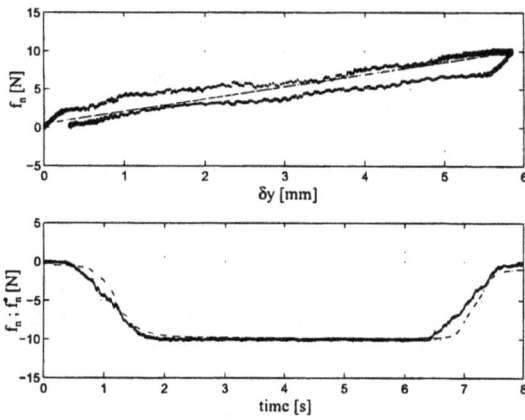

Fig. 2. Estimated environment stiffness $\hat{k}_e$ and contact forces. (Note: dashdot - estimated force $\hat{f}_n = \hat{k}_e \delta y$, solid - actual force f_n).

2.1 Control hardware

The control hardware consists of a PC Pentium 200 MHz, a servo-axis I/O board, linear power amplifiers and a 6-axis force/torque sensor. The power amplifiers are configured to operate as current amplifiers. In this functioning mode, the input control signal is a voltage in the range of ± 10 V with current ratings in the interval $[-3 , 3]$ A. The signals are processed through a low cost ISA-bus servo I/O board from SERVO TO GO, INC. The contact forces are measured by a JR[3] 6-axis force/torque sensor mounted at the end-effector of the arm. The force sensor system provide decoupled and digitally filtered data at a frequency rate of 8 KHz for each channel.

2.2 Real-time control software

In the last years, several open control architectures for robotic applications have been proposed (Jaritz and Spong 1996, Mandal and Payandeh 1995). In general these solutions rely on digital signal processor techniques (Lasky and Hsia 1994) or in expensive VME hardware running under the VxWorks operating system (Chiaverini et al. 1996, Kieffer and Yu 1999). This fact has motivated the development of a low cost open PC-based software architecture with kernel of management, supervision and control (Baptista et al. 1998). The real-time PC software tool was developed considering requirements such as low cost, flexibility for incorporation new technology and requirements and full in-group developed source code (Costescu and Dawson 1998).

The control software was developed in C++ language for the MS-DOS operating system. Real-time performance is guaranteed by reprogramming the Programmable Interval Timer (PIT) of the computer system (Kyle and Rabinowitz 1988). Thus, in the developed program the periodic interrupt generated by the servo-axis I/O board PIT is configured to the Interrupt Request 5 (IRQ 5), which activates the Interrupt Service Routine 13 (ISR 13). The ISR 13 is

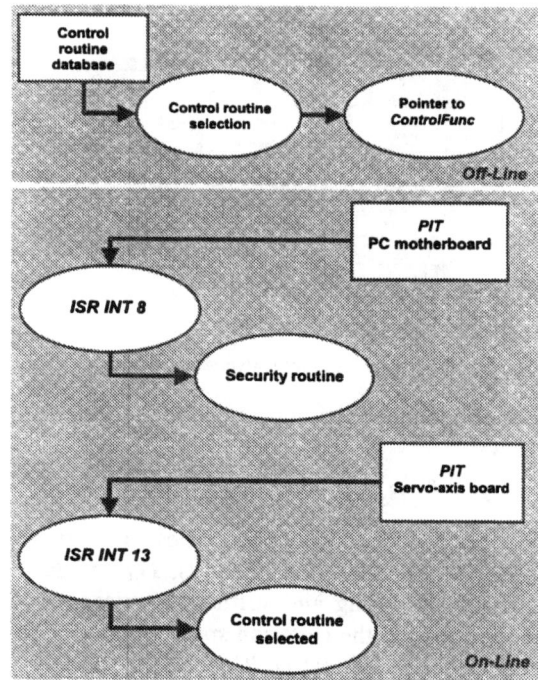

Fig. 3. Block diagram of the real-time operating procedure.

reserved for the control algorithm implementation at a sampling frequency F_s=1 KHz. The ISR 8 activated by the computer's IRQ 0 is redefined to execute the security routine at a frequency that is the double of the sampling frequency. In Fig. 3 the schematic diagram of the implemented real-time procedure is presented.

In order to avoid the accumulation of interrupt requests, the ISR 13 increments a flag at the beginning of the selected control routine execution. This flag is decremented at the end of the execution of the control procedure. The flag's content is checked by the security routine, which aborts the real-time procedure if its content is superior to the unity. In this situation, the failure is reported to the operator by the interface program.

In Fig. 4 the simplified software operating mode is presented. First, the robot parameters data files and other configuration parameters are loaded into the computer's memory. Secondly, the trajectory data file generated in MATLAB is executed by the selected control algorithm during the specified duration of time. Finally, the data obtained from the experiment is saved in an output file. The experimental data is further analyzed in MATLAB/SIMULINK.

3. FORCE CONTROL ALGORITHMS

In this section the impedance control approach is applied to obtain a force control scheme in non-rigid environments with force tracking capability. A simplified impedance control scheme can be obtained in quasi-static conditions, which means that velocities defined in the cartesian space and in the joint space, are assumed to be $\dot{x}$=0 and $\dot{q} \approx 0$, respectively. This

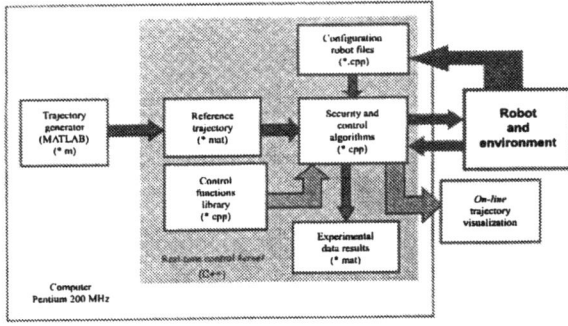

Fig. 4. Block diagram of the control software functioning mode.

approach leads to the following simplified impedance control law:

$$\tau = J(q)^T (B_d \dot{e} + K_d e) + g(q))$$ (2)

where $e = x_d - x$, $\dot{e} = \dot{x}_d - \dot{x}$ are cartesian position and velocity errors, respectively. In (2), B_d and K_d are the desired damping and stiffness matrices, respectively, defined in the cartesian space where $J(q)$ represents the Jacobian matrix. This control scheme is equivalent to a PD cartesian position control law with gravity compensation. It is possible to show that for a constant reference x_d and in the absence of contact forces, asymptotic stability is assured (de Witt et al. 1997). When the end-effector reaches the environment a contact force appears and it is not possible to assure asymptotic stability of x_d. In this case, a steady-state equilibrium situation is reached between deformation of the environment and position compliance of the controller, that is regulated by the inverse of matrix K_d. The stability analysis can be obtained considering the environment modeled as a generalized spring with stiffness K_e and surface coordinates x_e. Let's consider for simplicity the analysis only for the constrained direction. In this case, the environment modeled as a spring with stiffness k_e will generate a reaction force if $x > x_e$. Then, when the manipulator is in steady-state contact with the environment, the normal contact force is given by

$$f_n = k_e \delta x$$ (3)

where $\delta x = x - x_e$. Thus, the following steady-state balance condition is achieved

$$k_d(x_d - x) = k_e(x - x_e)$$ (4)

which leads to the equilibrium position (Singh and Popa 1995)

$$x_{eq} = \frac{k_d x_d + k_e x_e}{k_d + k_e}$$ (5)

and to the equilibrium force

$$f_{eq} = \frac{k_e k_d}{k_e + k_d}(x_d - x_e)$$ (6)

It is possible to show that the equilibrium point given in (5) of the closed-loop system is globally asymptotically stable (de Witt et al. 1997).

Considering the steady-state position and force given in (5) and (6), respectively, it is possible to apply a

desired contact force f_d on the environment, while simultaneously achieving the desired impedance by computing a virtual or reference position x_v in the constrained direction, given by:

$$x_v = \begin{cases} x_e + \dfrac{f_d}{k_d} & \text{if} \quad f_n = 0 \\ x_e + f_d \left(\dfrac{k_d \delta x + f_n}{k_d f_n} \right) & \text{if} \quad f_n \neq 0 \end{cases}$$ (7)

which is valid for contact and non-contact condition. Notice that it is possible to obtain $\dot{x}_v$ by numerical differentiation of x_v. However, due to the high frequency noise usually presented in the contact force measurements, $\dot{x}_v$ is assumed to be zero.

The control law (2) can be used to apply a desired contact force on the environment considering the virtual references given by the vector $x_t = [x_v \quad x_d]^T$. This approach enables the control law (2) with force tracking capability without explicit knowledge of the environment stiffness parameter.

From the impedance controller with force tracking (ICFT) described above a modified version of this algorithm is implemented in real-time, considering the Y-axis as the constrained direction (see Fig. 1 for details). The force controller can be described by the following steps in each sampling period T_s:

Step 1: calculation of the normal contact force $f_n(k)$.

Step 2: detection of the contact condition,

$$y_v(0) = y_e \quad \text{if} \quad f_n(k) < 0$$ (8)

Step 3: calculation of the virtual trajectory $y_v(k)$ in the constrained direction

$$y_v(k) = \begin{cases} y_v(k-1) - \dfrac{\delta f_d(k)}{k_d} \\ \quad \text{if} \quad f_n(k) \geq 0 \\ y_v(k-1) - \left(\dfrac{\delta f_d(k) + \kappa_f e_f(k)}{\hat{k}_e} \right) \\ \quad \text{if} \quad f_n(k) < 0 \end{cases}$$ (9)

where κ_f is a force error gain and

$$e_f(k) = f_d(k) - f_n(k)$$ (10)
$$\delta f_d(k) = f_d(k) - f_d(k-1)$$ (11)

Step 4: calculation of the control actions in the cartesian space

$$f(k) = K_d e(k) + T_s K_i I_p(k) - B_d \dot{x}(k)$$ (12)

where $e(k) = x_t(k) - x(k)$ are the virtual position errors and

$$I_p(k) = \frac{e(k) + e(k-1)}{2}$$ (13)

Step 5: calculation of the joint control actions

$$\tau(k) = J(q(k))^T f(k)$$ (14)

Notice that in (9), $\hat{k}_e$ given in Section 2 is used to calculate $y_v(k)$. This algorithm compensate environment geometric uncertainties, since the virtual trajectory is obtained without explicit knowledge of y_e. The term $\kappa_f e_f$ in (9) is used to compensate stiffness uncertainty and other environment non-modeled effects (see Fig. 2). An integral control action with gain K_i is included in (2) leading to (12) in order to improve force tracking performance. In Fig. 5 the schematic diagram of the force control scheme implemented in real-time is shown.

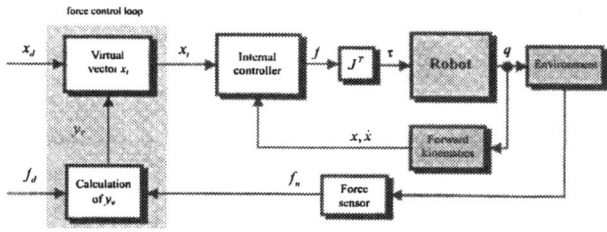

Fig. 5. Block scheme of impedance controller with force tracking.

Fig. 6. View of the 2-DOF robot manipulator executing a task on the contact surface.

The other control algorithm implemented in real-time is an hybrid force/position controller (HFPC) and can be described by the control laws for unconstrained space and constrained space, respectively given by:

$$f_p(k) = K_P e_x(k) + T_s K_I I_x(k) - K_D \dot{x}(k)$$
$$f_f(k) = f_d(k) + K_{P_f} e_f(k) + T_s K_{I_f} I_f(k) \quad (15)$$
$$- K_{D_f} \dot{y}(k)$$

where $e_x(k) = x_d(k) - x(k)$. Notice that velocity damping is included in the control law for constrained direction, as reported in Mandal and Payandeh (1995).

4. EXPERIMENTAL RESULTS

In this section experimental results obtained with the force control algorithms are presented. In Fig. 6 the 2-DOF manipulator executing the desired task on the contact surface is presented.

The control algorithms are implemented in the cartesian space with a sampling period T_s=1 ms. The internal controller gains in (12), experimentally tuned to give a good tracking response are presented in Table 2.

Table 2 Force control gains.

Controllers	ICFT			HFPC		
Axis	K_d	K_i	B_d	K_P	K_I	K_D
X−axis	2000	3000	5	2000	3000	5
Y−axis	2000	3000	0	750	1000	0

Notice that gain B_d in the constrained direction for the impedance controller and gain K_D for the hybrid controller are set to zero due to the noise present in the cartesian velocity signal. Notice that κ_f=1.0 is also selected from experimental trials. The implemented force/position task is a linear trajectory with 100 mm

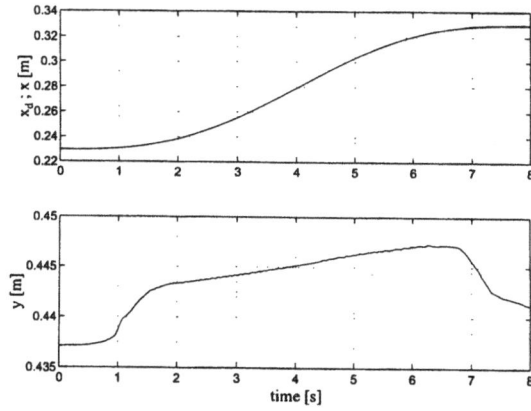

Fig. 7. ICFT: X-axis position evolution and Y-axis constrained position. (Notice the environment geometric uncertainty along this coordinate).

length between the initial position x_i=0.2295 m and the final position x_f=0.3295 m, with simultaneous application of a smooth 10 N step force profile along Y-axis, during 8 seconds. It is assumed that the arm is initially in contact with the environment. Moreover, geometric uncertainty in the contact surface of 5 mm at the end of the trajectory is intentionally imposed to the environment. Filtered force data (f_x, f_z) are obtained from the JR^3 DSP board considering a first order low-pass digital filter with 500 Hz of cut-off frequency in order to attenuate the force sensor noise. The results obtained with the proposed impedance control scheme (ICFT) are presented in Figs. 7 to 9. The position and contact force errors for the hybrid controller (HFPC) are presented in Fig. 10. The euclidian norm ($\| \cdot \|$) of position and force errors and the absolute maximum error values along the trajectory are presented in Table 3.

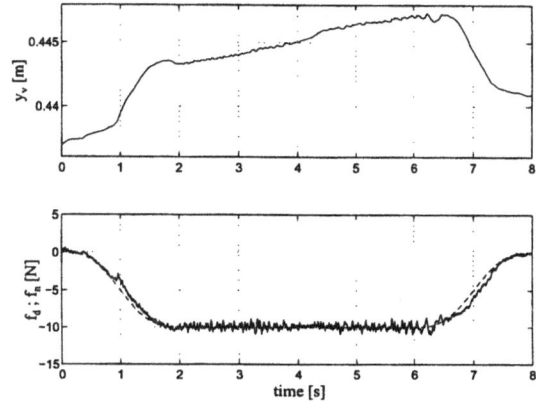

Fig. 8. ICFT: Y-axis virtual trajectory, desired and actual contact force (Note: solid-actual force ; dashdot-desired force).

Table 3 Euclidian norm of the position and force errors.

Controller	$\| e_x \|$ [m]	$\max(e_x)$ [mm]	$\| e_f \|$ [N]	$\max(e_f)$ [N]
ICFT	0.011	0.485	50.484	1.852
HFPC	0.011	0.485	97.982	3.321

The results presented for the impedance controller reveal a more accurate force tracking performance than the equivalent results obtained with the hybrid control

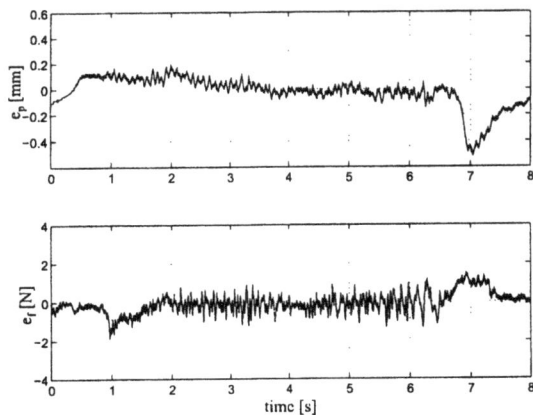

Fig. 9. ICFT: X-axis position errors and Y-axis contact force errors.

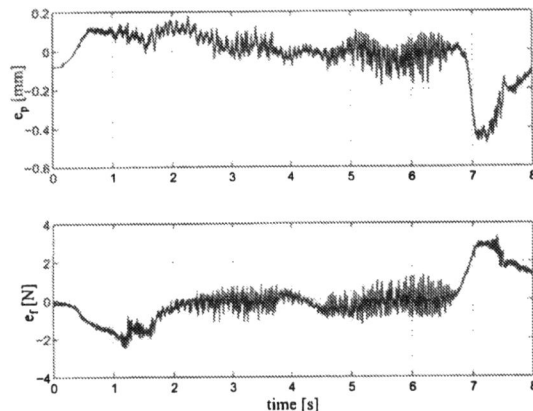

Fig. 10. HFPC: X-axis position errors and Y-axis contact force errors.

algorithm. The global results reveal the successful implementation of force control algorithms in real-time, especially for the impedance controller in the presence of non-rigid environments with uncertainties.

5. CONCLUSIONS

In this paper robot force control experiments in non-rigid environments were presented. The performance of an impedance controller with force tracking was evaluated considering an end-effector position and force tracking trajectory. The experimental results obtained for the impedance controller have shown a more accurate force tracking performance in comparison with hybrid force control results. Future research will concentrate on the improvement of a real-time software functionality and the experimental study of the impact situation on the contact surface.

6. REFERENCES

Baptista, L.F. (2000). Controlo adaptativo de posição e de força de robôs manipuladores. Phd thesis (in portuguese). Technical University of Lisbon, Department of Mechanical Engineering. Lisbon.

Baptista, L.F., J. Martins and J.M. Sá da Costa (1998). An experimental testbed for position and force control of robotic manipulators. In: *Proceedings of The 5th IEEE International Workshop on Advanced Motion Control AMC 98*. Coimbra, Portugal. pp. 222–227.

Chiaverini, S., B. Siciliano and L. Villani (1996). Parallel force/position control schemes with experiments on an industrial manipulator. In: *Proceedings of The IFAC 13th Triennial World Congress*. San Francisco, USA. pp. 25–30.

Costescu, N. and D.M. Dawson (1998). Qmotor 2.0-A PC based real-time multitasking graphical control environment. In: *Proceedings of The American Control Conference*. Philadelphia, USA. pp. 1266–1270.

de Witt, C.C., B. Siciliano and G. Bastin (1997). *Theory of robot control*. Comunications and Control Engineering. Springer-Verlag. UK.

Hogan, Neville (1985). Impedance control: an approach to manipulation: Part I-II-III. *Journal of Dynamic Systems, Measurement, and Control* **107**, 1–24.

Jaritz, A. and M. W. Spong (1996). An experimental comparison of robust control algorithms on a direct drive manipulator. *IEEE Transactions on Control Systems Technology* **4**, 614–626.

Kieffer, J. and K. Yu (1999). Robotic force/velocity control for following unknown contours of granular materials. *Control Engineering Practice* **7**, 1249–1256.

Kyle, J and C. Rabinowitz (1988). Hardware interrupt handlers. *MS-DOS Encyclopedia* pp. 409–427.

Lasky, T. A. and T. C. Hsia (1994). Application of a digital signal processor in compliant cartesian control of an industrial manipulator. In: *Proceedings of the American Control Conference*.

Mandal, N. and S. Payandeh (1995). Control strategies for robotic contact tasks: An experimental study. *Journal of Robotic Systems* **12(1)**, 67–92.

Raibert, M. and J. Craig (1981). Hybrid position/force control of manipulators. *Journal of Dynamic Systems, Measurement, and Control* **102**, 126–133.

Siciliano, B. (1998). Control in robotics: Open problems and future directions. In: *Proceedings of The 1998 IEEE Conference on Control Applications*. Vol. 1. Trieste, Italy. pp. 81–85.

Singh, S. K. and D. O. Popa (1995). An analysis of some fundamental problems in adaptive control of force and impedance behavior: Theory and experiments. *IEEE Transactions on Robotics and Automation* **11**, 912–921.

Zeng, G. and A. Hemami (1997). An overview of robot force control. *Robotica* **15**, 473–482.

DICOS: A REAL-TIME DISTRIBUTED INDUSTRIAL CONTROL SYSTEM FOR EMBEDDED APPLICATIONS

J.C. Campelo, P. Yuste, P.J. Gil, J.J. Serrano

Fault Tolerant Systems Group
Department of Computer Engineering
Technical University of Valencia
Camino de Vera s/n, 46022-Valencia, Spain
e-mail: {jcampelo, pyuste, pgil, jserrano}@disca.upv.es

Abstract: The Fault Tolerant Systems Group of the Technical University of Valencia has developed the DICOS system. This paper describes DICOS (**D**istributed **I**ndustrial **CO**ntrol **S**ystem) nodes. The architecture of DICOS nodes and the error detection mechanisms used are presented. These mechanisms are based on the internal capabilities of the 16-bit microcontroller used and control flow checking and deadlines control with the aid of a second 8-bit microcontroller. *Copyright © 2000 IFAC*

Keywords: embedded system, distributed computer control system, reliability, safety, deadline

1. INTRODUCTION

Real time distributed industrial control systems are becoming one of the most important areas in embedded control applications research. In this sense, control and supervision of those systems is accomplished through the co-operation of different nodes interconnected through industrial local area networks. In these systems, control and supervision elements are based on a microcontroller, in the simplest as well as in the most complex form (e.g. hard real time systems).

In systems that could involve certain risk for the process to control or the users, is necessary to stud and improve dependability. When a system is able to continue its work in presence of failures is called a **fault tolerant system** (LIS, 1996). Furthermore, a system is **safe** (LIS, 1996) if, when a failure is produced and it can not continue correct operation, the system stops in a well-known state which does not cause risk to the process. These aspects justif

the incorporation of different mechanisms in order to detect and correct as many as possible of the failures that may appear in such systems. Also, when designing real-time systems, mechanisms to detect and resolve missing deadlines are also needed.

From dependability point of view, one of the pursued objectives in the design phase of a distributed industrial control system is to obtain the **fail-silent** capability (LIS, 1996). So even in the case a node of the distributed system presents a wrong operation due to some fault, this must not transcend to the rest of the system. To obtain this behaviour hardware (component redundancy) or software mechanisms can be used.

From real-time point of view, although a system has a correct schedulability, transient faults can be fatal to meet deadlines. Reconfiguration techniques, recovery points, re-execution of tasks imply some time the impossibility to meet deadlines. So mechanisms in order to detect these faults and to determine if deadlines will be missed are necessary.

DICOS (Distributed Industrial COntrol System) is a fault tolerant real-time distributed industrial control system. The DICOS node architecture is based on two microcontrollers, one for the execution of the application code and one for communication tasks acting, also, as watchdog processor of the first one to achieve a better error coverage and detect missed deadlines. Both microcontrollers are connected through a dual port memory.

This paper is organised as follows: After the introduction, similar systems developed in other research centres are commented. In section 3 DICO nodes and their error detection mechanisms ar presented. In last section, conclusions are commented and future work is outlined.

2. RELATED WORK

In the bibliography related to this topic other systems with similar characteristic and similar objectives can be found. The nearest architecture to our objectives is DACAPO from the University of Göteborg (Rostamzadeh *et al.*, 1995; Rostamzadeh and Torin, 1995a; Rostamzadeh and Torin, 1995b): *Dependable distributed computer Architecture for Control Applications with Periodic Operation*. It is considered a fault tolerant distributed system with real-time characteristics. Its application is as a control and supervision system in automotive environments, so the system is composed by a small number of nodes communicated through a fault-tolerant network.

Another example is found in the University of Vienna. Its name is MAR : *Maintainable Real Time System* (Karlsson *et al.*, 1995; Kopetz *et al.*, 1990). Just as the previous one it is a fault tolerant distributed system with real-time characteristics. Also, just as DACAPO, it is a time-driven system and is based on replica of the process elements to obtain what is called FTU (*Fault Tolerant Unit*).

Others examples are DELTA-4 from LAAS (Arlat *et al.*, 1990), a distributed system with fault-tolerant characteristics, EDR (*Extended Distributed Recovery Block*) architecture (Hecht *et al.*, 1989), DUSBER (*DUplex System with Backward Error Recovery*) from Essen University (Grans, 1998) and ARTS *Adaptable Real-Time System* (Baba *et al.*, 1996).

Although those architectures have interesting characteristics from dependability and real-time points of view, nor DACAPO, DUSBER or ARTS have a real implementation (they are very complex architectures or are now under study). MARS have some prototypes but it is a massive-redundant architecture, with a high cost and nowaday its

communication protocol -TTP: *Time triggered protocol*- (Kopetz *et al.*, 1997) is under validation.

With DICOS system to use low-cost nodes to achieve a high reliability and safety system in real-time distributed embedded applications is proposed.

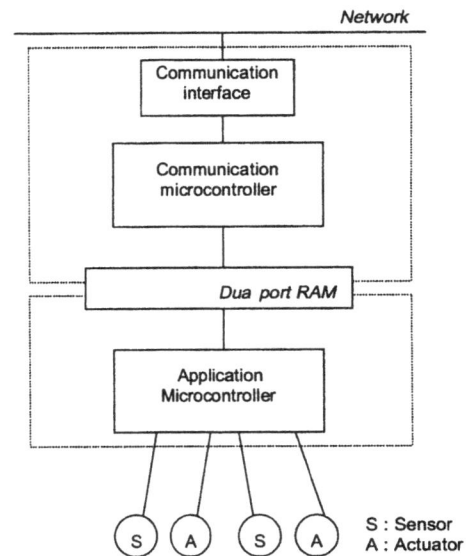

Fig. 1. Basic architecture

All these systems share with the one proposed here the basic principles of operation, with little differences from each other in the communication channel or links between elements and their purpose. Thus, as aspects common to all, the following parts (Fig. 1) can be distinguished:

- There is a microcontroller devoted to communication tasks.
- Another microcontroller is devoted to the execution of the application.
- The interface among both microcontrollers is based on a dual-port RAM or FIFO memory

Our fault-tolerant distributed system is based on a set of nodes connected through two local area networks based on the CAN (Controller area network) protocol (Bosch, 1991).

3. DICOS

In DICOS system, two different node architectures can be selected as a function of the systems dependability requirements. The Controller Area Network (CAN) is used as the communication channel among nodes. The most relevant characteristics of these node architectures are emphasised in the next paragraphs

3.1 DICOS Nodes Architecture

Architecture I

The objective of the first architecture is to obtain improvements in reliability and safety with a lo cost. Due to this, the basic node components are used in a single way to improve these parameters. Fig. 2 depicts the proposed architecture. It is composed of two communication networks, two microcontrollers, one for communication tasks (8-bit microcontroller) and another for the execution of the control algorithms (16-bit microcontroller), and one dual-port RAM.

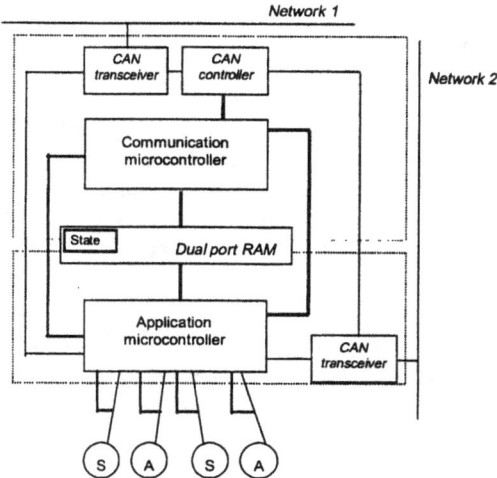

Fig. 2. Architecture I

Commonly in industrial applications different nodes can accomplish similar functions. This is possible because they can be connected to the same set of sensors and actuators. So, an important aspect is to store the application state of each one in the dual-port RAM and update it periodically. This way, if the application microcontroller fails, the communication microcontroller can send the state to another node that could continue its tasks (if it is connected to the same set of sensors and actuators). For this mechanism to be useful, it is important to obtain high error coverage, mainly for the application microcontroller (the most complex component).

In case of failure of the dual-port RAM, the microcontrollers can continue the correct functioning due to the direct connection through parallel ports or local serial channels. (In this case the node state must be stored in the local memory of both microcontrollers).

In order to obtain high error coverage, the communication microcontroller fulfils, also, functions of watchdog processor. The communication microcontroller can detect control flow errors in the other processor and inhibit possible outputs to the actuators and possible error messages

to the network (this is known as fail-silent behaviour).

Also, DICOS system proposes to use the *watchdog* processor for another new function: to analyse the tasks execution time and to detect if some task misses its deadline. In order to achieve this behaviour th watchdog processor uses the signatures that th application microcontroller sends periodically through the dual-port RAM to the watchdog. The application microcontroller, as function of its execution flow, sends a signature that identifies the exact execution point. If the watchdog processor, that has the information about the correct application flow, does not receive this signature before a time-out occurs, it will detect an error in the application. Also, if the signature is not correct (it is not the next expected signature) an error is also detected. In order to use this mechanism to detect missing deadlines th watchdog processor only has to know the signatures sequence and the deadlines between signatures (or some of them). So, the watchdog processor can kno when a task begins, when it ends and the time employed. With all this information, when an error is detected by one of the error-detection mechanisms, the watchdog processor can detect if a task will miss its deadline (in function of the recovery strateg needed) and to execute the operations needed to offe a safe-result. Obviously the watchdog processor has control over the application microcontroller in order to command the actions to do in those cases.

From the practical point of view the main node components are:

- 16-bit microcontroller for the execution of the control algorithms. Siemens C167CR (with built-in CAN controller, 9 I/O ports, A/D converter, 144 pins, ...) is used. The sensors and actuators are connected to it. This microcontroller accesses to 256 KB of stati RAM and 256 KB of Flash memory in our actual implementation.

- 8-bit microcontroller devoted to the communications and watchdog. In this case, the Intel 251 is used (8051 improvement). This microcontroller accesses to 64 KB of static RAM and 128 KB of Flash memory in our implementation.

- Dual port RAM (16K×8) to pass information between both microcontrollers and to store the node state.

- Two communication networks based on the CAN protocol.

- A serial RS-232 is included for both microcontrollers in order to make local serial channels possible in each node (if they are needed) and also, to allow software updates via RS-232 for both microcontrollers.

More information about this architecture and its theoretical dependabilit Markov model can be found in (Campelo, *et al.*, 1997; Campelo, *et al.*, 1998).

Architecture II

The second node architecture offers better reliabilit (Fig. 3). It modifies the previous one with th addition of a second 16-bit microcontroller for the application (spare), so the system can recover from a permanent error. If the communication microcontroller detects a permanent error in the main application microcontroller, it will stop it and run the spare.

Also, when it is not (physically) possible for sensors and actuators to be connected to several nodes, a structure more reliable is needed. So this architecture continues its function in the presence of a permanent fault in one of the application microcontrollers and tolerates, as the previous one, transient faults.

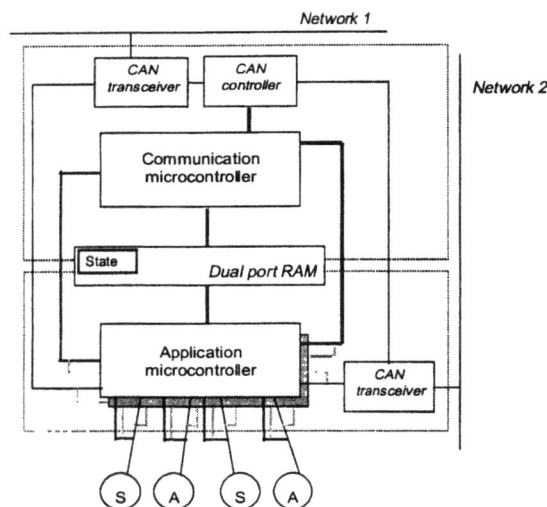

Fig. 3. Architecture II

As the first architecture, more information and the theoretical dependability Markov model of this architecture and a comparison with other node structures can be found in (Campelo *et al.*,. 1997; Campelo *et al.*,. 1998). In Fig. 4 a photograph of a DICOS architecture-I node can be seen.

Error Detection Mechanisms

The error detection mechanisms included in DICOS nodes can be classified in internal and external. Those which are built-in in the microcontroller hardware will be internal and the external will be the mechanisms included in the watchdog processor and software mechanisms in the application microcontroller. These mechanisms are:

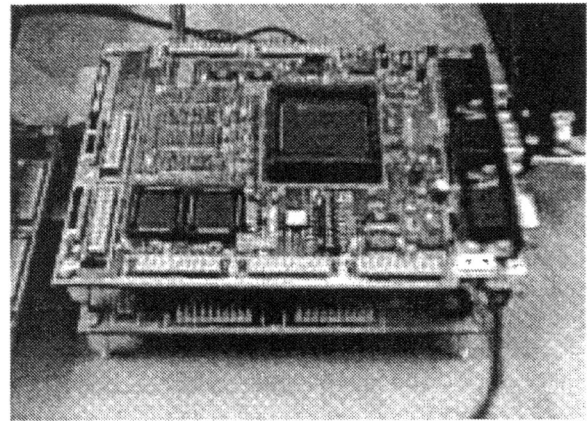

Fig. 4. DICOS node

- ***Internal mechanisms:***
  - Stack overflo : whenever the stack pointer is decremented to a value which is less than the value in the stack overflo register the microcontroller will enter th stack overflow trap routine.

  - Stack underflow: whenever the stack pointer is incremented to a value which is greater than the value in the stack underflow register, the microcontroller enters the stack underflow trap routine.

  - Undefined operation code: when the instruction currently decoded by the microcontroller does not contain a valid C167 operation code, it enters in the undefined operation code trap routine.

  - Protection fault: whenever one of the special protected instructions is executed where the operation code of that instruction is not repeated twice in the second word of the instruction and the byte following the operation code is not the complement of the operation code, the microcontroller enters in the protection fault trap routine. Examples of protected instructions are th *software reset, power down* or *idle mode* instructions.

  - Illegal word operand access: whenever a word operand read or write access is attempted to an odd byte address the microcontroller enters in the illegal word operand access trap routine.

  - Illegal instruction access: whenever a branch is made to an odd byte address the microcontroller enters in the illegal instruction access trap routine.

  - Illegal external bus access: whenever the microcontroller requests   an external instruction fetch, data read or data write, and no external bus configuration has been specified, it enters in the illegal bus access trap routine.

- ***External mechanisms:***

  - Control flow error The communication microcontroller, acting as watchdog processor, can detect an incorrect flow of the application if this microcontroller sends a signature after a timeout occurs o if this signature it is not the expected.

  - Deadline error: The communication microcontroller, acting as watchdog processor, can detect missing deadlines, using the same signatures the application microcontroller sends.

  - Fault tolerant software mechanisms : DICOS system can support different software techniques to improve error detection capabilities. Actually double task execution is used, but other schemes are possible.

4. CONCLUSIONS

In this paper a new distributed system for embedded real-time applications is presented. This system has as main characteristic its high reliability and safet achieved with low cost. It is obtained due to the use of the communication microcontroller as watchdog processor. As a novelty, the watchdog processor, besides control the execution flow of the application can take into account the execution time of tasks. So, this processor can detect missing deadlines.

As future work, an experimental validation of thi architecture will be made. In order to obtain the error coverage and latency time of errors, different fault injectors will be use. An interesting study will be the analysis of the fault duration in the proportion of missed deadlines.

5. ACKNOWLEDGEMENTS

This work is supported by the Spanish Comisión Interministerial de Ciencia y Tecnología under project CICYT TAP99-0443-C05-02.

6. REFERENCES

Arlat, J., M Aguera, Y. Crouzet, J.C. Fabr , E. Martins, and D Powell (1990). Experimental evaluation of the fault tolerance of an atomic multicast system. In: *IEEE Transactions on Reliability*, **39**, nº 4, pp. 455-466.

Baba, M.D., H Ekiz, A. Kutlu and E.T. Powner (1996). Toward adaptable distributed real-time computer systems. In: *Proceedings 3rd International Workshop on Real Time Computing Systems and Applications (RTCSA'96)*, pp. 170-175. Seoul, Korea.

Bosch (1991) *CAN Specification Version 2.0*. Robert Bosch GmbH

Campelo, J.C., F Rodríguez, J.J. Serrano and P.J. Gil (1997). Dependability evaluation of fault tolerant architectures in distributed industrial control systems. In: *Proceedings 2nd IEEE Workshop on Factory Communication Systems*, pp. 193-200. Barcelona, Spain.

Campelo, J.C., A Rubio, F. Rodríguez and J.J. Serrano (1998). Fault tolerance in distributed industrial control systems. In: *Proceedings Western Multiconference on Computer Simulation*, pp. 87-92. San Diego, USA.

Grans, K. (1998). DUSBER: a new fault tolerant technique combining duplication and recovery In: *Proceedings Ninth European Workshop on Dependable Computing, (EWDC-9)*, pp.61-65. Gdansk.

Hecht, M., J. Agron and S. Hochhauser (1989). A distributed fault tolerant architecture for nuclear reactor control and safety functions. In: *Proceedings 10th Real Time Systems Symposium*, IEEE Computer Society Press, pp-214-221.

Karlsson, J., P. Folkesson, J. Arlat, Y. Crouzet, G. Leber and J. Reisinger (1995). Application of three physical fault injection techniques to the experimental assessment of the MARS architecture. In: *Proceedings Fifth IFIP Working Conference on Dependable Computing for Critical Applications (DCCA-5)* , pp. 150-161. Urbana-Champaign, Illinois, EEUU

Kopetz, H., R Hexel, A. Krüger, et al. (1997). A prototype implementation of a TTP/C controller. In: *Proceedings Society of Automotive Engineers. SAE Congress and Exhibition*, Detroit, MI, EEUU. SAE paper number 970296.

Kopetz, H., H Kantz, G. Grünsteidl, P. Puschner and J. Reisinger (1990). Tolerating transient faults in MARS. In: *Proceedings 20th International Symposium on Fault Tolerant Computing (FTCS-20)*, pp. 466-473. Newcastle upon Tyne.

LIS (1996). *Guide de la sûreté de fonctionnement*. Laboratoire d'Ingenierie de la Sûreté de fonctionnement (LIS). Cépaduès – Éditions.

Rostamzadeh, B., H Lonn, R. Snedsbol and J. Torin, (1995). DACAPO: a distributed computer architecture for safety-critical control application. In: *Proceedings of IEEE Int. Symposium on intelligent vehicles*, pp. 376-381. Detroit, USA.

Rostamzadeh, B. and J. Torin (1995a). Design principles of fail-operational/fail-silent modular node in DACAPO. In: *Proceedings of the International Conference on Electrical Engineering (ICEE)*, Tehran, Iran.

Rostamzadeh, B. and J. Torin (1995b). A study of four configurations for design of a fault-tolerant computation unit in DACAPO, *Technical Report 222*. Department of Computer Engineering, Chalmers, Sweden.

APPLICATION-LEVEL TIME-OUT SUPPORT FOR REAL-TIME EMBEDDED SYSTEMS

Vincenzo De Florio, Geert Deconinck, Rudy Lauwereins

Katholieke Universiteit Leuven
Electrical Engineering Department, ACCA Division,
Kard. Mercierlaan 94, B-3001 Leuven, Belgium

Abstract: A common requirement for many applications and services like, for example, membership protocols, is the availability of a class of functions for managing timeouts, i.e., objects that schedule an event, typically a function call, to be generated after a given amount of time. This paper describes an application-level time-out management system that exploits multiple alarm execution threads in order to reduce alarm execution congestion and consequent run-time violations. Through some experimental results it is shown how and in which cases this system can be made able to fulfill real-time requirements in spite of this congestion. *Copyright © 2000 IFAC*

Keywords: Embedded systems, real-time systems, software tools, alarm systems, parallel processing.

1. INTRODUCTION

A class of C functions implementing an application-level time-out management system is herein presented and analyzed. The tool, called TOM (time-out manager), has been developed in the framework of the ESPRIT projects "EFTOS" (Embedded Fault-Tolerant Supercomputing) (Deconinck et al., 1997) and "TIRAN" (tailorable fault tolerance frameworks for embedded applications) (Botti et al., 1999). In this context, a time-out is an object which postpones a function call by a certain number of local clock ticks. The class essentially manages a list of these objects, ordered by clock-ticks-to-expiration. When the specified amount of time elapses for the top of the list, its function—let us call it "alarm"—is executed and the object is either thrown out of the list or renewed (in this second case, a time-out is said to be "cyclic"). A special thread monitors and manages one or more such lists, checking for the expiration of the entries

in the time-out list. This is done periodically at a fixed period which constitutes a trade-off between performance and the upper bound for function call delay. The mechanism also exploits multiple alarm execution threads in order to reduce the congestion due to concurrent alarm execution and the consequent run-time violations.

A promising model for solving problems like, for instance, dynamic membership (Cristian and Schmuck, 1995) in distributed systems has been recently introduced (Cristian and Fetzer, 1999). The availability of a class of functions for managing time-outs is a fundamental requirement of that model.

Section 2 presents the application programmer's interface and the design of the TOM system. Real-time issues are dealt with in Sect. 3. Section 4 closes this work with a brief summary.

2. A TIME-OUT MANAGEMENT SYSTEM

This section describes the architecture, the application programmer's interface (i.e., the client-side

---

[1] Partially supported by the ESPRIT-IV project 28620 "TIRAN". Geert Deconinck is a Postdoctoral Fellow of the Fund for Scientific Research - Flanders (Belgium) (FWO).

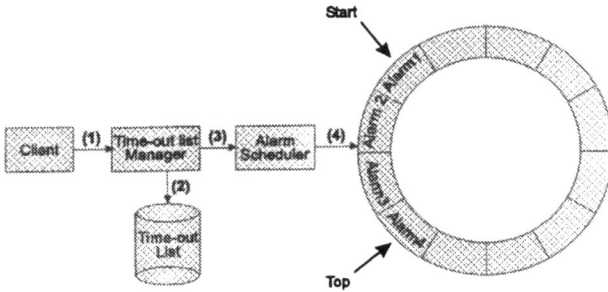

Fig. 1. Architecture of the time-out management system.

view), as well as the server-side protocol of the TOM class of C functions. TOM can be regarded as a client-server application such that the client issues requests according to a well-defined protocol, while a server thread fulfills these requests registering, updating, modifying, purging entries of the time-out list, also executing the corresponding alarms. This server side is totally transparent to the user module (the client).

2.1 The Architecture of the TOM System

Figure 1 shows TOM's architecture. In

(1), the client process sends requests to the time-out list manager via the API to be described in Sect. 2.2; in

(2), the time-out list manager accordingly updates the time-out list with the server-side protocol to be described in Sect. 2.3;

(3) each time a time-out reaches its deadline, a request for execution of the corresponding alarm is sent to a task called alarm scheduler ($\mathcal{AS}$);

(4) this latter allocates an alarm request to the first available thread out of those in a circular list of alarm threads ($\mathcal{AT}$'s), possibly waiting until such a thread becomes available.

The availability of $\mathcal{AS}$ and $\mathcal{AT}$'s can have positive consequences on fulfilling real-time requirements. These aspects are dealt with in Sect. 3.

2.2 Client-Side Protocol

The TOM class appears to the user as a couple of new types and some function calls. The first of these types comes into action the moment the user starts using the time-out management service. Specifically, to declare a time-out manager, the user needs to define a pointer to a TOM object. This is more or less like defining a FILE pointer in standard C: TOM *tom.

Just as, within the standard C class "FILE", function fopen attaches an object to that pointer and opens a connection with the file system, likewise

function tom_init defines a time-out manager object, and sets up a connection with TOM's server-side:

int alarm(TOM *);   tom = tom_init(alarm);

(function alarm is the default function to be called when a time-out expires. A class-specific variable stores this function pointer. This default value can be changed afterwards setting another, object-specific, variable). Function tom_init is node-specific; that is, the first time it is called on a node, a custom thread is spawned on that node. That thread is the actual time-out manager, that is, TOM's server-side.

At this point the user is allowed to define time-outs. This is done via type timeout_t and function tom_declare; an example follows:

timeout_t t;
tom_declare(&t,TOM_CYCLIC, TOM_ENABLE, TID, DEADLINE).

In the above, time-out t is defined as:

- a cyclic time-out (renewed on expiration; as opposed to TOM_NON_CYCLIC, i.e., purged on expiration),
- enabled (only enabled time-outs "fire", i.e., call their alarm on expiration; an alarm is disabled with TOM_DISABLE),
- with a deadline of DEADLINE clock ticks before expiration.

Furthermore, time-out t is identified by integer TID.

Once defined, a time-out can be submitted to the time-out manager for insertion in its running list of time-outs—this will be explained in more detail in Sect. 2.3. From the user viewpoint, this is managed by calling function tom_insert.

After successful insertion (to be tested via the return value of tom_insert), and in case no further control is specified, an enabled time-out will trigger the call of the default alarm function after the specified deadline. This behaviour would be cyclic if the original time-out is defined as such. Further control is still possible though. For instance, a time-out can be temporarily suspended while in the time-out list via function tom_disable and (re)-enabled via function tom_enable.

Furthermore, via similar functions, the user is allowed to specify a function other than the default one to be called upon expiration of the deadline (function tom_set_action), as well as to specify new deadline values (function tom_set_deadline). Other functionality of our tool includes the capability of deleting a time-out from the list (tom_delete) as well as the ability of *renewing* a time-out— which substantially means removing a time-out and inserting it with the original values set

Table 1. An example of usage of the TOM class.

```
1.   /* declarations */
     TOM *tom; timeout_t t1, t2;
     int my_alarm(TOM*), another_alarm(TOM*);
2.   /* definitions */
     tom ← tom_init(my_alarm);
     tom_declare(&t1, TOM_NON_CYCLIC, TOM_ENABLE,
                 TIMEOUT1, DEADLINE1);
     tom_declare(&t2, TOM_CYCLIC, TOM_DISABLE,
                 TIMEOUT2, DEADLINE2);
     tom_set_action(&t2, another_alarm);
3.   /* insertion */
     tom_insert(tom, &t1),
     tom_insert(tom, &t2);
4.   /* control */
     tom_enable(tom, &t2);
     tom_set_deadline(&t2, NEW_DEADLINE2);
     tom_renew(tom, &t2);
     tom_delete(tom, &t1);
5.   /* deactivation */
     tom_close(tom);
```

again (tom_renew). Function **tom_close** terminates the time-out management service also performing garbage collection. Table 1 summarizes the above client-side protocol with an example.

2.3 Server-Side Protocol

The server-side protocol is run by a component called time-out list manager (TOLM). TOLM basically checks every **TOM_CYCLE** whether

- there are any incoming requests for manipulating the time-out list; if so, it deals with those requests;
- there are time-outs whose deadlines are reached. If so, it manages the execution of the corresponding alarm.

This section describes the server-side protocol of TOLM, i.e., the way TOLM manages the list of time-outs.

Each time-out **t** is characterized by its *deadline*, viz. t.deadline, a positive integer representing the number of clock ticks that must separate the time of insertion or renewal from the scheduled time of alarm. This field can only be set by functions tom_declare and tom_set_deadline. Each time-out t holds also a field, t.running, initially set to t.deadline.

Each time-out list object, say **tom**, hosts a variable representing the origin of the time axis. This variable, **tom.starting_time**, is related in particular to the time-out at the top of the time-out list— the idea is that the top of the list be the only entry whose running field needs to be compared with current time in order to verify the occurrence of the time-out-expired event. For the time-outs behind the top one, that field represents relative

values, viz., distances from expiration time of the closest, preceding time-out. In other words, the overall time-out list management aims at isolating a "closest to expiration" time-out, or head time-out, that is the one and only time-out to be tracked for expiration, and at preserving the coherence of a list of "relative time-outs".

Let us call TimeNow the system function returning the current value of the clock register. In an ordered, coherent time-out list, residual time *for the head time-out* t, is given by

$$\text{t.running} - (\text{TimeNow} - \text{tom.starting\_time}), \quad (1)$$

that is, residual time minus time already passed by. Let us call quantity (1) as r_1, or head residual. For time-out n, $n > 1$, that is for the time-out located $n - 1$ entries "after" the top block, let us define

$$r_n = r_1 + \sum_{i=2}^{n} \text{t}_i.\text{running} \quad (2)$$

as the n-th residual, or residual time for time-out at entry n. If there are m entries in the time-out list, let us define $r_j = 0$ for any $j > m$.

It is now possible to formally define the fundamental operations on a time-out list: insertion and deletion of an entry.

2.3.1. Insertion.
Basically there are three possible insertions, namely on top, in the middle, and at the end of the list.

2.3.1.1. Insertion on top.
In this case a new time-out object, say t, has to be inserted on top, such that $t.\text{deadline} < r_1$, or whose deadline is less than the head residual. Let us call u the current top of the list. Then the following operations need to be carried out:

$$\begin{cases} t.\text{running} \leftarrow t.\text{deadline} + \text{TimeNow} - \\ \qquad\qquad \text{tom.starting\_time} \\ u.\text{running} \leftarrow r_1 - t.\text{deadline}. \end{cases}$$

Note that the first operation is needed in order to verify relation

$$t.\text{running} - (\text{TimeNow} - \text{tom.starting\_time})$$
$$= t.\text{deadline},$$

while the second operation aims at turning the absolute value kept in the **running** field of the "old" head of the list into a value relative to the one stored in the corresponding field of the "new" top of the list.

2.3.1.2. *Insertion in the middle.* In this case a time-out t such that, for some j,

$$r_j \le t.\text{deadline} < r_{j+1},$$

is to be inserted. Let us call u time-out $j+1$. (Note that both t and u exist by hypothesis). Then the following operations need to be carried out:

$$\begin{cases} t.\text{running} \leftarrow t.\text{deadline} - r_j \\ u.\text{running} \leftarrow u.\text{running} - t.\text{running}. \end{cases}$$

Observation 1. Note how, both in the case of insertion on top and in that of insertion in the middle of the list, time interval $[0, r_m]$ has not changed its length. It has, however, been further subdivided, and is now to be referred to as $[0, r_{m+1}]$.

2.3.1.3. *Insertion at the end.* Let us suppose the time-out list consists of $m > 0$ items, and that a time-out t has to be inserted, with $t.\text{deadline} \ge r_m$. In this case the item is simply appended and initialized so that

$$t.\text{running} \leftarrow t.\text{deadline} - r_m.$$

Observation 2. Note how insertion at the end of the list is the only way to prolong the range of action from a certain $[0, r_m]$ to a larger period $[0, r_{m+1}]$.

2.3.2. *Deletion.* The other basic management operation on the time-out list is deletion. Three types of deletions can be found: deletion from the top, from the middle, and from the end of the list.

2.3.2.1. *Deletion from the top.* A singleton list is a trivial case. Let us suppose there are at least two items in the list. Let us call t the top of the list and u the next element, the one that will be promoted to top of the list. From its definition it is known that

$$r_2 = u.\text{running} + r_1$$
$$= u.\text{running} + t.\text{running} - $$
$$(\text{TimeNow} - \text{tom.starting\_time}). \quad (3)$$

By (1), the bracketed quantity is elapsed time, so the amount of absolute time units that separate current time from the expiration time is given by $u.\text{running} + t.\text{running}$. In order to "behead" the list t needs to be updated as follows:

$$u.\text{running} \leftarrow u.\text{running} + t.\text{running}.$$

2.3.2.2. *Deletion from the middle.* Let us suppose that there are two consecutive time-outs in the list, t followed by u, such that t is not the top

of the list. Again, before actually removing t from the list, the following step is required:

$$u.\text{running} \leftarrow u.\text{running} + t.\text{running}.$$

2.3.2.3. *Deletion from the end.* Deletion from the end means deleting an entry which is not referenced by any further item in the list. Physical deletion can be performed with no need for any updating. Only the interval of action is shortened.

Observation 3. The variable **tom.starting_time** is never touched when deleting from or inserting entries into a time-out list, except when inserting the first element, when it is set to the current value returned by TimeNow.

3. REAL-TIME SUPPORT

What has been reported so far applies in the ideal situation such that

- all management operations occur instantaneously, that is, inducing no real-time delay,
- time-out detection delay is negligible,
- other threads/processes running on the same processing node have negligible influence on the performance of the manager,
- alarm execution delay is negligible.

This is in general far from being true. Management tasks do impose a non-negligible overhead, time-out checks occur once every **TOM_CYCLE** clock ticks (one may reduce this value, though this is detrimental to performance, because this increases the frequency of checking, which implies more context switches and more CPU cycles assigned to TOM), overloading the process with threads results in a smaller time slice per thread, and, last but not least, the alarm function is a big source of non determinism due to the fact that it is defined by the user. In our experience, for instance, the alarm function is often related to some communication task—typically the alarm triggers the transfer of a message, possibly across the network. Frequently this happens through synchronous links and results in blocking the sender until the receiver is ready to receive and the transfer has been made.

In conclusion, there is a non negligible delay between the time at which r_1 becomes zero and the time that event is managed; such delay can be expressed as

$$(\text{TimeNow} - \text{tom.starting\_time}) - t.\text{running}.$$

This quantity must be immediately propagated to those entries who follow the top of the list. This is done by determining the integer $j \ge 1$ such that

$$r_j < 0 \quad \wedge \quad r_{j+1} \ge 0, \quad (4)$$

where this time r_1 can also be negative. The time-out management thread needs therefore first to check whether r_1 is less than zero; if so, it must calculate index j such that (4) is verified; and finally command the execution of all the corresponding alarm functions.

Finally, if the list is not empty, that thread must adjust the corresponding **running** field as follows: let t be time-out $j + 1$; then

$$t.\text{running} \leftarrow t.\text{running} + r_j.$$

Clearly the above mechanism only works well if there is a way to keep under control the congestion that is due to alarm execution. The rest of this section describes a mechanism with this aim.

Alarm execution is managed via the mechanism shown in Fig. 1, points (3) and (4), i.e., through the $\mathcal{AS}$, which gathers all alarm execution requests and forwards them to the next entry of a circular list of threads. As already said, alarms often imply communications, therefore using a list of concurrent threads might in principle result in better performance, should the underlying system offer means for managing I/O in parallel. In general, if the alarms do not compete "too much" for the same resources at the same times, this scheme should allow a better exploitation of the available resources as well as a higher probability of controlling alarm congestion. The present Section deals with the estimation of the capability of this mechanism to control alarm congestion under different levels of congestion and in two opposite cases of alarm interference, i.e., no competition and full competition.

In order to estimate the average run-time delay imposed by TOM, the following experiment has been performed: 1000 non-cyclic time-outs have been generated pseudo-randomly with a deadline uniformly distributed in time interval $]0, T]$. Each time a time-out triggered an alarm, the difference between expected and real time of execution of the alarm has been computed. Let us call δ this value. TOM's period, i.e., **TOM_CYCLE** was set to 50000 clock ticks, 1 clock tick being 1μs on the target machines. The experiment was performed on a single node of a Parsytec PowerXplorer, using a PowerPC 601 at 66MHz, and has been repeated on a Parsytec CC system, with a PowerPC 604 at 133MHz. More or less the same results have been observed on both machines (the clock frequency only influences alarm list management times, which is a small percentage of processing time).

In order to measure the alarm congestion capability of TOM, the experiment has been repeated configuring TOM with no threads in the circular list of $\mathcal{AT}$'s (alarm execution managed by TOLM)

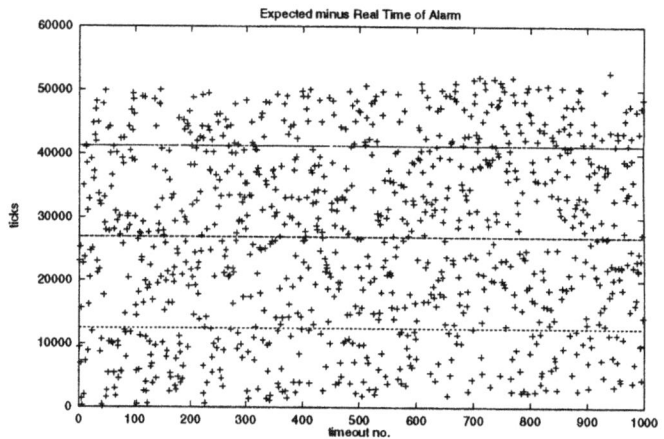

Fig. 2. 1000 time-outs are uniformly generated in $]0, 100]$s. In this case $\delta \approx 50\mu$s and $\tau = 0$. The maximum value is 52787, minimum is 157. The average is 26892.74, standard deviation is 14376.89; 20 of the 1000 time-outs exceeded **TOM_CYCLE**.

and later adding more and more threads in that list. Let us call τ the number of threads in the list.

Furthermore, the following overheads have been artificially imposed on the alarm functions: nearly no overhead (just that of calling a function and copying a 20-byte message—this is the typical delay used in most of our applications, lasting approximately 50μs on a DEC Alpha board using the TEX nanokernel (Anonymous, 1997b),—10ms, 100ms, and so on. This overhead has been imposed either by loading the alarms with some purely computation-oriented tasks, in order to make them compete for the same resource (being the only CPU in the system) or by executing a function (**TimeWait**) that puts the calling thread in the wait state for the specified amount of time—this way alarms do not compete at all with each other.

Figure 2 summarizes the results of the first experiment. The experiments show that minimal (effective) overhead is imposed on the alarms of 1000 time-outs whose deadline is generated pseudo-randomly in $]0, 100$s]. TOLM directly executes alarms in this case ($\tau = 0$).

3.1 Best case scenario.

Another experiment aimed at evaluating what happens when δ is increased while keeping τ equal to zero and there is no competion among the alarms. The results have been also summarized in Table 2 which reports the number of items that resulted in a real-time violation (alarm delay greater than **TOM_CYCLE** clock ticks) as well as other statistics. Particularly meaningful is the largest violation, which, in the case of a delay of

Table 2. The table reports on experiments aiming at measuring alarm execution congestion. The first column represents the duration of each alarm.

delay	items	average	stdev	max
$\approx 50\mu s$	20	50967	762	52782
1ms	34	51282	976	53263
5ms	77	53078	2774	64328
10ms	132	57606	6918	88183
20ms	238	65640	14572	124535

Table 3. A run with $\delta = 20$ms and $0 < \tau \le 5$. μ and σ are resp. the average and std. deviation of the 1000 outcomes. γ is the number of time-outs for which real time of alarm minus expected time of alarm was > TOM_CYCLE (50000 clock ticks). μ' and σ' are resp. the average and std. deviation of the γ time-outs. "max" represents the largest outcome, i.e., the worst run-time violation.

τ	μ	σ	γ	max	μ'	σ'
0	35963	22683	264	130993	65483	14160
1	30201	18447	108	108991	64676	15648
2	27557	14472	46	63659	52228	2927
3	27287	14370	45	53934	51281	1015
4	27494	14058	47	53372	51526	922
5	27423	14080	46	54077	51357	1079

Table 4. A run with $\delta = 100$ms and $0 < \tau \le 5$.

τ	μ	σ	γ	max	μ'	σ'
0	3692159	2966039	974	12277216	3790058	2943376
1	33674	27688	140	264882	83245	38052
2	28177	16802	54	146684	66738	20706
3	27622	14277	45	100260	52872	7359
4	27435	14087	44	53712	51376	1001
5	27476	14107	44	53992	51444	1009

20ms, was equal to 124535 clock ticks, or 2.49 times TOM_CYCLE.

In particular, when the duration of the alarm became larger than TOM_CYCLE (the checking quantum of the time-out list manager), an ever increasing violation of real-time requirements due to alarm execution congestion has been observed.

It has been experimentally found that the presence of the $\mathcal{AS}$ and of the circular list of threads has a great consequence on alarm congestion control: Table 3 for instance summarizes the results of increasing τ from 0 to 5 when δ is set to $20\mu s$. In particular when τ is equal to 2, the number of "run-time violations", i.e., the number of time-outs for which the true time of execution of the alarm minus the expected time of execution of the alarm is greater than TOM_CYCLE, drops from 264 to 46 items, while with $\tau = 3$ the worst case violation was equal to 53934 clock ticks, or just 1.07868 times TOM_CYCLE—a violation of something less than 8%. Table 4 shows a similar behaviour when δ is larger than TOM_CYCLE.

Table 5. A run with $\delta = 20$ms and $0 < \tau \le 2$. This time $\mathcal{AT}$'s do compete for a unique system resource, so adding threads does not improve the behaviour.

τ	μ	σ	γ	max	μ'	σ'
0	35252	21667	226	126931	66224	13713
1	35533	21584	248	127403	64378	13803
2	35735	21607	250	125910	64681	13306

3.2 *Worst case scenario.*

The worst case is when all $\mathcal{AT}$'s compete for the same set of resources at the same time. An easy way to accomplish this has been to let each alarm function perform a pure integer computation task so that each thread competes with all the others for the single CPU of our systems. Given a fixed $\delta = 20$ms, τ has been increased and the corresponding real-time violations have been measured. The results have been summarized in Table 5. Adding threads did not produce any useful result in this case. Note how, adopting a parallel system, one could automatically and transparently schedule the parallel execution of alarms on the available processors to further enhance the ability of TOM to fight alarm execution congestion.

4. CONCLUSION

A software system for the management of time-outs at application-level has been presented. It has been shown how exploiting an asynchronous mechanism for the execution of alarms can reduce and, in some cases, solve the problem of alarm congestion even under heavy load conditions—long lasting alarm functions. The tool, currently available for Windows and Parsytec EPX machines and on a DEC Alpha board, is compliant to the timed-asynchronous distributed system model (Cristian and Fetzer, 1999). As such, it can serve as a basic tool for developing dependable embedded services compliant to that model.

5. REFERENCES

Anonymous (1997b). *TEX User Manual*. TXT Ingegneria Informatica. Milano, Italy.

Botti, O. et al. (1999). TIRAN: Flexible and portable fault tolerance solutions for cost effective dependable applications. In: *Proc. of EuroPar'99, LNCS* Vol. 1685. pp. 1166–1170.

Cristian, F. and C. Fetzer (1999). The timed asynchronous distributed system model. *IEEE Trans. P.D.S.* **10**(6), 642–657.

Cristian, F. and F. Schmuck (1995). Agreeing on processor group membership in asynchronous distributed systems. Technical Report CSE95-428. UCSD.

Deconinck, G., et al. (1997). Industrial embedded HPC applications. *Supercomputer* **XIII**(69), 23–44.

TOWARDS UNIFIED COMPOSITIONAL DESIGN OF CONTROL SYSTEMS

Jüri Vain, Marko Kääramees

Institute of Cybernetics, Tallinn Technical University

Abstract: In this paper an approach is developed for systematic construction of control system design specifications. Our approach is based on the notion of generic component, which serves as a reference model for design refinement and allows us to represent hybrid dynamics and dependability properties of components in the uniform methodological framework. Applicability of this approach is demonstrated on a water temperature control system design using PVS system. *Copyright© 2000 IFAC*

Keywords: design specification, compositional verification, generic component, correctness, hybrid dynamics, dependability.

1. INTRODUCTION

Compositional specification and verification methods are increasingly applied for design of real-time programs (de Roever, 1998). Their main advantage in case of large parallel programs is the opportunity to avoid the state explosion problem. Therefore compositional methods combined with data abstraction techniques have a good prospect to become an effective weapon for attacking real industrial-scale design tasks. An extensive effort has been also made to build proof assistance tools relying on these techniques (Berghammer and Lakhnech, 1998). Still, quite often the most time-consuming part in tool supported verification is not the search of proofs itself but definition of proof obligations which is dependent on particular application, specification style and methodological framework of design.

One solution to that problem discussed in this paper is a construction of proof obligations from requirements and design specifications using semantics-based specification schemas which are recursively applicable throughout many design phases. As a constructive guide for getting such schemas certain reference models can be used like RM-ODP (The Reference Model of Open Distributed Processing) for distributed systems, CIM OSA (Open System Architecture for CIM) for computer integrated manufacturing systems etc.

In design of control systems some specific aspects must be represented in the reference model. First, for verification of behavioral properties the hybrid dynamics has to be specified, and second, if safety critical applications are considered, dependability properties have to be related to the specification of functionality.

The reference model introduced in this paper for compositional specification and verification of control systems is called a *generic component*. The generic component is an abstraction of typical control system components such as sensors, controllers, actuators and their compositions. It can be used in design steps where hybrid behavior and fault-tolerance are design concerns and have to be specified explicitly.

The paper consists of the following parts. In Section 2, a generic component model is defined. Section 3 explains how to construct tractable sets of abstract states for systems modeled with generic component. In Section 4, possible refinement scenarios of a component hybrid dynamics are discussed. Section 5 introduces a structure for specification of dependability properties. Section 6 illustrates the whole approach with a simple control system design example.

2. GENERIC COMPONENT

The generic component model is represented using two-level abstraction (see figure 1). The higher level model carries only the information necessary for component interface specification. The lower level model, being structural refinement of the higher level model, introduces subcomponents of fixed type which are necessary for separation of different design concerns. For simplicity the case is observed where the multiplicity of subcomponents of the same type is reduced to one. Types of subcomponents are the following:

- Functional (*F*-) subcomponent – models basic functionality of the component;

- Local environment (*LE*-) subcomponent – models specific to that component influence of the environment which induces faults and degradation of the component;

- Corrector (*CR*-) subcomponent (optional) – models fault-tolerance of the component, e.g., correction of faults and compensation of component degradation if this is a design concern.

Note: Subcomponent types introduced above are only for semantic distinction of subcomponents. Syntactically all subcomponents are indistinguishable and can be specified and refined (along further design steps) exactly as the component of higher level model.

A component description is started by definition of its state space. Due to compositionality the only observable state variables of the upper level model are component inputs and outputs. In general, the specification of a component *C* is triple

$$spec\ C \equiv < I, O; R(I,O) > \qquad (1)$$

where
I, O – specification of observable inputs, outputs

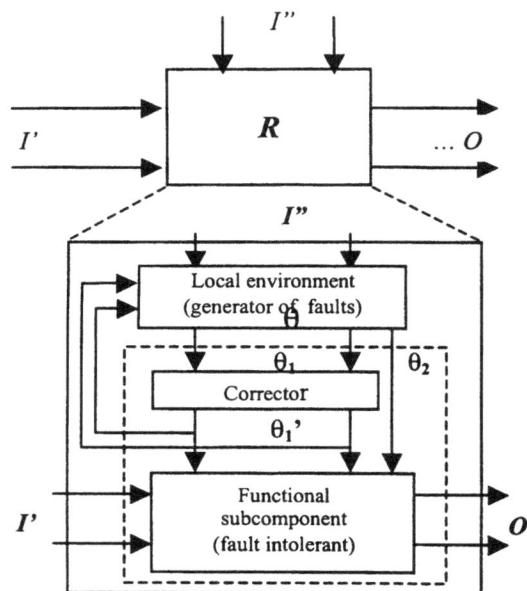

Fig. 1. Generic component model

(signals, disturbances, etc) and their types;
$R\ (I,O)$ – set of (static and dynamic) input-output relations.

The set of inputs is split into two subsets: **I'**-modeling functional inputs and **I''**-modeling impacts from the component's environment which cause faults and degradation.

An important modeling assumption is independence of all inputs. If some of the inputs have qualitative co-effects, then the introduction of an extra component is required with these inputs and with the output, which models their co-effects.

To achieve better modularity of the specification, the input-output relation is split by outputs into a set of simpler relations:

$$R(I,O) \equiv \bigcup_{j=1}^{m} R_j(I_1,.., I_n; O_j)$$

where $\#I = n, \qquad \#O = m$.

In the following sections elements of the specification (1) are examined in detail.

3. CONSTRUCTION OF ABSTRACT STATES

Control system components interact with real-world processes. To transform the infinite state space of components to a finite one, the abstract states are constructed using partition refinement (Spelberg, *et al.*, 1998) on domains of input and output variables.

The aim of partitioning is to achieve the minimum number of symbolic states without loss of important state information. For traditional control system components (sensors, actuators, controllers) at least two kinds of characteristics can be given e.g.,

- domain and range of component transfer function;

- expected lifetime depending on operation conditions.

It is demonstrated in the following how to use them as state partition criteria. Let $\mathbf{I}_i$ denote the set of all physically possible values of input I_i. By the first criteria, two input regions can be distinguished. Domain of transfer function defines those input values ($I_i \in \mathbf{I}_i^R$) which guarantee that the output value O_j depending on it is in the expected range ($O_j \in \mathbf{O}_j^R$), i.e., satisfies the specification of transfer function. If the value of i-th input is out of the domain ($I_i \in \mathbf{I}_i^U$, where $\mathbf{I}_i^U = \mathbf{I} \setminus \mathbf{I}_i^R$), nothing can be said about the component output. This provides a partition of the input domain usually into three parts $\mathbf{I}_i^{U1}, \mathbf{I}_i^R, \mathbf{I}_i^{U2}$ (where $\mathbf{I}_i = \mathbf{I}_i^{U1} \cup \mathbf{I}_i^R \cup \mathbf{I}_i^{U2}$ and $\mathbf{I}_i^{U1} \cap \mathbf{I}_i^R \cap \mathbf{I}_i^{U2} = \emptyset$) (see Figure 2).

Similarly, using second partition criteria the input domain is split into regions of the values that cause

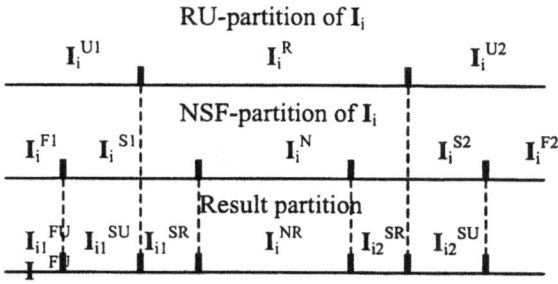

Fig. 2. Construction of combined partitions

qualitatively different degradation of the component. Thus, instead of expected lifetime its dual–approximate degradation rate is used. Here the following input regions can be distinguished: normal mode ($I_i \in \mathbf{I}_i^N$) - input values do not cause any observable degradation of the component; stress mode ($I_i \in \mathbf{I}_i^S$) - the input values cause forced degradation of component quality characteristics; and the failure mode ($I_i \in \mathbf{I}_i^F$) - the component totally looses its functionality.

Note: The state splitting may define not necessarily convex intervals, e.g., as shown in figure 2. In both sides of convex region $\mathbf{I}_i^N$ are sub-intervals $\mathbf{I}_i^{S1}$ and $\mathbf{I}_i^{S2}$ of $\mathbf{I}_i^S$.

If for definition of abstract states many partition criteria are used simultaneously, combined partitions have to be constructed as non-overlapping intersections of parts of basic partitions (see Figure 2).

Abstract states corresponding to parts of combined partitions of a input I_i can be defined now as comprehension sets (I_i; g_i^l), where for each part $\mathbf{I}_i^l$ there is a characteristic formula g_i^l of the form:

$$\bigwedge_{k=1}^{\kappa} g_i^{l_k}, \text{ where}$$

κ - number of partitions
l_k – index of the part of k-th partition which forms the l-th part of combined partition, and

$$g_i^{l_k} \equiv \exists_1 \mathbf{I}_i^{l_k} \in Partition_k : I_i \in \mathbf{I}_i^{l_k}.$$

Assuming that the same partitions are applied for all inputs, the number of abstract states is

$$(\prod_k^{\kappa} \#Partition_k)^m \qquad \text{where}$$

$\#Partition_k$ - number of parts in $Partition_k$
m – number of inputs for which all κ partitions are applied.

To avoid exponential growth of the number of abstract states, characteristic formulas can be weakened under certain assumptions. For instance, an ordering relation $\prec$ within those partitions is defined, which are used for definition of combined abstract states. Intuitively speaking, the relation $\prec$

determines which set of input values has causally stronger influence to output behavior. For example, the order $I^N \prec I^S \prec I^F$ within NSF-partition means that whenever at least one input has a value in failure region then the behavior of the component is qualitatively determined by that input because I^F is the most influential value region of the given partition. Similarly for RU-partition relation $I^R \prec I^U$ holds. Generalizing this observation to all partitions for which the relation $\prec$ can be defined, a weaker characteristic formula for l-th combined abstract state can be formulated as:

$$\bigwedge_{k=1}^{\kappa} g_k^l, \qquad \text{where}$$

$$g_k^l \equiv \left\{ \begin{matrix} \exists\, I_i, \forall I_j \in I, I_i \neq I_i, \exists \mathbf{I}_k^l, \exists \mathbf{I}_k^l \in \\ Partition_k, I_i \in \mathbf{I}_k^l, I_j \in \mathbf{I}_k^l \end{matrix} \right\} : \neg(\mathbf{I}_k^l \prec \mathbf{I}_k^l).$$

In the sequel those states are called *dominated abstract states* and it can be shown that the number of those states does not depend on the number of inputs anymore.

4. SPECIFICATION OF HYBRID DYNAMICS

Dominated abstract states constructed in section 3 define the set of abstract states where in each state exactly one (different from others) input-output relation holds. On the other hand, all these abstract states are non-overlapping. Thus, the characteristic formula of each l-th abstract state entails an input-output relation R_l:

$$(\bigwedge_k^{\kappa} g_k^l) \Rightarrow R_l(I, O).$$

In other words, the set of concrete states that satisfies the characteristic formula of a l-th abstract state, constitutes the domain of a relation R_l. For further refinement of the input-output relations dynamics (either continuous, discrete or hybrid) has to be introduced. General specification formula for hybrid dynamics of a component is

$$Q_0 \wedge \bigwedge_i Q_i \wedge \bigwedge_j T_j \qquad \text{where} \qquad (2)$$

Q_0 – specification of the initial state
Q_i – specification of a i-th phase (with continuous dynamics)
T_j – specification of j-th transition.

As far as discrete transitions T_j can be specified using standard methods, such as Hoare triples, assumption-commitment pairs, etc., only on phase specifications are focused here.

In general, a phase specification formula has the form:

$$Q_i \equiv g_i \Rightarrow q_i \qquad \text{where}$$

g_i – phase stay condition
q_i – specification of continuous dynamics during the phase

For those abstract states where only continuous dynamics occurs the phase stay condition coincides with the characteristic formula of that abstract state. The specification of continuous dynamics of the phase may be constructed trough several refinement steps. Starting with a very rough description, when minimum information about component behavior is available, one can give a specification $\varphi^I \Rightarrow \psi^O$ expressing just the fact that "if the input of the component satisfies condition φ^I, the output value will satisfy condition ψ^O". For components, such as electronic devices, the input-output relation is often given by transfer function f_m and an output error δ_m. This allows one to specify a non-deterministic relation between the set I of inputs and an output O_m where non-determinism is parameterized by an error estimate δ_m:

$$O_m(I) \in [f_m(I) - \delta_m; f_m(I) + \delta_m].$$

In the simplest case where the output is proportional to the input I_i by some coefficient a_{im}, the transfer function for an output O_m is approximated by linear function

$$f_m(I_i) = f(0) \pm a_{im} \cdot I_i \qquad \text{where}$$

$$f_m(0) = O_m^{R\pm}$$

$$a_{im} = \left| \frac{O_m^{R-} - O_m^{R+}}{I_i^{R-} - I_i^{R+}} \right|$$

$$dom(f_m) = [I_i^{R-}; I_i^{R+}]$$
$$ran(f_m) = [O_m^{R-}; O_m^{R+}].$$

Non-determinism is reduced by introducing the time variable t and delay τ_{im} into transfer function e.g.,

$$f_m(I_i, t) = f(I_i, 0) \pm a_{im} \cdot I_i(t - \tau_{im}).$$

For further reduction of non-determinism, more complicated functions or differential equations may refine the transfer function but in many engineering applications the linear approximation suffices.

5. SPECIFICATION OF DEPENDABILITY PROPERTIES

When modeling dependability properties together with functional properties it needs to be clarified whether the elements of the functional model carry all necessary information and can it be used for proper characterization of component dependability properties.

The generic component model allows one to define a set Θ of parameters that characterize component dependability. These quality characteristics are parameters of input-output relation such as output error δ_m, delay τ and bounds of input partitions, e.g., bounds of normal and stress region characterize sensitivity to disturbances and faults. Another important aspect of modeling dependability is the description of dynamics of quality characteristics and factors influencing their dynamics. External influences are represented in generic component model as the subset Γ' of inputs I. Still some sources of faults become explicit only after implementation decisions are made about the component. The local environment (LE) subcomponent (see figure 1) is defined to allocate these influences to a dedicated part of the model. "Locality" of the environment means that this influence is encapsulated into that particular component and it is not necessary to define new inputs to any upper level component (it would violate also compositionality principle). Outputs of the LE-subcomponent are dependability characteristics Θ and their dynamics is specified as input-output relation $R_{LE}(\Gamma', \Theta)$ of LE-subcomponent.

For modeling fault-tolerance the corrector (CR-) subcomponent is introduced. Its inputs are defined as a subset Θ_1 of Θ ($\Theta = \Theta_1 \cup \Theta_2$) influenced by a LE-subcomponent and its outputs are elements of Θ_1' that are affected by compensation and recovery procedures. The effect of the corrector is specified as a relation $R_{CR}(\Theta_1, \Theta_1')$.

Practically no real component is completely fault-tolerant which means that the functionality of the component is always affected by environment in some extent. Therefore, the set of functional (F-) subcomponent inputs I_F must include all these impacts, i.e. $I_F = \Gamma \cup \Theta_1' \cup \Theta_2$. The output of F-subcomponent coincides with the set of outputs O of upper level component.

To summarize the role of generic component model, it is important to note that it gives a guided way to express the input-output dependencies of a component in the form of a composition of simpler relations where different design concerns are explicitly related to predefined subcomponents. Recursive usage of given refinement strategy provides a specification of regular structure.

6. CASE STUDY

A water temperature control system is considered in the case study. Plant to be controlled is a water supply system for a drum boiler (Davidson, 1990) where the goal of the control is to keep the temperature of water flowing into the boiler within the range of $70 \pm 4 \; C^\circ$. The regulator has two incoming water pipes – one for hot water and the other for cold water. The control system (see figure 3) consists of a temperature sensor measuring the temperature of the water running into the boiler;

$I_{SENS.}$ / $I_{SENS.i}$	Unit	$I_{SENSOR}^{T}(0)$ / (a_T^N / a_T^S)	$I_{SENSOR}^{N}(0)$ / (a_N^N / a_N^S)	$I_{SENSOR}^{S}(0)$ / (a_S^N / a_S^S)
T	°C / °C/s	[0;100] / $(\pm10^{-7}/\pm10^{-6})$	[-30;130] / (0/0)	[-50;-30[,]130;180] / (0/0)
T'	°C/s	[-1;1] / (0/0)	$[-\infty;\infty]$ / (0/0)	-
τ_s	s / s/s	[0;1.5] / $(\pm10^{-7}/\pm10^{-6})$	$[-\infty;\infty]$ / (0/0)	-
δ_ω	Hz / Hz/s	[0;20] / $(\pm2\cdot10^{-7}/\pm3\cdot10^{-7})$	[0;30] / (0/0)	]30;60] / (0/0)

Table 1. Initial values and degradation coefficients of input regions

programmable logic controller (PLC) and an actuator which is a motor driven hot and cold water mixing valve. It is assumed that the standard analysis for control design is done and transfer functions of components are identified.

6.1. Requirements and assumptions

User requirements say that the control system must keep the temperature of the water running into the boiler always within the range 66 – 74 °C.

To have a feasible design also some engineering constraints must be met. For instance, due to limited number of allowed switching cycles of the valve motor it is important to avoid over-regulation and to minimise the motor use. A design constraint to controller is introduced saying "if T_{cold} and T_{hot} have been stable within their Reliable regions at least t_{stable} = 2 sec then control activities must be completed and the motor switched off".

Characteristics of the plant are the temperatures of incoming hot water and cold water (T_h and T_c, respectively) and the temperature T of mixed water flowing into the boiler. It is assumed that temperature T_h varies within the range 80-90 C° and temperature T_c within the range 20-40 C°. The pressure in both inflow pipes is constant and balanced, but the maximum total temperature change rate of both cold and hot water together, may be at most 0.5 C°/s.

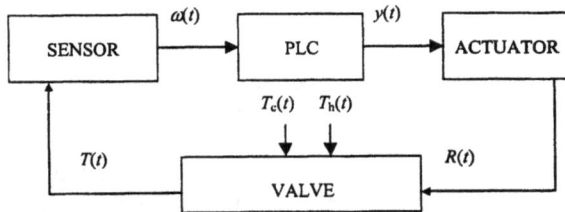

Fig. 3 Temperature control system

6.2. Components

Temperature sensor transforms the measured temperature T to frequency ω

$\omega(t) = 25 * T(t - \tau_s) + 2830$, where τ_s = 0.5 sec. Frequency 4630 Hz corresponds to 72 C° and 4530 Hz to 68 C°. Based on sensor output readings the PLC computes in each 0.1 sec a new control signal and switches "ON" one of two output relays providing corresponding PLC output values 1 or -1. Values –1 and 1 denote voltages with different polarity driving the rotation direction of the valve motor. Value 0 denotes the situation where both output relays are "OFF" and the motor stops. The transfer function of the PLC is

$$y(t) = \begin{cases} 1, & 4630 < \omega(t - \tau_c) \\ 0, & 4530 < \omega(t - \tau_c) < 4630 \\ -1, & \omega(t - \tau_c) < 4530 \end{cases}$$

with τ_c = 0.1 s.

Maximum time of turning the valve's position by motor between two extreme positions is 14 s. The first derivative of valve's relative position $R \in [0,1]$ is $\dot{R}(t) = k * y(t)$, where k =1/14 s^{-1}. For simplicity it is assumed that mixing cold and hot water in the valve is instantaneous and temperatures are related as in the following equation

$$T(t) = R(t) * T_h(t) + (1 - R(t)) * T_c(t).$$

Initially the component specification is given using a concise tabular form consisting of the specification of component inputs, outputs, their partitions, dynamics of partitions and I/O relations. Given forms are automatically translated to a set of first order logic formulas of type (2) and verified in PVS system towards requirements using formulas of type (3). Table 1 is an example of specification of the temperature sensor SMT160-30-18 'Smartec', with inputs: T - measured temperature, T' – derivative of T, τ_s – time lag of the sensor, δ_ω - measuring error. The output is ω - the signal frequency with reliable region: O_ω^T = [2830; 5330] Hz.

6.3. Verification

For mechanical verification of design correctness, the theory of hybrid dynamics is created in the proof assistant PVS (Owre, 1992). The main objects of the theory are real valued functions of time for continuous behavior and time dependent predicates for specification of discrete dynamics. Arithmetical and logical operations used in verification are defined in terms of those functions and predicates. Figure 4 shows a fraction of the theory.

The specification of our example system is translated into the form of a specialized theory where the general hybrid systems theory is imported into it. A fragment of the specialized theory is shown in Fig 5.

```
hybridcalc      : THEORY
BEGIN
Value           : TYPE = real
Time            : TYPE = real
SFunct          : TYPE = [Time -> Value]
SPred           : TYPE = pred[Time]
Interval        : TYPE = setof[Time]
I               : VAR Interval
+ : [SFunct, SFunct -> SFunct] =
        (LAMBDA f, g: (LAMBDA t: f(t) + g(t)));
<= : [SFunct, SFunct -> SPred ] =
        (LAMBDA f, g: (LAMBDA t: f(t) <= g(t)));
G(P)      : bool = (FORALL t: P(t));    %Globally
dur(P, I) : bool = (FORALL t: I(t) IMPLIES P(t));
d         : [SFunct -> SFunct]      %Derivative
END hybridcalc
```

Fig. 4. A fragment of the hybrid systems theory in PVS

```
heating  : THEORY
BEGIN
IMPORTING hybridcalc
% Specification
S1       : bool = G((T >= 66) AND (T <= 74))
S2       : bool = FORALL t1:
    (dur((d(Tc)=0 AND d(Th)=0), cx(t1)) IMPLIES
    EXISTS t2: t2 > t1 AND dur(d(T)=0, cx(t2)))
% Motor specification
r        : SFunct
k        : real = 1/14
Sm: bool = G(((r >= 0)AND(r <= 1)) IMPLIES
    d(r) = (LAMBDA t: k * y(t)))
END heating
```

Fig. 5. A fragment of the control system specification in PVS

6.4 Fault tolerant system design

About the dependability properties of components it is known that the expected lifetime of the valve's motor is $2 \cdot 10^5$ switching cycles and the lifetime of PLC output relays 10^5 switching cycles. The sensor is in direct contact with warm water and due to the precipitation its thermal capacity and consequently its time lag increases with rate at most 1 sec/year. Another degradation effect of the sensor is the drift of its output frequency 20 Hz/year which needs correction at least once a year.

Having proved functional and timing correctness of a design step assuming ideal conditions (in the sense of reliability), our next design goal is to guarantee fault-tolerant operation of the design with more realistic (degrading and unreliable) components. As far as the noticeable degradation of sensor only is assumed (see table 1) specifications of other components remain the same.

Similarly to component specifications, requirement specification for whole composition must be refined due to fault tolerance constraints. Dependability requirements must be a part of user requirement specification. In the example it is assumed that the required dependability characteristic is $\mathfrak{S}$ - the expected after service fault free operation period. The dependability constraints are added to control the system correctness formula by restricting it temporally by operator "*during* $\mathfrak{S}$". It means that for proving correctness of the dependable design it must be shown that at least during $\mathfrak{S}$ after last service the system meets its requirement specification:

$$(C_{SENSOR} \wedge C_{PLC} \wedge C_{ACTUATOR} \Rightarrow C_{CONTROL}) \; during \; \mathfrak{S}. \quad (3)$$

Verification of fault tolerant design. To simplify verification let us make a natural assumption that degradation is an irreversible and monotonic process. It suffices to prove design correctness using values of degrading parameters at time instant $t_0 + \mathfrak{S}$ (these calculations are made using degradation rate parameters a_T^N, a_T^S, see table 1). The formula (3) can be proved by proving its simplified version without "during" operator with calculated values at $t_0 + \mathfrak{S}$. Proofs with PVS show that the given design is not correct. To achieve the correct implementation of $C_{CONTROL}$ the PLC design is changed increasing its insensitivity to the longer sensor delays, e.g., design where the PLC activates the motor only during 0.5 s and then blocks for 2 s before being ready for the next control action, can be proved correct w.r.t. formula (3).

REFERENCES

Berghammer, R., Y. Laknech (Ed) (1998). *Proc. of Int. Workshop on Tool Support for System Specification Development and Verification.* Univ. of Kiel, Germany.

de Roever, W.-P. (1998) The need for compositional proof systems: a survey. In: *LNCS* **1536**, pp. 1-22.

Davidson, E. (Ed) (1990). Benchmark Problems for Control System Design. *Report of IFAC Theory Committee.*

Owre, S., J. Rushby, and N. Shankar (1992). PVS: a prototype verification system. In: *LNCS* **607**, pp. 748-752.

Spelberg, R., H.Toetenel, M.Ammerlaan, (1998). Partition refinement in Real-Time Model Checking. In: *LNCS* **1486**, pp. 143-157.

Robust and Efficient Software for Control Problems:
The SLICOT library.

I. Blanquer, V. Hernández, A. Vidal, E. Arias

Universidad Politécnica de Valencia, Dpto. de Sistemas Informáticos y
Computación (DSIC). Camino de Vera s/n, Valencia 46022.
{iblanque, vhernand, avidal, earias}@dsic.upv.es

Abstract. This article describes the library SLICOT (Software Library In COntrol Theory) [40] [7] [43] [2] that provides Fortran 77 implementation and also MATLAB [1] and SciLAB [39] interfaces for numerical algorithms for computations in systems and control theory. The library is build up on an efficient kernel of basic numerical linear algebra subroutines and provides tools for solving matrix equations, model reduction, subspace identification, robust control and non-linear control. SLICOT is freely available and it has been tested on many industrial problems. *Copyright © 2000 IFAC*

Keywords. Computer-aided control systems design, Parallel algorithms, Numerical algorithms, Robust control, Model reduction, Nonlinear control systems, Identification.

1. Introduction

In the frame of the NICONET [28] [29] [6] [5] [38] [41] (Numerics In COntrol NETwork, BRITE EURAM BRRT-CT97-5040) an open-source numeric software library called SLICOT is being designed. The objective is to provide efficient and robust numerical tools for Control Theory. SLICOT comprises previous work of the Working Group of Software (WGS), the RASP library [9] and other developments. Currently SLICOT provides more than 300 routines.

The NICONET project is aimed at the following objectives:

1. To stimulate the co-operation among control and numerics experts.

2. To increase the functionality of SLICOT.

3. To design and implement interfaces for linking SLICOT to user-friendly CACSD packages, such as MATLAB or SciLAB.

4. To provide with benchmarks based on real industrial problems that could accurately test the suitability of SLICOT.

5. To promote and freely distribute the SLICOT library.

The partners of the NICONET project are 9 Universities (UCL-CESAME, Katholieke Univ. Leuven, Eindhoven Univ. of Technology, Delft Univ. of Technology, Leicester Univ., Univ. Bremen, TU Chemnitz, Umeå Univ. and Univ. Politécnica de Valencia), 2 Research Centres (DLR and INRIA) and six industrial companies (IPCOS, LMS International, TBZ-Pariv, SFIM Industries, NAG, OMRON Spain).

The routines that comprises the SLICOT library have been developed by many different authors. The collaboration is open and there exist implementation guidelines [47] [46].

The NICONET project is divided into five main tasks, covering the areas:

I: Basic Numerical Tools for Control
II: Model Reduction
III: Subspace Identification
IV: Robust Control
V: Nonlinear Systems in Robotics

In each one of these tasks, different routines have been developed addressing to different numerical problems. A detailed description of the taks is shown in section 3. Section 2 contains a short history of the development of SLICOT, section 4 describes the progress with the interface to CACSD packages, such as MATLAB and SciLAB and section 5 describes the parallel version of SLICOT. Finally, performance results and application examples are shown on sections 6 and 7.

2. Short History of SLICOT

The reasons for developing the SLICOT software library are mainly three:

The limitations of CACSD packages, such as MATLAB, ANDECS, EASY5, SciLAB, XMath. User-friendly packages are of main interest, although their performance is low. As an example, non-blocked algorithms prevents MATLAB from reaching better performance.

The lack of robustness of available software (e.g. the control Toolbox of MATLAB) and the

---

[1] MATLAB is a registered trademark of Mathworks Inc.

need of controllers for complex processes, (flight control, satellite positioning) demands accurate and robust tools.

The need of structure preserving algorithms. Exploiting any structural information leads to increasing efficiency and reducing requirements Improving accuracy and reliability.

The development of efficient, reliable, and portable numerical software is challenging and time-consuming and requires joining expertise in theory, numerics, computers and engineering.

Several efforts have been initiated in the past to develop control libraries. The Scandinavian control library, and the Swiss control library AUTLIB were the first ones. The SLICE, from Kingston Polytec. and improved and distributed by NAG is another example.

In the 70s, the Working Group of Software (WGS) was founded involving academia and industries. The WGS developed SYCOT which was integrated with the SLICE library, creating the first version of SLICOT. The association with the DLR promoted the integration of RASP, leading to the SLICOT 3.0 release, the starting point for the NICONET project.

3. The SLICOT 4.0 Library

The current release of the SLICOT library is divided into several chapters, each devoted to a global area. The library contains two categories of routines: *Fully Documented Driver Routines* and *Supporting Routines*, which are provided for software developers. Depending on the routine, SLICOT uses four different formats for linear time-invariant systems: State-space, Polynomial Matrix, Rational Matrix or Time Response.

To increase the robustness, performance and portability, the computational kernel of SLICOT was migrated to LAPACK and BLAS-3 All the software required for SLICOT is public-domain, as well as SLICOT itself.

The following sections describe the modules of SLICOT. Two new chapters devoted to Adaptive Control and Identification are planned to be added in the 2nd half of NICONET.

3.1   Analysis Routines

Currently, SLICOT provides routines for models in the standard and generalised state-space representations including continuous and discrete time versions. This chapter includes routines for reduction to canonical and quasi-canonical forms [27] and computation of reduced models [9] [4] [3], system norms [8], dual systems, poles, and interconnections of subsystems.

3.2   Data Analysis

This section includes routines for the computation of specific characteristics [37] of a signal or data set: Convolution or deconvolution of two real signals, sine transform or cosine transform of a real signal, Discrete Fourier transform of complex and real signals, anti-aliasing window applied to a real signal. Future releases of the library will include statistical properties, trend removal, z-transforms, prediction or filter design.

3.3   Filtering

This chapter includes routines performing certain filter operations or designing specific Kalman filters for discrete-time systems described by state-space models. Both time-varying and time-invariant square root covariance and information filters are included. On the other side, a routine fast recursive least-squares filtering is also provided.

3.4   Mathematical Routines

This chapter contains some numerical routines not included in LAPACK or BLAS libraries [8], [30], which are interesting for control theory problems. The routines can be grouped into:

Routines for solving Linear Equations and Least Squares of structured matrices (triangular or complex upper Hessemberg), or providing more information as SVD decompositions, error estimations, or condition numbers.

Routines providing tools for computing Eigenvalues and Eigenvectors of structured matrices, with rank estimation and error bounds.

Decompositions and Transformations, such as RQ, QR, LQ, singular subspaces, partial diagonalisation of bi-diagonal matrices, square-reduced form of a Hamiltonian matrix.

Matrix Functions Routines, computing matrix exponential with accuracy estimate, Polynomial and Rational Function Manipulation.

Operations on Polynomial Matrices.

3.5   Non-linear Control Systems

Routines for solving Ordinary Differential Equations and Algebraic Differential Equations are being included in this chapter. A common SLICOT-like interface to several well-known integrator packages is provided [41] [10].

Routines for Optimisation and solving Non-linear Systems of Equations will be included in the 2nd half of the NICONET project.

3.6   Synthesis Routines

This chapter includes routines for the design of a system with some desired behaviour [8] and following some prescribed rules. Routines for the solution of standard and generalised Sylvester and Lyapunov and Riccati equations are also provided. Typical state-space design techniques, e.g., observer design pole assignment, feedback design, deadbeat control, optimal regulation, etc., are based on the solution of such equations. SLICOT also provides routines for polynomial and rational matrix synthesis and Optimal and sub-optimal H_2 and $H_?$ controller design [19] [1] [18] [33] [20] [34] [14].

3.7 *Transformation Routines*

This chapter contains routines for transforming one representation of a linear time-invariant system to another. More specifically, routines for transforming state-space, polynomial, rational, or time response representations, to any other representation are included.

4. CACSD Interfacing

Each one of the tasks of NICONET comprises the selection, standarisation and collation of FORTRAN numerical routines. These routines will be integrated in the SLICOT library and also made availabe for using through out CACSD packages, such as MATLAB or SciLAB [45] [11] [8] [20] [41].

In order to maintain the efficiency of SLICOT, gateway routines in 'mex' module files have been implemented for interacting with Matlab. These 'mex' files are called through out Matlab source code functions, obtaining better results either in performance or accuracy [44]. A similar work has been done with SciLAB.

The CACSD platforms provide the user with many tools that ease the development of numerical codes. SLICOT provides the packages with better accuracy and performance (e.g. SLICOT uses LAPACK and BLAS-3 instead of LINPACK as MATLAB does, or the exploitation of the structure of the matrices).

Currently, 11 mex libraries have been developed in NICONET:

ARESOL. Solvers for the continuous and discrete-time Ricatti equation, expressed in different formats and using different [45].

GENLEQ. Solvers for the generalised, continuous-time and discrete-time Lyapunov and Sylvester equations [11].

LINMEQ. Solvers for the standard, continuous-time or discrete-time Lyapunov and Sylvester equations [11].

SYSCOM. Computation of the observable, and controlable forms and Minimal Realizations. [8].

SYSTRA. Computation of matrix balancing and transformations (Schur form, triangularization). [8].

SYSRED. Model reduction using different approximations (square-root, balancing-free, Hankel Norm). [9].

CONHIN. H_2 and $H_?$ design of controllers for continuos-time systems.[20].

DISHIN. H_2 and $H_?$ design of controllers for discrete-time systems. [20].

HINORM. Computation of the $H_?$ norm of a continuous-time stable system [20].

ODESOLVER. Ordinary Differential Equations Solver. This package includes the standard solvers ODEPACK (LSODE and LSODES), RADAU5, DASSL and DASPK [41].

DAESOLVER. Differential Algebraic Equations Solver. This package is analogous to the previous one. It includes the ODEPACK (LSODI and LSOIBT), RADAU5, DASSL, DASPK and GELDA. The advantages of both packages lie on the use of a single interface, and thus a single syntax, and the use of MATLAB files for the definition of the system functions. The use of a single interface permits to change the solver without impact on the code, even in the user-defined functions, and the us of MATLAB allows to use the FORTRAN packages using only MATLAB files [41].

5. The PSLICOT Library

A Parallel Version of SLICOT is in progress [22] [31]. The objective of this library is to be able to perform efficiently the solution of large-scale or real-time problems. The next release of the SLICOT library will include parallel versions of basic tools and Model Reduction routines. The library will be oriented to Distributed Memory computers, including clusters of PCs and will make use of standard message passing libraries, such as BLACS and MPI.

The parallel library will comprise parallel versions of the routines of the sequential library. Therefore, similar implementation guidelines were defined [22].

5.1 *Parallel Basic Numerical Tools*

As SLICOT uses LAPACK and BLAS, PSLICOT will use ScaLAPCK and PBLAS. PSLICOT will include routines not available on ScaLAPACK or PBLAS which are required as components for the rest of the PSLICOT library. These are mainly the routines of the Mathematical chapter of SLICOT.

5.2 *High-Order Systems Model Reduction*

The PSLICOT library will include routines for model reduction based on the Bartels-Stewart and Hammarling's methods [23] [12] and routines based on the matrix sign function [31].

6. Advantages of SLICOT

The advantages of SLICOT lie mainly on the high performance and accuracy. Many experiments have been performed to compare SLICOT with other systems [44].

For the validation of SLICOT, a large benchmark battery has been provided [26] [17] [15] [13] [42]. Along with this battery, several industrial cases have been obtained for promoting the results on the control engineering industry.

Performance

This section shows several examples of the gain in performance of SLICOT routines with respect to MATLAB.

Figure 1 displays comparative results for the SLICOT Fast Fourier transform routine DG01MD (for complex sequences), and the corresponding MATLAB function fft. Random sequences X of length n were generated. Besides better efficiency, the accuracy of SLICOT routine shows an improvement which can be two orders of magnitude.

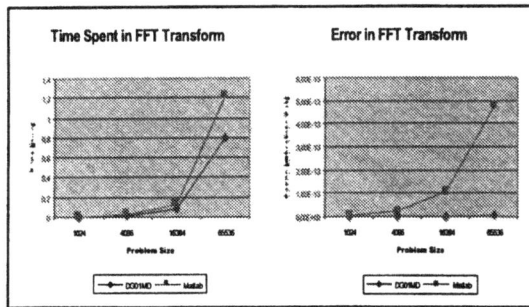

Fig. 1. Comparison with the Fast Fourier Transformation.

Figure 2 gives performance results for SLICOT's general Lyapunov solver SB03MD, which solves both continuous- and discrete-time equations, and MATLAB function lyap. The SLICOT routine also provides an estimate of the condition number is returned. The data used was generated directly, so the solutions were not known. Therefore, only relative residuals are given in the figure.

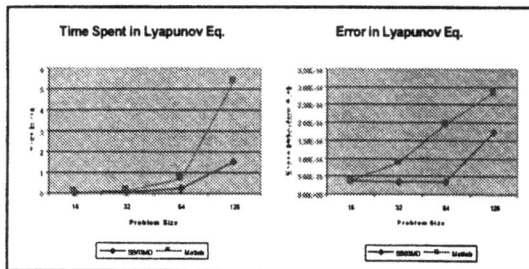

Fig. 2. Comparison between SB03MD and MATLAB results.

Figure 3 displays timing results for the SLICOT Hamiltonian eigenvalue solver based on the square-reduced form (MB04ZD and MB03SD)

and MATLAB function eig. There is no MATLAB function exploiting the Hamiltonian structure. Moreover, for random matrices, Matlab often gave eigenvalues very close to the imaginary axis, and which do not appear in pairs lambda, and -lambda, as the structure ensures.

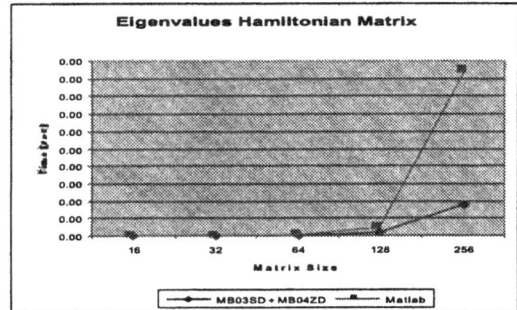

Fig. 3. Comparison between MB04ZD + MB03SD and MATLAB results

Accuracy

In the previous section, several comparisons in term of accuracy were performed. This section shows another examples of the accuracy and robustness of SLICOT routines with respect to MATLAB.

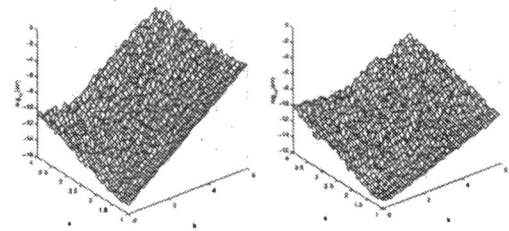

Fig. 4. A comparison on the Error of riccshur (left) and ricc (right).

Figure 4 shows the results of an extensive check of the accuracy of the Ricatti solver [35] [18] [36]. In this experiment, 2.000 Ricatti Equations were generated with known solution. The system generator, described in [18] used to parameters (s and k) to fill the matrix coefficients. An increase on the s coefficient leads to an increase of the condition number, but an increase on the k coefficient does not. In such way, the error of the routine should not increase with k. In both riccshur (MATLAB) and ricc (SLICOT) the error increases, thus showing some lack of robustness. However, the increase on the case of ricc is slower and of less magnitude, which shows a better behaviour. Moreover, an error estimation is provided by ricc, with very good results, and thus, the outcome of this routine does not lead to misunderstanding.

7. Industrial Applications

The benefits of using SLICOT lie either on the high-performance or high accuracy of its routines. Many problems that require a real-time behaviour can be implemented using SLICOT, and many other can be possible due to the accuracy of the routines. This section presents some industrial problems [24] that were solved using SLICOT.

In first place, there are three examples in which SLICOT is used for real-time control. First one is automatic driving of tractors in large fields and second and third ones consist in $H_?$ control design. The first $H_?$ controller is designed for helicopter flight control and the second one is a $H_?$ control design for a distillation column.

Finally, four examples of the application of SLICOT's model reduction routines are shown. In the examples, it has been possible to work with models of dimensions tens of times lower without problems on accuracy.

Automatic Tractor Driving

The GPS Lab at the Department of Aeronautics and Astronautics of Stanford University (USA) has developed a control mechanism for automatic steering of a farm tractor provided of a GPS. The controller has been implemented using SLICOT [32] for solving in real-time a set of Riccati equations and controlling the exact path to be followed.

Helicopter Control

The University of Leicester, is developing robust controllers for helicopters collaborating with other institutions. The design of flight controllers for helicopters is a difficult task due to the instability of the craft, complexity of the interactions among elements and particularly, the model uncertainties and disturbances. A controller for a Bell 205 helicopter was designed with the SLICOT $H_?$ routines.

Distillation Columns Control

This example shows the controller design for a distillation column using $H_?$ optimisation. The model is originally implemented using SpeedUp (Aspen Technology Inc.) and improved with Aspen Custom Modeller. The nonlinear system has 82 states. It is linearised at a set point and then reduced to a model of 17 states. There are 13 inputs, among which the most effective 6 are selected for the control system design. The system has 5 measurable outputs to be controlled. The synthesis of a controller uses the SLICOT synthesis [18] for obtaining a sub-optimal controller. SLICOT also provides condition numbers and accuracy estimates.

Model Reduction

The Model Reduction routines of SLICOT have been applied successfully to reduce different large industrial application models [4].

Microwaves control

This model includes the control of both a microwave heating and forced convection (air)

heating sources. Optimal control of a combination oven will give a higher quality end-product with a more uniform temperature distribution and a better cook-quality. The model was reduced in just 25 minutes from a dimension of 441 to 29, with a similar behaviour.

Aircraft control

The linearized aircraft model of ATTAS (Advanced Technology Testing Aircraft System) from DLR, describes the linearized rigid body dynamics of the plane during the landing approach. The total order of the model is 55. The 15th order approximation obtained fits almost exactly the original model both in terms of step and Nyquist frequency responses.

A CD-player finite element model.

This is a 120-th order single-input single-output system which describes the dynamics between the lens actuator and radial arm position of a portable compact disc player. Due to constraints on the size of the controller, an equivalent reduced model with order at most 15 is desired. The Balance & Truncate SLICOT routine gave a good 10-th order approximation.

A GEC ALSTHOM gasifier model

The model includes all significant effects; e.g., drying of coal and limestone, pyrolysis and volatilisation of coal, the gasification process itself and elutriation of fines. Three linearized non-minimal models of order 25 were obtained. Numerical difficulties with respect to using these models have been Rep.ed. The reduced order models of state dimension 16 obtained, are almost identical to the original ones, whose numerical difficulties do not appear on the reduced ones since scaling is done by default in all model reduction routines of SLICOT.

8. References

[1] A. A. Stoorvogel. Numerical problems in robust, H-inf optimal control. NICONET Rep. 1999-13.
[2] A. van den Boom, S. Van Huffel. Developments around the Freeware Standard Control Library SLICOT. Proc. CACSD'96 Symp, Dearborn, MI, pp. 473-476, 1996.
[3] A. Varga. Numerical Methods, Software Tools for Model Reduction. Proc. 1st MATHMOD Conf., Vienna, pp. 226-230, 1994.
[4] A. Varga. Task II.A.1 - Selection of Model Reduction Routines. SLICOT Work Note 1998-2.
[5] A. v. d. B., S. V. Huffel, P. Benner. NICONET: a Network for Numerically reliable Software in CACSD. Journal A, vol.38, no. 3, pp. 20-21, 1997
[6] A. v. d. Boom, S. V. Huffel. The numerics in control network NICONET. Proc. 17-th Benelux Meeting on Systems, Control, Mierlo, the Netherlands, March 4-6, p. 186, 1998
[7] Ad van den Boom, V. Sima, Peter Benner, Sabine Van Huffel. The freeware numerical subroutine library SLICOT for systems, control theory. Proc. 17-th Benelux Meeting on Systems, Control, Mierlo, the Netherlands, March 4-6, p. 185, 1998
[8] Andras Varga, Paul Van Dooren. Task I.A - Basic software tools for standard, generalized state-space

systems, transfer matrix factorizations . SLICOT Working Note 1999-17: Dec. 1999.

[9] Andras Varga. Model reduction routines for SLICOT. NICONET Rep. 1999-8: June 1999.

[10] Andras Varga. Standardization of Interface for Nonlinear Systems Software in SLICOT.. SLICOT Working Note 1998-4, June 1998.

[11] Andras Varga. Task I.A.1 - Selection of Basic Software Tools for Standard, Generalized State-Space Systems, Transfer Matrix Factorizations.. SLICOT Working Note 1998-3, June 1998.

[12] D. Guerrero, V. Hernández, J. E. Román, A. M. Vidal Parallel Algorithms for the Cholesky Factor of Generalized Lyapunov Equations. 5th IFAC Workshop on Alg, Archs. for Real-Time Control (AARTC'98), Cancun, Mexico, April 15-17, 1998

[13] D. Kressner, V. Mehrmann, T. Penzl. CTDSX - a Collection of benchmark examples for state-space realizations of time-invariant continuous-time systems. SLICOT Working Note 1998-9.

[14] D.-W. Gu, P. Hr. Petkov, M. M. Konstantinov. Direct formulae for the H-infinity sub-optimal central controller.. NICONET Rep. 1998-7.

[15] Daniel Kressner, Volker Mehrmann, Thilo Penzl. CTLEX - A collection of benchmark examples for continuous-time Lyapunuv equations. SLICOT Working Note 1999-6: June 1999.

[16] Daniel Kressner, Volker Mehrmann, Thilo Penzl. DTDSX - a Collection of benchmark examples for state-space realizations of time-invariant discrete-time systems. SLICOT Work Note 1998-10: Nov. 1998, June 1999.

[17] Daniel Kressner, Volker Mehrmann, Thilo Penzl. DTLEX - A collection of benchmark examples for discrete-time Lyapunuv equations. SLICOT Working Note 1999-7: June 1999.

[18] Da-Wei Gu, Petko Hr. Petkov, Mihail Konstantinov. H-inf, H2 optimization toolbox in SLICOT. SLICOT Working Note 1999-12.

[19] Da-Wei Gu, Petko Hr. Petkov, Mihail Konstantinov. H-inf. loop shaping design procedure routines in SLICOT. NICONET Rep. 1999-15.

[20] Da-Wei Gu, Petko Hr. Petkov, Mihail M. Konstantinov. An introduction to H-infinity optimisation designs. NICONET Rep. 1999-4.

[21] F. Alvarruiz, V.Hernández, P. A. Ruiz, A. M. Vidal. Algoritmos Paralelos para la Obtención de Realizaciones Minimales de Sistemas Dinámicos Lineales. VIII Jornadas de Paralel. Cáceres, Sep. 97

[22] I. Blanquer, D. Guerrero, V. Hernandez, E. Quintana-Orti, P. Ruiz. Parallel-SLICOT Implementation, Documentation Standards. SLICOT Working Note 1998-1: Sep. 1998.

[23] I. Blanquer, H. Claramunt, V. Hernández, A. M. Vidal. Algoritmos Secuenciales y Paralelos para la Resolución de la Ecuación de Lyapunov Generalizada por el Método de Bartels-Stewart. Encuentro de Análisis Matricial y Aplicaciones (EAMA-97), Sevilla, Septiembre 1997

[24] Industrial Examples home page http:///www.win.tue.nl/niconet/NIC2/examples.html

[25] J. Abels, P. Benner. DAREX: A Collection of Benchmark Examples for Discrete-Time Algebraic Riccati Equations. SLICOT Working Note 1999-16.

[26] Jörn Abels, Peter Benner. CAREX --- A Collection of Benchmark Examples for Continuous-Time Algebraic Riccati Equations (Version 2.0). SLICOT Working Note 1999-14: Dec. 1999.

[27] Konstantinov, M.M., Petkov, P.Hr., Christov, N.D. Orthogonal Invariants, Canonical Forms for Linear Controllable Systems. Proc. 8th IFAC World Congress, Kyoto, 1, pp. 49-54, 1981.

[28] NICONET home page http://www.win.tue.nl/niconet/niconet.htm

[29] P. Benner, V. Mehrmann, V. Sima, S. V. Huffel, A. Varga. SLICOT - A Subroutine Library in Systems, Control Theory. NICONET Rep. 97-3: (to appear in Applied, Computational Control, Signals, Circuits)

[30] Paul Van Dooren. Selection of basic software tools for structured matrix decompositions, perturbations. SLICOT Working Note 1999-9: June 1999.

[31] Peter Benner, Enrique S. Quintana-Orti, Gregorio Quintana-Orti. A portable subroutine library for solving linear control problems on distributed memory computers. NICONET Rep. 1999-1.

[32] Peter Benner, Heike Fassbender. SLICOT drives tractors!. NICONET Rep. 1999-2: Jan. 1999.

[33] Petko Hr. Petkov, Da-Wei Gu, Mihail M. Konstantinov. Fortran 77 routines for H-infinity, H2 design of discrete-time linear control systems. NICONET Rep. 1999-5: May 1999.

[34] Petko Hr. Petkov, Da-Wei Gu, Mihail M. Konstantinov. Fortran 77 routines for H-infinity, H-2 design of continuous-time linear control systems. NICONET Rep. 1998-8: Sep. 1998.

[35] Petko Petkov, Da-Wei Gu, Mihail M. Konstantinov, Volker Mehrmann . Condition, Error Estimates in the Solution of Lyapunov, Riccati Equations). SLICOT Working Note 2000-1: Jan. 2000.

[36] Petko Petkov, Mihail Konstantinov, Da-Wei Gu, Volker Mehrmann. Numerical solution of matrix Riccati equations: a comparison of six solvers. NICONET Rep. 1999-10: Aug. 1999.

[37] Rabiner, L.R., Rader, C.M. Digital Signal Processing. IEEE Press, 1972.

[38] Sabine Van Huffel, Ad J. W. van den Boom. NICONET: network for performant numerical software development in control engineering. Proc. 7th IFAC Symposium on Computer-Aided Control Systems Design, Ghent, Belgium, April 28-30, 1997, paper 95.

[39] SciLAB home page http://www-roq.inria.fr/scilab/scilab.html

[40] SLICOT home page http://www.win.tue.nl/niconet/NIC2/slicot.html

[41] V. Hernández, I. Blanquer, A. Vidal, E. Arias. SLICOT: una Librería de Software Numérico Eficiente y Fiable para Problemas de Control con Interfaces MATLAB, Proc. of the III Congress of MATLAB users in Spain. pp. 143-152, 1999.

[42] V. Mehrmann, T. Penzl. Benchmark collections in SLICOT. SLICOT Working Note 1998-5.

[43] V. Sima, S. Van Huffel. The freeware numerical subroutine library SLICOT for systems, control theory. Abstracts ICCoS study day on Identification, Feb. 4, 1998, Vrije Universiteit Brussel, Brussels

[44] Vasile Sima, Peter Benner, Sabine Van Huffel, Andras Varga. Improving the efficiency, accuracy of the MATLAB control toolbox using SLICOT-based gateways.. Presented at MTNS98 Padova, Italy, July 1998, (to appear in Proc. of MTNS98).

[45] Volker Mehrmann, Vasile Sima, Andras Varga, Hongguo Xu. A MATLAB MEX-file environment of SLICOT. SLICOT Working Note 1999-11.

[46] WGS. Contributor's Kit. WGS Rep. 96-2, Aug. 1996; (revised version of WGS Rep. 94-1).

[47] WGS. SLICOT Implementation, Documentation Standards. WGS Rep. 96-1, Aug. 1996.

INTERVAL MODEL PREDICTIVE CONTROL

J.M. Bravo[1], C.G. Varet[1] and E.F. Camacho[2]

[1] *Departamento de Ingeniería Electrónica, Sistemas Informáticos y Automática*
Universidad de Huelva (Spain)
[2] *Departamento de Ingeniería de Sistemas y Automática*
Universidad de Sevilla (Spain)

Abstract: Model Predictive Control is one of the most popular control strategy in the process industry. One of the reason for this success can be attributed to the fact that constraints and uncertainties can be handled. There are many techniques based on interval mathematics that are used in a wide range of applications. These interval techniques can mean an important contribution to Model Predictive Control giving algorithms to achieve global optimization and constraint satisfaction. *Copyright © 2000 IFAC*

Keywords: Predictive Control, Intervals, Constraints, Nonlinear Control, Optimization,

1. INTRODUCTION

Model Predictive Control (MPC) is a very ample range of control methods developed around certain common ideas. A model is used to predict the future plant outputs. The elaboration of mathematical models of processes in real life requires simplifications to be adopted. In practice, no model capable of exactly describing a process exists. Therefore, no model can be considered to be complete without taking into account possible modeling errors or uncertainties (Camacho and Berenguel, 1997).

In practice, all processes are subjected to constraints. The control system normally operates close to the limits and constraints violations may occur. The control system has to anticipate constraints violations and correct them in a appropiate way.

If no constraints are present, the model process is linear and the cost function is quadratic, the resulting control law is easy to implement and requires little computation. However, its derivation is more complex than that of the classical PIDs controllers. If the process dynamic does not change, the derivation of the controller can be done

beforehand, but in the adaptive control case all the computation has to be carried out at every sampling time.

When constraints are considered, the control signal has to be computed using a numerical optimization algorithm and the amount of computation required is even higher. The design algorithm is based on a prior knowledge of the model and it is independent of it, but it is obvious that the benefits obtained will depend on the discrepancies existing between the real process and the model used.

This work proposes control algorithms that can be used with linear and nonlinear models, quadratic or non quadratic objective function, linear and nonlinear constraints and bounded uncertainties.

The plant to be controlled can be described by the following non linear state-space model:

$$x(k) = f(x(k-1), u(k), \theta(k))$$
$$y(k) = g(x(k))$$

$$(1)$$

Where $u(k)$ is a vector of inputs or manipulated variables, $x(k)$ is a vector of state variables, $\theta(k)$ is a vector of uncertainties and *y(k)* is a vector of controlled variables or outputs.

Section 2 shows Model Predictive Control formulation. In sections 3 and 4, interval techniques for modeling and nonlinear constraints solving are described. Section 5 proposes a new Model Predictive Control strategy: Interval Model Predictive Control (IMPC). In order to illustrate the algorithm an application to a nonlinear model of an evaporator is presented in section 6.

2. MODEL PREDICTIVE CONTROL

Model Predictive Control (MPC) does not designate a specific control strategy but a very ample range of control methods which make an explicit use of a model of the process to obtain the control signal by minimizing an objective function. The ideas appearing in greater or lesser degree in all the predictive control family are basically (Camacho and Bordons, 1999): Explicit use of a model to predict the process output at future time instants (horizon). Calculation of a control sequence minimizing an objective function. Receding strategy, so that at each instant the horizon is displaced towards the future, which involves the application of the first control signal of the sequence calculated at each step.

No model capable of exactly describing a process exists. Parametric uncertainties are uncertainties in the parameters of the model without changing its order. That is, the structure of the process is the same as the model's but with parameters differing from the real ones.

It is difficult to handle uncertainties in the process model because its formulation is complex in many cases. Interval mathematics can be used to model bounds uncertainties, it can reduce the complexity of this formulation.

In practice all processes are subject to constraints. Actuators have a limited range of action and a limited slew rate. The operating points of plants are determined to satisfy economic goals and lie at the intersection of certain constraints. There are many aspects that cause bounds in process variables.

The constraints acting on a process can originate from amplitude limits in the control signal, slew rate limits of the actuator, and limits on the output signals. However, other types of constraints may exist: band constraints, overshoot constraints, monotonic behavior, actuator non linearities and others (Camacho and Bordons, 1999).

It may occur that process or constraints are not linear, or cost function are not quadratic. In these cases global optimization algorithms using interval analysis can solve this problem.

3. INTERVAL TOOLS

Interval Mathematics has been used since the early 60s giving rise to the development of algorithms to achieve global optimization and constraints satisfaction. Interval Mathematics can be said to have begun with the appearance of R.E. Moore's book Interval Analysis (1966). Moore's work transformed this simple idea into a viable tool for error analysis. Instead of merely treating rounding errors, Moore extended the use of Interval Analysis to bound the effect of errors from all sources, including approximation errors and errors in data.

Since the appearance of Moore's book, over 1000 publications on interval analysis has appeared as journal articles and reports. Over two dozen books are devoted entirely or in part to the subject (Moore, 1966; Neumaier, 1990; Hansen, 1992; Kearfott, 1996).

The main idea used in Interval Analysis is to consider that a variable X does not take a real value but is instead define by and interval. That is:

$$X = [a,b] = \{x : a \leq x \leq b\} \quad x, a, b \in \Re$$

(2)

All the variables and constants used to model or control systems will take interval values. In this case is important to have a set of tools which allow us to work with the previous variables and constants. This set of tools is formed by Interval Arithmetic, Extended Relation Operators and Interval Functions.

3.1 Interval Arithmetic

Interval Arithmetic is the set of arithmetic operations defined over intervals. Definition of addition, subtraction, multiplication and division are showed below.

$$[a,b] + [c,d] = [a+c, b+d]$$
$$[a,b] - [c,d] = [a-d, b-c]$$
$$[a,b] \cdot [c,d] = [min(ac, ad, bc, bd), max(ac, ad, bc, bd)]$$
$$[a,b] / [c,d] = [1/d, 1/c] \cdot [a,b]$$

(3)

Extended and generalized operations has been developed (Hansen, 1992).

3.2 Extended Relation Operators

To provide operators to compare two intervals are needed. The semantic of these operators over real numbers is clear, but a definition is necessary over intervals:

$$[a,b]=[c,d] \Leftrightarrow [a,b] \cap [c,d] \neq \emptyset$$
$$[a,b]>[c,d] \Leftrightarrow b>c$$
$$[a,b]<[c,d] \Leftrightarrow a<d$$
$$[a,b]\geq[c,d] \Leftrightarrow b \geq c$$
$$[a,b]\leq[c,d] \Leftrightarrow a \leq d$$

$$(4)$$

This definition favors that all the solutions of the problem stay in the space of study, however there may be definitions more appropriate for others applications. To get this robustness, the definition, has been carried out following this criterion:

$$X op Y \Leftrightarrow \exists x \in X, \exists y \in Y \mid x op y$$

$$(5)$$

3.3 Interval Functions

An Interval Function is an interval extension of a real function. There are several ways of extending (Hansen, 1992). Natural Interval Extension has been used in this work. Other kind of extensions and their influence in the control will be studied in later works.

The Natural Interval Extension F of a real function f is formed as the following: Each constant k of f is substituted by the interval [k,k]. Each real variable x of f is substituted by an interval variable X. Each real operator op is substituted by an interval operator OP

4. BRANCH & BOUND ALGORITHMS

The Branch and Bound algorithms are tools used to solve a lot of problems. Its methodology consists on dividing a certain search space into smaller spaces where the search is easier. The solution of the global problem will be the best solution of the smaller problems. The best advantage of these algorithms is the guarantee of offering a global solution against local methods.

4.1 Traditional definition

Traditional works (Mitten, 1970) define the important points that characterize the Branch and Bound algorithms:

Problem to solve and search space.

Division Rule; define how to divide the search space.

Bound Rule; define a limit to solve the problem in a certain space.

Selection Rule; allow to select the following space to work with.

Suppression Rule; define what spaces can be suppressed because of its lack of solution.

4.2 Predictive Control formulation

In order to solve a problem of Model Predictive Control, it should be taken into account the following points:

Problem to solve and search space; in MPC the problem is to find the set of control signals that minimize a cost function, so the search space is composed by the control signals domains.

Division Rule; a simple bisection will be used to divide the present search space.

Bound Rule; interval Arithmetic allow to evaluate the cost function in intervals obtaining an upper and lower value.

Selection Rule; select the space, heuristically, which has the higher probability of containing an optimal signal control.

Suppression Rule; two different criteria will be considered: Verify if the present search space may have a better solution that the one already obtained. If it is not possible, it will be dismissed. And verify if the present search space may satisfy the constraints imposed to control and in negative case, to dismiss this space.

5. INTERVAL MODEL PREDICTIVE CONTROL

The present work studies the viability to make an interval extension of the predictive control which permits to solve problems that are difficult for traditional algorithms. For example, non linear models and constraints and modeling uncertainties through interval values.

5.1 Problem Formulation

Consider the plant to be controlled described by (1).The problem to be solve by the interval model predictive controller at each sampling time may be stated as follows:

$$\min_{u} J(u(k), y(k), \omega(k), \theta(k))$$

Subject to

$$C_1(u(k)), C_2(y(k)), C_3(\theta(k)) \qquad (6)$$

J defines an objective function over a finite control horizon, $w(k)$ defines the set point sequence and $C_i(k)$ are sets of non linear constraints.

The key idea of the algorithm is that all variables and constants are given an interval value instead of a real value. That is, uncertainties both in model parameters and errors are given the interval value defined by the minimum and maximum value that the uncertainty can take. In general, the uncertainties, when bounded, can be described by a set of non linear constraints.

The problem is to determine u(k), in an interval mathematics form such that it minimizes J while satisfying the sets of constraints $C_i(k)$. Notice that the decision variables u(k), θ(k) and w(k) are exogenous variables and y(k) is predicted from the model equations and the values taken by the other variables. But this prediction is made using interval arithmetic.

Three steps are needed to define the problem of the predictive control in the interval domain field: First make the interval extension of the cost function and the model constraints, according to the steps described above. Second design an algorithm to resolve the sets of constraints. Third design an algorithm to minimize the cost function.

5.2 Algorithm to resolve the sets of constraints

There are a lot of works (Kearfott, 1996) about the constraints resolution based on Branch and Bound algorithms and interval techniques. In this work two main algorithms have been developed: First an algorithm of constraint satisfaction allows to know if a certain solution space satisfies the constraints imposed to the control. Those algorithms consist on replacing the interval domains of the solution space in the variables and verifying through relational operators, whether these constraints are satisfied or not. Second an algorithm to bound the solution space in charge of eliminating those parts of the space that do not satisfy constraints, returning a surrounding bound of the minimum space that satisfies them. In control, it is interesting to found a bound of the space adjusted enough to satisfy those constraints. The algorithm adjust made in that bound will depend on the required precision, the available time to make the calculations, and the overestimation that interval arithmetic introduces.

Algorithm to bound the space that satisfies constraints.

```
X = InitialSpace
C = Constraints
if NotSatisfyConstraints(C,X) ToReturn ListEmpty
List=Insert (X)
While Precision reached
        X = FirstElement(List)
        List = List -X
        (X1,X2) = SplitSpace(X)
        if SatisfyConstraints(C,X1) List=List + X1
        if SatisfyConstraints(C,X2) List=List + X2
EndWhile
Return UnionSpace(List)
```

5.3 Algorithm to minimize functions

The development of global optimization algorithms has been a great success in interval analysis (Hansen, 1992; Kearfott 1996). Those algorithms allow to find the global minimum of functions subject to constraints. These functions and constraints can be linear or non linear and differentiable and non differentiable. The more information the functions and the constraints have, the higher the convergence speed will be. For this reason, the minimization of differentiable functions will permit methods with a higher degree of convergence.

This work has developed a valid algorithm for non linear and non differentiable functions and constraints. The algorithm has as an input an initial search space, a cost function and a set of constraints. Specifying the predictive control problem, we found:

Initial search space; it is the set domain of the present and future inputs that compound the predictive control. The dimension of that space will depend on the number of input variables and horizon control and prediction used.

Cost function; its minimization permits to calculate the optimal control signals.

Input constraints; constraints imposed to control.

A first step in the algorithm consists on bounding the search space to the constraints as much as the interval techniques allow. This very first part of the algorithm returns a subspace that bound externally the defined space by the constraints. Next, the subspace is inserted in an ordered list used by the algorithm to keep the different subspaces created. The following step will be to execute a loop that will be repeating while a certain precision is reached. This precision will be calculated by the subspaces width. The loop starts taking the first element of the list. The mid point is calculated, and the cost function is evaluated with these values. A minimal bound is obtained. If that bound is better than the ones obtained previously, the bound is taken. Following, the space is split into two subspaces. The following operations are made with these two subspaces: Verify that constraints are satisfying, and in a negative case, to dismiss them; evaluate the cost function obtaining an upper and lower bound for these subspaces and verify that the lower bound does not exceed the present minimum (if it does, it will be dismissed) and to insert it in an ordered way in the list. The list can be ordered according to the best upper or best lower bound.

Once the precision is obtained by achieving the subspaces width wished, the algorithm ends returning the first of the list.

Algorithm to minimize cost functions

X = InitialSpace
C = Constraints
J = FunctionToMinimize
List = Insert(BoundSpaceSatisfyConstraintns(C,X))
While PresicionNonReached
    X = FirstElement(List)
    Minimum = MidPointTest(X)
    List = List - X
    (X1,X2) = SplitSpace(X)
    if SatisfyConstraint(C,X1)
        Evaluate J with X1
        if LowBound(J) > Minimo
            List = InsertOrdered(X1)
        Endif
    Endif
    if SatisfyConstraint(C,X2)
        Evaluate J with X2
        if LowBound(J) > Minimo
            List = InsertOrdered(X1)
        Endif
    Endif
EndWhile
Return FirstElement(List)

More efficient algorithms have been developed, however it is needed the use of the gradient (Van Hentenryck, 1997).

6. SIMULATION

An evaporator has been chosen as a testing bed for interval model predictive control. The results presented in this section have been obtained by simulation on a non linear model of the process (Newell and Lee, 1989).

The system dynamics is mainly dictated by the differential equations modeling the mass balance. In the solute the mass balance can expressed by the differential equation:

$$M \frac{dX_2}{dt} = F_1 X_1 - F_2 X_2$$

$$(7)$$

Where M is a constant that defines the total quantity of liquid in the evaporator, F_1 is the feed flowrate, X_1 is the feed composition, F_2 is the product flowrate and X_2 is the product composition. F_2 is the manipulated variable. X_2 is the process variable. F_1, X_1 are disturbances.

Taking a sampling time of one minute, the non linear discrete model used by the interval controller is:

$$X_2(k) = X_2(k-1) + \frac{(F_1(k) \cdot X_1(k) - F_2(k) \cdot X_2(k))}{20}$$

$$(8)$$

Four results of applying the interval controller are presented. Figure 1 shows the system response and the control signal without disturbances and without constraints. Figure 2 shows the system response and the control signal with a constraint over X_2. Figure 3 shows the system response and the control signal with a disturbance of ±10% in X_1. Figure 4 shows the system response and the control signal with the same disturbances but using a interval model where these disturbances are considered.

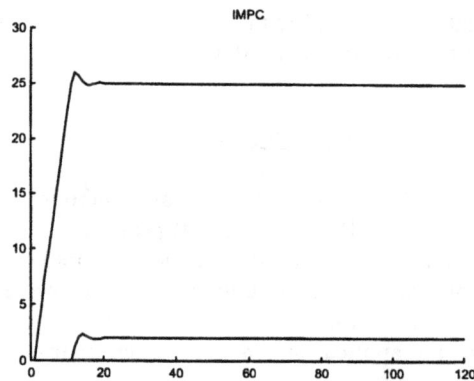

Fig. 1. Interval control without constraints, without disturbances.

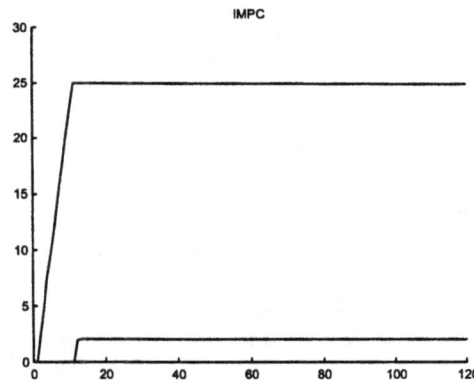

Fig. 2. Interval Control with $X_2 \leq 25$, without disturbances

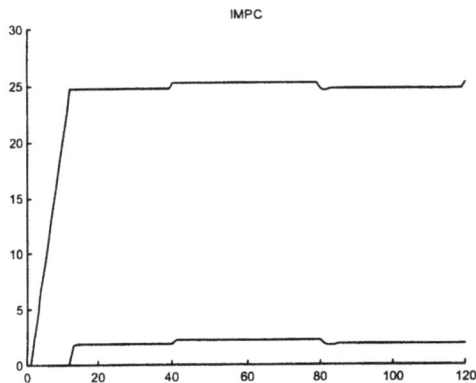

Fig. 3. Interval Control with disturbances not considered by the model.

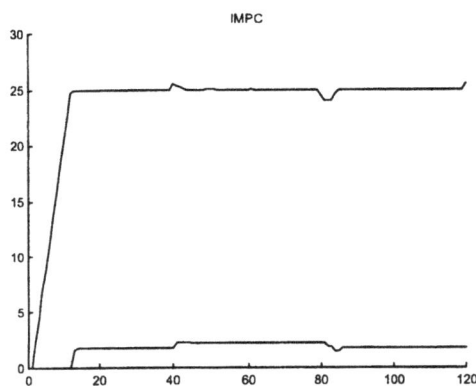

Fig. 4. Interval Control with disturbances considered by the model.

7. CONCLUSIONS

In this work, a set of interval methods for application in a MPC framework is proposed. These methods allow to use nonlinear models, nonlinear constraints and bounds uncertainties. However, there are some limitations: High computational cost, so its application in real time control is difficult if the process is fast; high storage cost (to handle spaces generated by branch & bound algorithms) and function overestimation in Interval mathematics.

Hardware development more and more fast make possible the application of interval methods.

REFERENCES

Camacho, E.F. and Bordons, C. (1999). *Model Predictive Control*. Springer-Verlag.

Camacho, E.F. and Berenguel, M. (1997). Robust Adaptive Model Predictive Control of a Solar Power Plant with Bounded Uncertainties, *International Journal of Adaptive Control and Signal Processing*, **Vol. 11**, pp 311-325.

Hansen, E. (1992). *Global Optimization Using Interval Analysis*. Marcel Dekker, Inc. New York.

Kearfott, R.B. (1996) *Rigorous Global Search: Continuous Problems*. Kluwer Academic Publishers.

Mitten, L.G. (1970). Branch and Bound methods: general formulation and properties. *Operation Research* 18:24,34.

Moore, R.E. (1966). *Interval Analysis*. Prentice Hall.

Neumaier, A. (1990) *Interval Methods for Systems of Equations*. Cambridge University Press, Cambridge

Newell, R.B. and Lee, P.L. (1989*). Applied Process Control. A case Study*. Prentice Hall

Van Hentenryck, P. (1997) *Numerica: A Modeling Language for Global Optimization*. The MIT Press, Cambridge, Massachussetts.

OPTIMISATION ALGORITHMS FOR INPUT/OUTPUT STRUCTURE SELECTION OF LARGE MULTIVARIABLE CONTROL SYSTEMS

M. Zhang* and L.F. Yeung[†]

*FernUniversität, Faculty of Electrical Engineering, 58084 Hagen, Germany,
min.zhang@fernuni-hagen.de
[†] City University of Hong Kong, Dept. of Electronic Engineering, Kowloon, Hong Kong,
eelyeung@cityu.edu.hk

Abstract: Often a subset of input/output variables is sufficient to control a multivariable system with acceptable performance degradation, facilitating real time control of such a system. Here, combinatorial optimisation techniques are developed to reduce the input/output structure of large multivariable control systems, which are based on the controllability and the observability indices. Two fast optimisation procedures are presented and compared; one employs Integer Programming, the other one a Genetic Algorithm. *Copyright © 2000 IFAC*

Keywords: Multivariable systems, input/output structure selection, combinatorial optimisation, integer programming, genetic algorithm.

1. INTRODUCTION

For the control of a processing plant, the selection of a good control structure is at least as important as the design of a controller (Engell, 1997). The choice of the control structure has a much stronger influence on the resulting performance than the design of the control algorithm. The selection of a control structure should be based on some notion of controllability and observability. Input/output controllability and observability are the ability to achieve acceptable control performance. They are affected by the selection of inputs and outputs, but cannot otherwise be changed by the control engineer. Therefore, the selection of the input/output structure is always beyond controller design. The best set of manipulated and controlled variables is one for which a plant is optimally controllable. Thus, a simpler and more effective controller can be designed instead of a complex controller.

For economic reasons, control engineers always prefer to use knowledge on process dynamics and control to help to design plants being easy and safe to control, rather than to try controlling ill-designed plants. When applying a certain design technique, in most cases it is assumed that the manipulated and controlled variables, i.e., the plant inputs and outputs, have already been chosen. In fact, an important part of multivariable design, especially in the process industries, is the choice of those variables which are most appropriate to control, and the choice of those variables which can be manipulated most effectively. Therefore, the input/output structure problem is an important issue to be addressed. This is also due to the realisation that many physical systems can in fact be viewed as interconnections of several smaller subsystems. Such kind of structural property makes it easy to apply decentralised control in real plants.

It is always seen that, although the number of output and control variables may be large, only a few of the output variables usually need to be controlled, and only a few control variables significantly influence the overall performance of a system. Experience shows that subsets of all inputs and outputs often suffice to control systems within reasonable performance and stability margins, thus facilitating their real time control. For this reason, the number of effective variables can be reduced to a more manageable dimension. There are many criteria that can be used for selecting the measurements and manipulations, such as the cross-coupling indices, the relative gain array or the controllability and observability indices (Moore, 1981). The latter have successfully been used in model reduction (Hyland and Bernstein, 1985) for the last two decades, especially in model order reduction.

In this paper, we develop a model input/output structure reduction technique based on an index related to system controllability/observability. Many strategies can be employed to find the best subsets of system input and output variables based on some system controllability/observability indices, for instance, graph theory, integer programming or genetic algorithms.

This project was supported by the RGC of Hong Kong, grant no. 9040173

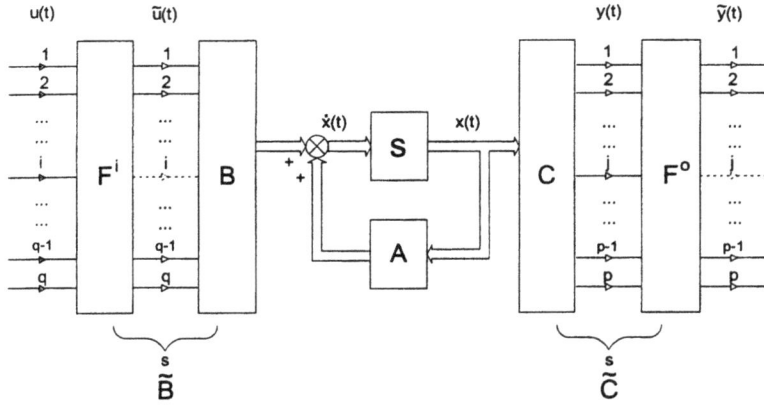

Fig. 1. The input/output connection matrix

For a control system, the concept of the input/output connection matrix having binary elements, only, is introduced, and so is the input/output weighting matrix. Owing to its particular structure, a genetic algorithm (GA) can be applied to start a multipoint optimisation search. In order to find the smallest subset of inputs and outputs meeting the constraint that the controllability and observability measures of the reduced system remain within pre-set limits, a GA combined with a linear search scheme is developed. By further investigating the controllability/ observability measures, an integer programming (IP) based algorithm is introduced, which requires large memory space for calculation, but needs less objective function calls compared to the GA procedure.

2. INPUT/OUTPUT STRUCTURE REDUCTION AND QUALITY MEASURES

Consider a linear time-invariant system $A \in R^{n \times n}, B \in R^{n \times q}, C \in R^{p \times n}$ given by

$$
\begin{aligned}
\dot{x} &= Ax + Bu \\
y &= Cx
\end{aligned}
\tag{1}
$$

The realisation (A, B, C) is assumed to be minimal. In the development, we only consider asymptotically stable systems, i.e., the spectrum of A is a subset of the open LHP. Define the input (respectively output) connection matrix F^i (respectively F^o) as a $q \times q$ (respectively $p \times p$) diagonal matrix with either 1 or 0 along the diagonal.

The rth input (respectively rth output) is connected when the rth diagonal entry of F^i (respectively F^o) is 1, otherwise it takes a value equal to 0 (Fig. 1). We can easily see that $F^i \in S_{\tilde{q},q}$ and $F^o \in S_{\tilde{p},p}$, where

$$
\begin{aligned}
S_{r,n} := \{ \text{diag}(s_1, s_2, ..., s_n) : s_i = 0 \text{ or } 1, \\
\text{with total number of 1 equal to } r \}
\end{aligned}
\tag{2}
$$

To measure the error induced by reducing the input and output simultaneously, the impulse response error

$$
e(t) = \begin{pmatrix} C & -F^o C \end{pmatrix} \exp \begin{pmatrix} At & 0 \\ 0 & At \end{pmatrix} \begin{pmatrix} B \\ BF^i \end{pmatrix}
\tag{3}
$$

is considered. Define the objective function based on L_2 measure on $e(t)$ given by

$$
\begin{aligned}
\tilde{J}^{io} &= \text{trace} \left(\int_o^\infty e(t) e^T(t) dt \right) \tag{4} \\
&= \text{trace} \left(\begin{pmatrix} C & -F^o C \end{pmatrix} R \begin{pmatrix} C^T \\ -C^T F^o \end{pmatrix} \right)
\end{aligned}
$$

where $R \geq 0$ is the unique solution of

$$
\begin{pmatrix} A & 0 \\ 0 & A \end{pmatrix} R + R \begin{pmatrix} A^T & 0 \\ 0 & A^T \end{pmatrix} + \begin{pmatrix} B \\ BF^i \end{pmatrix} \begin{pmatrix} B^T & F^i B^T \end{pmatrix} = 0
\tag{5}
$$

Thus, R is given by

$$
R = \begin{pmatrix} R_1 & R_2 \\ R_2 & R_2 \end{pmatrix}
\tag{6}
$$

where $R_1 = R_1^T > 0$ (since (A, B, C) is minimal) and $R_2 = R_2^T \geq 0$ are unique solutions of

$$
\begin{aligned}
AR_1 + R_1 A^T + BB^T &= 0 \\
AR_2 + R_2 A^T + BF^i B^T &= 0
\end{aligned}
\tag{7}
$$

It can be easily seen that $R_1 > R_2 \geq 0$. Hence

$$
\tilde{J}^{io} = \text{trace}(CR_1 C^T) - \text{trace}(F^o C R_2 C^T)
\tag{8}
$$

Since $\text{trace}(CR_1 C^T)$ is independent of both F^i and F^o, the minimisation of $\tilde{J}^{io}$ is equivalent to maximising

$$
J^{io} := \text{trace}(F^o C R_2 C^T)
\tag{9}
$$

As far as computation is concerned, we first calculate and store the matrices P_i by solving

$$
AP_i + P_i A^T + b_i b_i^T = 0
\tag{10}
$$

where b_i is the ith column of B, and obtain

$$
R_2 = \sum_{i \in \tilde{q}} P_i
$$

where $\tilde{q}$ represents the selected inputs according to diagonal values of F^i. Then

$$
\begin{aligned}
J^{io} &= \text{trace}(F^o C R_2 C^T) \tag{11} \\
&= \text{trace}(C^T F^o C R_2) \tag{12} \\
&= \sum_{j \in \tilde{q}} \text{trace}(C^T F^o C P_i) \tag{13}
\end{aligned}
$$

64

The maximum J_1^{io} can be found by selecting different F^i and F^o. Then the best subset of $\tilde{p}$ inputs and $\tilde{q}$ outputs of the entire q inputs and p outputs can be found by solving the following:-

$$\mathbf{P_1}: \quad \max_{F^i \in S_{\tilde{q},q}, \; F^o \in S_{\tilde{p},p}} J^{io} \quad (14)$$

If $\tilde{p}$ and $\tilde{q}$ denote the number of selected outputs and inputs, respectively, then the total number of possible combinations of input and output selections is given by $_pC_{\tilde{p}} \times _qC_{\tilde{q}}$. When there is no reduction in the inputs or outputs, respectively, it can be shown that J^{io} is reduced to respectively

$$J^o = \text{trace}(C^T F^o C P) = \sum_{\tilde{p}} \text{trace}(B^T Q_i B), \quad (15)$$

$$where \quad \begin{aligned} AP + PA^T + BB^T &= 0 \\ A^T Q_i + Q_i A + c_i^T c_i &= 0 \end{aligned}$$

and

$$J^i = \sum_{\tilde{q}} \text{trace}(C^T P_i C) = \text{trace}(B F^i B^T Q). \quad (16)$$

$$where \quad \begin{aligned} AP_i + P_i A^T + b_i b_i^T &= 0 \\ A^T Q + QA + C^T C &= 0 \end{aligned}$$

Due to the positive semidefiniteness of $B^T Q_i B$ and $C P_i C^T$, the maximisation of these objectives requires a mere sorting procedure on the values of $\text{trace}(B^T Q_i B)$ and $\text{trace}(C^T P_i C)$.

For the objective J^i, the best $\tilde{q}$ inputs can be found by maximising $\sum_{\tilde{q}} \text{trace}(C^T P_i C)$ of the selected $\tilde{q}$ inputs. This amounts to arrange $\text{trace}(C^T P_i C)$ in descending order, and choose the first $\tilde{q}$ inputs as the reduced system inputs. Similarly, for the objective J^o, the best $\tilde{p}$ outputs can be found by maximising $\sum_{\tilde{p}} \text{trace}(B^T Q_i B)$ of the selected $\tilde{p}$ outputs. This amounts to arrange $\text{trace}(B^T Q_i B)$ in descending order, and choose the first $\tilde{p}$ outputs as the reduced system outputs.

There are two special cases: input-state structure reduction and state-output structure reduction. When one considers minimising the input-to-state energy due to reduction of inputs only, the problem is a special case of J_1^i with $C = I$. When one considers minimising the state-to-output energy due to reduction of outputs only, the problem is a special case of J_1^o with $B = I$.

Also, note that

$$F^o C^T C = \sum_{\tilde{p}} c_j^T c_j \quad (17)$$

where c_j is the jth row of C, and $\tilde{p}$ represents the selected inputs according to diagonal values of F^o, the output connection matrix. Then the objective function (13) can be reformed to:-

$$J^{io} = \sum_{\tilde{q}} \sum_{\tilde{p}} \text{trace}(c_j^T c_j P_i) \quad (18)$$

Let us define a *cross-connection matrix* X, with its elements taking either '0' or '1' values, and that

$x_{ij} = 1$ implies that the *input-i* and *output-j* are selected, respectively. The total number of '1's in a row (column) is equal to the number of inputs (outputs) to be selected.

For instance, if we want to select a 2×2 input-output subset from a 3×3 system, the *cross-connection matrix* X may look like Fig. 2:-

	I_1		I_2
	0	0	0
r_1	1	0	1
r_2	1	0	1

Fig. 2. A 3×3 cross connection matrix

The 1-element inside the *cross-connection matrix* X shows us the connection status of the input and output. In this case, the selected inputs are the 2nd and the 3rd input, and the selected outputs are the 1st and the 3rd output of the original system.

The effectiveness of the ith input to control the jth output can be measured by the index w_{ij}:-

$$w_{ij} = trace(c_j^T c_j P_i) \quad (19)$$

Note that w_{ij} is based on (18) and W is called the *weighting matrix* here. Then the cost function of (18) can be re-written in terms of W as follows:-

$$J^{io} = \sum_{i=1}^{q} \sum_{j=1}^{p} w_{ij} x_{ij} \quad (20)$$

Thus this input/output reduction problem can be transformed into a combinatorial problem as follows:-

$$\mathbf{P_2}: \quad \max_X J^{io}(W, X) \quad (21)$$

Subject to:- $\quad \sum_{j=1}^{p} x_{ij} = r \quad i = 1, 2, \ldots, q$
$\quad\quad\quad\quad\quad \sum_{i=1}^{q} x_{ij} = l \quad j = 1, 2, \ldots, p$
Where $\quad x_{ij} = 1$ or 0;
$\quad\quad\quad r$ *is the desired reduced size of*
$\quad\quad\quad input group;$
$\quad\quad\quad l$ *is the desired reduced size of*
$\quad\quad\quad output group.$

Assume that we know how many inputs/outputs are wanted to remain to control/measure the whole plant, then the above formula can be considered as a transportation problem (Daellenbach and co-workers, 1983). Two procedures are developed to solve problem $\mathbf{P_1}$ (or $\mathbf{P_2}$). One is based on a branch-and-bound (B&B) method and the other one employs a genetic algorithm (GA).

3. INPUT/OUTPUT STRUCTURE REDUCTION BY A GENETIC ALGORITHM

According to Section 2, the diagonal elements of the input connection matrix F^i and the output connection matrix F^o are either '1' or '0'. The concatenation of each possible diagonal element of the connection matrices F^i and F^o can be considered as a chromosome (genotype). A genetic algorithm (GA) (Goldberg, 1989; Holland, 1992; Yeung and co-workers, 1996) can then be applied to start a multi-point optimisation search. Assume that we know how many inputs/outputs are wanted to remain to control/measure the whole plant, a simple input/output structure reduction genetic algorithm ($\mathbf{P_{GA}}$) can be developed to find the optimal group of inputs and outputs.

For a given linear system (A, B, C), we can now state a GA to solve $\mathbf{P_1}$:-

$(F^{I*}, F^{O*}, f^*) = \mathbf{P_{GA}}(\tilde{q}, \tilde{p}, q, p)$

$\tilde{q}$: number of inputs to be connected.
q : total number of inputs.
$\tilde{p}$: number of outputs to be connected.
p : total number of outputs.
P_c : probability of crossover.
P_m : probability of mutation.
ϵ : tolerance
f^* : best fitness value
F^{I*} : optimal input connection matrix
F^{O*} : optimal output connection matrix

1. Randomly generate the initial population represented by chromosomes having the form of binary strings with length $p + q$ such that there are $\tilde{p}$ 1-elements in positions $1, 2, ..., p$, and $\tilde{q}$ 1-elements in positions $p + 1, ..., p + q$ and all other entries being 0.

2. Scale objective function J^{IO} to fitness function form.

3. Calculate and assign the fitness function value to each chromosome of the initial population.

4. Set generation counter k to 0.

5. Start the GA cycle.

   5.1 $k = k + 1$

   5.2 Reproduce the population by evaluating the fitness function of individual chromosomes and randomly mate them.

   5.3 Generate new chromosomes by crossover and mutation according to the probabilities P_c and P_m.

   5.4 Make a judgement on the new chromosomes to verify whether they are out of range or not. If so, insert new randomly generated ones to replace the unsatisfactory ones.

   5.5 Calculate the fitness values of the chromosomes in the new population.

   5.6 For $k < 5$ go to step 5.

   5.7 For $k \geq 5$ calculate

   $$v_1 = \frac{\text{bestfit}(k) - \text{bestfit}(k-5)}{\text{bestfit}(k)},$$

   $$v_2 = \frac{\text{meanfit}(k) - \text{meanfit}(k-1)}{\text{meanfit}(k)}$$

6. if $(0 < v_1 < \epsilon)$ then Stop
   elseif $(0 < v_2 < \epsilon)$ then Stop
   else Go to step 4.

Results: Optimal connection matrices F^{I*} with $\tilde{q}$ inputs and F^{O*} with $\tilde{p}$ outputs, identified by the chromosome for which the maximum value f^* of the fitness function f is assumed in the population generated last, and the best fitness function value f^* according to F^{I*} and F^{O*}.

Comments

(1) Each chromosome represents one possible connection. For a certain structure, the ith input (respectively output) is considered controlling the plant if the ith diagonal element of the connection matrix F^I (respectively F^O) is '1'. Since J^{IO} is a monotonously decreasing function of the number of remaining inputs and outputs, we fix the number of '1s' in each chromosome.

(2) Since the number of '1s' in each chromosome is fixed, there exist extra restrictions on the process of crossovers and mutations. However, this may not cause essential difficulties as these processes are probabilistic.

(3) The probability of crossover (P_c) and the probability of mutation (P_m) have an influence on the convergence rate of the GA. A trial-and-error process may be used to find the fittest combination of P_c and P_m.

(4) The terminal condition ϵ should be appropriately chosen to avoid prolonged computation. To find the best terminal condition is also a trial-and-error process.

$\mathbf{P_{GA}}$ is a GA based procedure which can select the $\tilde{q}$ optimal connections to the system out of a total of q inputs, and $\tilde{p}$ optimal connections of the system out of a total of p outputs. To find the "smallest" subset of inputs and outputs such that the controllability and observability measures of the reduced system still remain within "acceptable" limits, a linear search scheme was used to find the best $\tilde{q}$ by solving a sequence of $\mathbf{P_{GA}}$ with $\tilde{q}$ set to $(q, q-1, q-2,...)$ in descending order and $\tilde{p}$ set to $(p, p-1, p-2, ...)$. The search is stopped when the controllability measure has reached a pre-set threshold. In general, we observed that the GA approach is reasonably fast for systems with modest size. For very large systems, we prefer the following Integer Programming approach which is faster but requires large memory space.

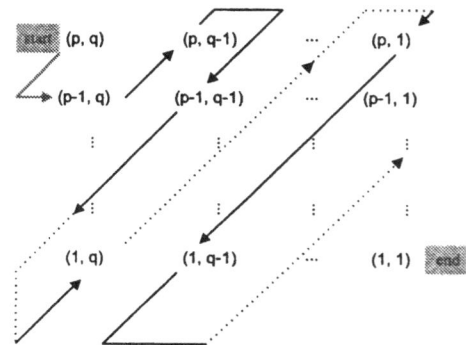

Fig. 3. Sequence of linear search

4. INPUT/OUTPUT STRUCTURE REDUCTION BY INTEGER PROGRAMMING

Algorithms for integer programming (**IP**) problems are usually based on algorithms for general linear programming problems, together with some procedure for constructing additional constraints which exclude non-integer solutions. In this paper, we use a standard **IP** routine from NAG to solve the problem $\mathbf{P_2}$. This subroutine is based on a branch and bound (B&B) method (Taha, 1987). The **IP** problem we are going to solve is:-

$$\mathbf{P_{IP}}: \quad \max_x F(x) \quad (22)$$

$$Subject\ to:- \quad \sum_{j=1}^{n} a_{ij} x_j \left\{ \begin{array}{c} = \\ \leq \\ \geq \end{array} \right\} b_i,$$

$$i = 1, 2, \ldots, m$$

$$Where \quad F(x) \triangleq \sum_j c_j x_j; \ j = 1, 2, \ldots, n$$

$$x_j = 0\ or\ 1; \ j = 1, 2, \ldots, n$$

Then $\mathbf{P_2}$ can be re-formulated into standard form:-

Step 1 Reform the 2-dimensional *cross-connection matrix* $X \in N^{m \times n}$ and the *weighting matrix* $W \in R^{m \times n}$ into a 1-dimensional one, i.e., let

$$\bar{W}(j + (i-1)*n) = W(i,j) \quad (23)$$
$$\bar{X}(j + (i-1)*n) = X(i,j)$$

where $\bar{W} \in R^{1 \times (m*n)}$ and $\bar{X} \in N^{1 \times (m*n)}$.

Step 2 Expand the *cross-connection matrix* $\bar{X} \in N^{1 \times (m*n)}$ to $\hat{X} \in N^{1 \times (m*n+m+n)}$:-

$$\hat{X} = \{\hat{x}_1, \hat{x}_2, \ldots, \hat{x}_{m \times n}, |\ y_x, \ldots, y_m, |\ z_1, \ldots z_n\}$$

where $Y \in N^{1 \times m}$ and $Z \in N^{1 \times n}$ are the supplementary variables, and

$$\hat{x}_j = either\ 0\ or\ 1, \quad j = 1, 2, \ldots, m*n$$
$$y_i = either\ 0\ or\ 1 \quad i = 1, 2, \ldots, m$$
$$z_i = either\ 0\ or\ 1, \quad i = 1, 2, \ldots, n$$

Step 3 Simultaneously expand the *weighting matrix* $\bar{W} \in R^{1 \times (m*n)}$ to $\hat{W} \in R^{1 \times (m*n+m+n)}$:-

$$\hat{W} = \{\bar{W} \mid \mathbf{0}\} \quad (24)$$

Step 4 Then the input/output structure reduction problem ($\mathbf{P_{IP}}$) can be solved by maximisation.

$$\mathbf{P_{IP}} \quad \max_x F(x) \quad (25)$$

$$where \quad F(x) = \sum_j \hat{w}_j \hat{x}_j$$
$$\hat{W} \in R^{1 \times (m*n)}\ and$$
$$\bar{X}^1 \in N^{1 \times (m*n+m+n)}$$

$$\sum_{i=1}^{m} y_i = r$$
$$\sum_{j=1}^{n} z_j = l$$
$$\sum_{j=1}^{n} x_{kj} - y_k * l = 0 \quad k=1,\ldots,m$$
$$\sum_{i=1}^{m} x_{ik} - z_k * r = 0 \quad k=1,\ldots,n$$

A test example for solving this input/output structure reduction problem is shown in Section 5.

5. EXAMPLES

Consider the following plant (A,B,C):-

$$A = \begin{bmatrix} -3.2 & -2.1 & -915.5 & 0.51 & 134.2 \\ 1.6 & -5.9 & -281.6 & 0.2 & 57.0 \\ 0.0169 & -0.026 & -10.03 & 0.008 & 0.58 \\ 0 & 0 & 0 & -10.0 & 0 \\ -2.16 & 6.86 & 740.5 & 1.19 & -171.5 \end{bmatrix}$$
$$(26)$$

$$B^T = \begin{bmatrix} 0.2 & 0.01 & 0.41 & 0.83 & 0.5 \\ 0.19 & 0.74 & 0.84 & 0.01 & 0.7 \\ 9.50 & 7.62 & 6.15 & 4.05 & 0.57 \\ 2.31 & 4.56 & 7.91 & 9.35 & 3.52 \\ 6.06 & 0.18 & 9.21 & 9.16 & 8.13 \\ 0.60 & 0.44 & 0.52 & 0.68 & 0.42 \\ 4.85 & 8.21 & 7.38 & 4.10 & 0.09 \\ 8.91 & 4.44 & 1.76 & 8.93 & 1.38 \\ 0.27 & 0.93 & 0.2 & 0.37 & 0.3 \\ 0.19 & 0.46 & 0.67 & 0.83 & 0.18 \end{bmatrix}$$
$$(27)$$

$$C = \begin{bmatrix} 1.93 & 6.97 & 4.96 & 6.6 & 7.27 \\ 0.7 & 0.79 & 0.97 & 0.13 & 0.66 \\ 0.54 & 0.95 & 0.27 & 0.01 & 0.28 \\ 6.82 & 3.78 & 8.99 & 3.41 & 3.09 \\ 0.44 & 0.52 & 0.25 & 0.89 & 0.46 \\ 3.02 & 8.60 & 8.21 & 2.89 & 8.38 \\ 0.69 & 0.88 & 0.87 & 0.19 & 0.06 \\ 5.41 & 8.53 & 6.44 & 3.41 & 5.68 \\ 1.50 & 5.93 & 8.17 & 5.34 & 3.70 \\ 0.62 & 0.17 & 0.73 & 0.29 & 0.98 \end{bmatrix}$$
$$(28)$$

The best 5-input/output subset to control the system is pre-determined to be $\{3,4,5,7,8\}$ (for input) and $\{1,4,6,8,9\}$ (for output), respectively.

Case 1: The result of $\mathbf{P_{GA}}$

By using $\mathbf{P_{GA}}$, the desired reduced size is chosen to be 5. We apply the traditional values of the probabilities of crossover and mutation, $P_c = 0.65$, $P_m = 0.005$. Since this example is not large, in order to save calculation time, the population size is set to 30. As termination condition the tolerance $\epsilon = 0.001$ is selected.

The program is terminated after the 52th generation. The $\mathbf{P_{GA}}$ program has called the objective function $1560 = 30 \times 52$ times. In fact, it requires 63504 ($_{10}C_5 \times _{10}C_5$) times for computing all possible combinations. The optimal input group and the remaining output group are found as expected.

In order to determine the effectiveness and efficiency of $\mathbf{P_{GA}}$, 9 sets of different combinations of the remaining input group/output group were investigated using the same example, as well as the same population size and termination condition. For all these different combination sets the $\mathbf{P_{GA}}$ program converged after around 50 generations. Table 1 compares the number of objective function calls. It can easily be seen that the GA is preferable for problems which require a lot of calculations: $(\bar{q}, \bar{p})$ is the combination of different remaining input and output sizes, $_q C_{\bar{q}} \times _p C_{\bar{p}}$ indicates the number of objective function calls required by exhaustive search, $\mathbf{P_{GA}}$ is the number of objective function calls by the GA, and $\frac{_qC_{\tilde{q}}\times_pC_{\tilde{p}}}{\mathbf{P_{GA}}}$ is the ratio of the objective function calls by these two algorithms. The table shows that there is a strong correlation between the problem size expressed by $_qC_{\tilde{q}}\times_pC_{\tilde{p}}$ and the reduction in computational effort achieved by using the GA as expressed by the ratio.

Table 1 Number of objective function calls

$(\tilde{q},\tilde{p})$	$_qC_{\tilde{q}}\times_pC_{\tilde{p}}$	$\mathbf{P_{GA}}$	$\frac{_qC_{\tilde{q}}\times_pC_{\tilde{p}}}{\mathbf{P_{GA}}}$
(5,5)	63504	56×30	40.7
(5,6)	52920	41×30	43.0
(4,6)	44100	45×30	32.7
(5,7)	30240	49×30	20.6
(7,5)	30240	52×30	18.0
(3,4)	25200	47×30	17.9
(7,6)	25200	52×30	16.2
(5,8)	11340	58×30	6.5
(8,6)	9450	51×30	6.2
(9,5)	2520	53×30	1.6

Case 2: The result of $\mathbf{P_{IP}}$

The weighting matrix W of this example system, which is used as coefficients for the NAG IP subroutine, is computed to be as follows.

The first five columns of the weighting matrix are:-

$$W(1:5) = \begin{bmatrix} 263 & 16.8 & 13.6 & 1261.4 & 6.5 \\ 1088.7 & 69.3 & 55.5 & 5339.4 & 28.2 \\ \underline{54790} & 7530 & 2830 & \underline{271500} & 1410,6 \\ 98870 & 6330 & 5110 & \underline{480810} & 2500 \\ 138440 & 8660 & 7040 & \underline{652390} & 3440 \\ 392 & 25.3 & 20.4 & 1934.7 & 10 \\ 83280 & 5360 & 4300 & \underline{411240} & 2150 \\ \underline{3206} & 204 & 161 & \underline{15813} & 77 \\ 36.2 & 2.5 & 1.9 & 211.4 & 0.99 \\ 706.7 & 45.4 & 36.6 & 3453.2 & 17.9 \end{bmatrix}$$
(29)

The second five columns of the weighting matrix are:-

$$W(6:10) = \begin{bmatrix} 533 & 18.6 & 1195.6 & 180.1 & 8.1 \\ 2172.1 & 77 & 4948 & 747 & 34.4 \\ \underline{110080} & 3920 & \underline{251050} & \underline{37540} & 1750 \\ \underline{199410} & 7030 & \underline{450900} & \underline{67820} & 3090 \\ \underline{277730} & 9620 & \underline{621060} & \underline{95100} & 4180 \\ 791.1 & 28.1 & 1797.7 & 268.2 & 12.5 \\ \underline{167240} & 5950 & \underline{381150} & \underline{57120} & 2650 \\ \underline{6129} & 224 & \underline{14020} & \underline{2127} & 104 \\ 71.2 & 2.8 & 171.1 & 24 & 1.4 \\ 1426.9 & 50.4 & 3231.6 & 484.6 & 22.2 \end{bmatrix}$$
(30)

The computed subset for the selected inputs is $\{3, 4, 5, 7, 8\}$ and the output subset is $\{1, 4, 6, 8, 9\}$, respectively, which is as expected. Note that the sum of all the underlined elements in W is the optimal cost.

The algorithm $\mathbf{P_{IP}}$ was written in FORTRAN, with the W matrix obtained in Matlab. For this (10×10) case, the B&B program had only searched for 2 nodes. The objective function calls needed are much less than the objective function calls that $\mathbf{P_{GA}}$ needed. Some other examples were calculated to test the correctness and the efficiency of $\mathbf{P_{IP}}$ as well.

6. CONCLUSION

In this paper, we have developed two algorithms for input/output structure reduction based on some system controllability/observability indices. From the example we can see that $\mathbf{P_{GA}}$ is easy to use and more flexible, the additional constraints can be catered for easily. On the other hand, $\mathbf{P_{IP}}$ is much faster but requires large memory space for calculation. However, the algorithms presented here can only be applied for large linear systems. In general process control, systems are normally described by sets of non-linear differential equations, which requires further investigation.

7. ACKNOWLEDGMENT

The authors would like to thank Dr. Daniel Ho and Dr. James Lam for their valuable comments and helps.

8. REFERENCES

Daellenbach, H.G., J.A. George and D.C. McNickle (1983). *Introduction to operations research techniques.* Allyn & Bacon Inc.

Engell, S. (1997). *Controllability analysis and control structure selection.* IFAC Symposium on New Trends in Design of Control Systems, Smolenice.

Goldberg, D. (1989). *Genetic algorithms in search, optimization and machine learning.* Addison-Wesley.

Holland, J.H. (1992). *Adaptation in natural and artificial systems.* MIT Press, Cambridge, MA.

Hyland, D.C. and D.S. Bernstein (1985). The optimal projection equations for model reduction and the relationship among the methods of Wilson, Skelton and Moore. *IEEE Trans. on Automat. Ctrl.*, Vol. AC-30, 1201–1211.

B.C. Moore (1981). Principal Component Analysis in Linear Systems: Controllability, Observability, and Model. *IEEE Trans. on Automat. Ctrl.*, Vol. AC-26, 17–31.

H.A. Taha (1987). *Operations Research: An Introduction*, Macmillan Publishing Co., New York.

Yeung, L.F., M. Zhang, J. Lam and D. Ho (1996). Input/Output Structure Reduction for Multivariable Control Systems by Genetic Algorithm. *Proc. 4th ICARCV*, Singapore. pp. 2119–2123.

CONTROL OF POWER SYSTEM USING ADAPTIVE FUZZY CONTROLLER

Oscar Calvo

LEICI, Facultad de Ingeniería, Universidad Nacional de La Plata
CICpBA, Argentina
Department of Physics, University of the Balearic Islands
Palma de Mallorca, Spain
oscar@galiota.uib.es

ABSTRACT: This paper presents a control system for an asynchronous AC drive based on an adaptive fuzzy identifier and a neuro-fuzzy controller. The system is able to adapt itself to changes in load or power line by changing the rules of the neurofuzzy controller. A model-reference approach is used to impose transient closed loop performance. The architecture chosen for the controller is based on an ANFIS structure (Jang 1993) that utilizes a hybrid mechanism of adjust combining LSE with back propagation and descend methods. The system is used to control an asynchronous tri-phase AC motor. *Copyright © 2000 IFAC*

Keywords: Fuzzy Control. Adaptive Control, Motor Control. Power Control

1. INTRODUCTION

Recently, AI techniques have been applied to control systems, especially since the introduction of the Fuzzy Sets (Zadeh 1965) and the development of the Fuzzy Control Systems. These systems are adopted by the control engineering community to solve problems with different levels of complexity. The most popular and one of the simples implementation of a fuzzy controller are the Fuzzy versions of standard PI, PD and PID controllers. They started by adding expert rules based on closed loop response: overshoot, transient response, etc., evolving to more complex versions like the PID with expert tuning. They are used mainly in industrial processes when the human expert put his knowledge into rules. These systems are usually multivariable and poorly modeled, but the behavior against changes in the control variables is understood.

When there is no expert capable of describing these input-output relations and the system is highly non-linear and time dependent, the optimum setup of these controller may be a long and costly process. In these cases, adaptive controllers come to the rescue, adjusting the rules automatically based in cost

functions. Within this framework a popular structure is the neuro-fuzzy controller, where the knowledge can still be described linguistically but the rules are adjusted using training algorithms developed for the Neural Networks. The expert can still contribute defining the initial rules if necessary but if the system changes or behaves nonlinearly the system adapts itself to the changes providing always the optimum response. A backpropagation technique can be used to train the Knowledge Base. In this direction Jang's works in his PhD thesis and later in his papers made a big contribution developing his ANFIS structure (Jang 1992, 1993).

2. NEUROFUZZY SYSTEMS

2.1 ANFIS Architecture

ANFIS (Adaptive Network-based Fuzzy Inference System) is a very well known scheme that allows the approximation of almost any non-linear function with a rule-based controller (Consoli 1995). Its parameters are adjusted by training input-output patterns using backpropagation techniques developed for Neural Networks (Narendra 1990). A short description will be done for clarity of the paper.

The system can be described by a set of rules of the Takagi-Sugeno (TKS) type. For instance, in a system with two rules:

Rule 1: IF x IS A_1 AND y IS B_1 THEN   (1)
$$f1=p_1x+q_1y+r_1$$
Rule 2: IF x IS A_2 AND y IS B_2 THEN
$$f2=p_2x+q_2y+r_2$$

Figure 1 represents graphically the previous rulebase. The net is integrated by fixed nodes represented by squares and adaptive nodes indicated by circles.

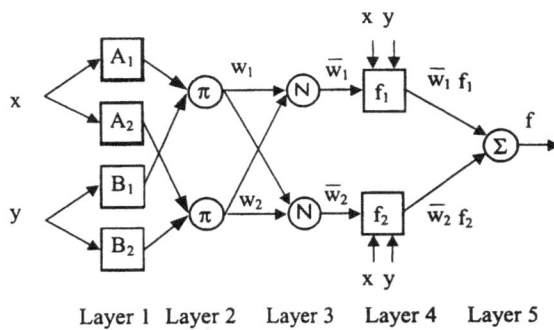

Layer 1   Layer 2   Layer 3   Layer 4   Layer 5

Fig. 1. Graphical representation of an ANFIS structure with two rules.

In this representation of the fuzzy system, x is the input to node i, μ_{Ai} is the membership of x to set Ai. Typical memberships used in ANFIS are:

$$\mu_{Ai} = \frac{1}{1 + \left(\left(\frac{x - c_i}{a_i} \right)^2 \right)^{b_i}} \quad (2)$$

where a_i, b_i y c_i are the adaptable parameters.

Also:

$$w_i = \mu A_i(x) \times \mu B_i(y) \quad i=1,2. \quad (3)$$

and

$$\overline{w}_i = \frac{w_i}{\sum_{i=1}^{n} w_i} \quad (4)$$

$$O_4^j = \overline{w}_i f_i = \overline{w}_i (p_i x + q_i y + r_i) \quad (5)$$

Finally, the output

$$O_5^i = Output = \sum_{i=1}^{n} \overline{w}_i f_i \quad (6)$$

The adaptation mechanism is performed with a mixed technique that uses the Backpropagation algorithm for the centers and widths of membership functions and the Recursive Least Square Estimator

(RLSE) for the Takagi Sugeno coefficients. This technique assures a global minimum and speeds-up the searching process necessary for Real Time Control.

3. CONTROL SYSTEM ARCHITECTURE

A standard scheme used to control variable non-linear systems is the Reference Model Controller, MRAC (Åstrom 1995). In this method the output of the plant is compared with the output of a model whose behavior the plant must imitate. The difference between both is used to adjust the controller.

This work proposes the use of this technique with an adaptive neurofuzzy system as a controller, adjusting its parameters by backpropagation. To use this algorithm, the output error has to propagate through the inverse model of the plant. Unfortunately in many cases the model of the plant is not known or the inverse Jacobian is difficult to compute. Since in our case, the AC model is variable and not known very well, we discard the analytical solution. The proposed solution is to identify the plant using another ANFIS structure as an estimator (Method I). A scheme of the proposed solution is shown in figure 2.

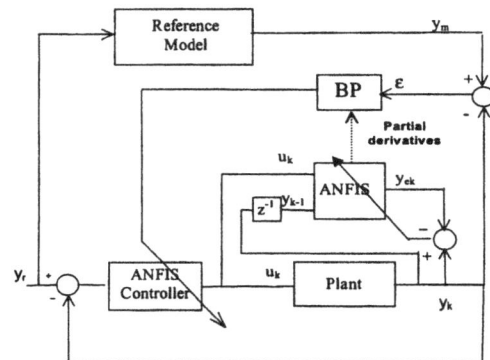

Fig. 2. Reference Model Control with controller and Estimator implemented with ANFIS

The model of the plant given by the ANFIS identifier is used to propagate the output error using the gradient descend to the Neurofuzzy controller with a similar structure to the one used by the identifier.

Another scheme presented in this work replaces the ANFIS identifier by a Fuzzy Inverse Model (FIM) originally proposed by Layne and Passino for the rudder control of ships (Layne 1992a, b). The Inverse model is based on the observations of the dynamics of the plant+reference-model and the corrective actions to be applied to the controller. There is a nonlinear mapping between the tracking error ε (between output of controlled process and model)

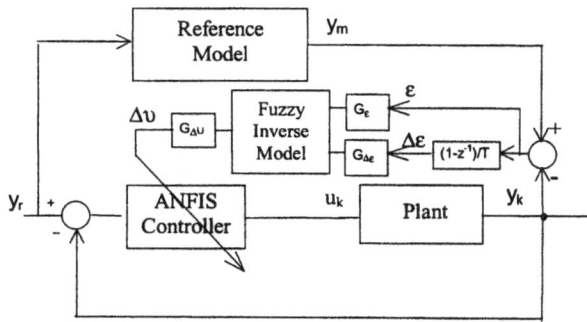

Fig. 3. MRAC with ANFIS controller and FIM
Identifier (Method II).

and its derivative $\Delta\varepsilon$ the necessary changes in the
control signal to bring these errors to zero ($\Delta\upsilon$).

In the scheme presented here, the FIM is used to
estimate the shift of the control signal $\Delta\upsilon$ from the
optimum situation (having ε and $\Delta\varepsilon$ equal to zero all
the time), and the RLSE algorithm is used to adjust
the output consequents of the TKS controller,
minimizing that error ($\Delta\upsilon$). A scheme of the
proposed system is seen in figure 3.

The Universes of discourse for input and output
variables are normalized into the [-1,1] interval hence
the gains G_ε and $G_{\delta\varepsilon}$ are adjusted so that the sampled
values are mapped into that interval. Similarly, the
gains G_υ and $G_{\Delta\upsilon}$ are adjusted so that υ y $\Delta\upsilon$ take the
proper values. Table I shows the rulebase to build the
25 rules FIM. The rule outputs indicate the error in
the control signal $\Delta\upsilon$, based on the tracking error ε
and its derivative $\Delta\varepsilon$.

Table I Rule Base for the Fuzzy Inverse Model
(FIM).

$\varepsilon / \Delta\varepsilon$	NB	NS	Z	PS	PB
NB	NB	NB	NB	NS	Z
NS	NB	NB	NS	Z	PS
Z	NB	NS	Z	PS	PB
PS	NS	Z	PS	PB	PB
PB	Z	PS	PB	PB	PB

4. APPLICATION TO AC DRIVES

The system described previously was utilized to
control a three Phase Induction Motor (Bose, 1986).
This work presents the results obtained by
simulations and the experimental setup is underway.

4.1. Dynamical Model of a Three Phase Induction Motor

The traditional model of an induction motor is based
on the conventional phase model that is only valid for
stationary operations. When the transient behavior

and its closed loop response are desired, a dynamical
model of the motor has to be used (Carrera, 1996).
Using a transformation from two to three phases for
the currents in both stator, rotor and voltages in the
stator, the drive can be seen as a two phase's
machine. Also, stator and rotor equations have to be
referenced to a common reference framework,
rotating at angular frequency ω_x. Differential
equations describing dynamical behavior of induction
motor for arbitrary voltage and current waveforms
are:

$$\overline{v}_s = rs \cdot i_s + ls \cdot \frac{d\overline{i}_s}{dt} + lm \cdot \frac{d\overline{i}_r^s}{dt} \qquad (8)$$

$$0 = rr \cdot i_r + lr \cdot \frac{d\overline{i}_r}{dt} + lm \cdot \frac{d\overline{i}_s^r}{dt} \qquad (9)$$

$$J \cdot \frac{d\omega_m}{dt} = P \cdot \frac{lm}{lr} \cdot (\phi_{\alpha r} \cdot i_{\beta s} - \phi_{\beta r} \cdot i_{\alpha s}) - T_L - B \cdot \omega_m \qquad (10)$$

$$\omega_m = \frac{d\varepsilon}{dt} \qquad (11)$$

Equation (8) represents stator voltage as a function of
stator and rotor currents referenced to stator
(Stationary coordinates, $\omega_x=0$), (9) represents rotoric
voltage in rotor coordinates and (11) and (12)
represent the mechanical dynamics of the motor and
load. It can be seen that in stationary state, the airgap
flux is proportional to the ratio Vs/ω_e, and a speed
control of constant Vs/ω_e, is proposed. This type of
control allows handling quick transient responses of
the converter since the motor operates in the constant
torque region.

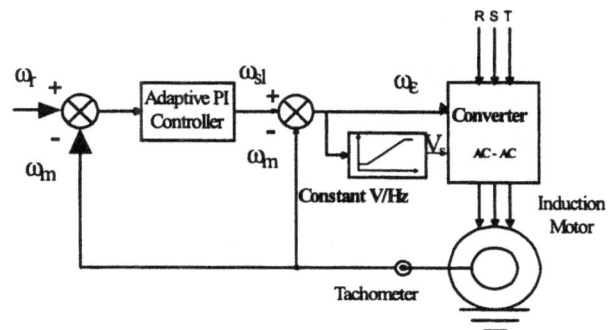

Fig. 4. Control Constant V/Hz control with slip
regulation.

Table II Motor Parameters

lm	45.68 mHy
ls	49.15 mHy
lr	47.88 mHy
rs	1.03 Ω
rr	0.51 Ω
P	1 par
J	0.006 kg.m^2
B	0.0033
T$_L$	0.2 N.m

The system is completed with a slip control. The
reference speed is compared with real speed given by

the tachometer and the adaptive PI controller integrates the error. The slip command signal ω_{s1} is added to the mechanical speed deriving the control actions in both frequency ω_s and voltage V_s keeping their ratio constant. A scheme of this controller can be seen in figure 4. The motor parameters are shown in Table II.

4.2. Reference Model

Based on experimental observations of the open loop response of the drive, the Reference model was chosen accordingly as a second order system:

$$T(s) = \frac{\omega_n^2}{s^2 + 2 \cdot \xi \cdot \omega_n \cdot s + \omega_n^2} \quad (12)$$

with

$$\xi = 0.95 \ and \ \omega_n = 100 \ ^{rad}\!/_{seg}$$

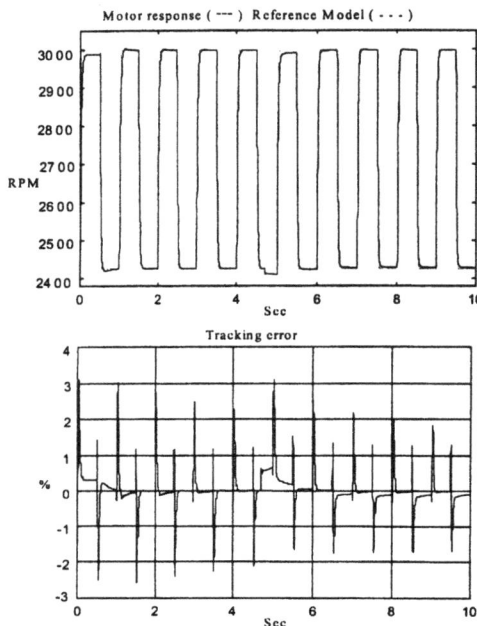

Fig. 5. (Top) Responses of Reference Model and Motor to squared speed reference. (Bottom) Tracking error. Both cases using method I.

5. RESULTS

The two methods described previously to adjust the rules of the fuzzy controller were tested to control the motor. Also, for comparison different rulebases were tried: one with five linguistic labels for each input variable, leading to a 25 rules knowledge base, and the other with three labels and 9 rules, giving similar results.

To compare both methods a square wave was applied as the speed reference in the two cases. Also, after 4.7 sec, the mechanical load was increased ten times to analyze the load perturbation rejection.

Figure 5 shows the results when the control scheme shown in figure 2 (Method I) is used. Both controller

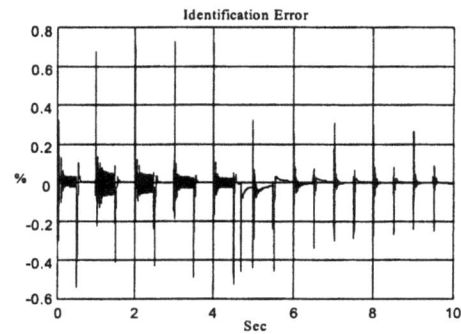

Fig. 6. Plant Identification error.

and identifier were ANFIS structures with the same number of rules. Bottom part of figure 5 shows the tracking error in speed between real speed of the motor and the Reference Model.

Figure 6 shows the identification error defined as the difference between the motor speed with the identifier for the same input. The identifier converges rapidly and the errors are within a 1%. The same analysis was performed for the strategy shown in figure 3 (Method II), using the fuzzy inverse model to propagate the error derivatives.

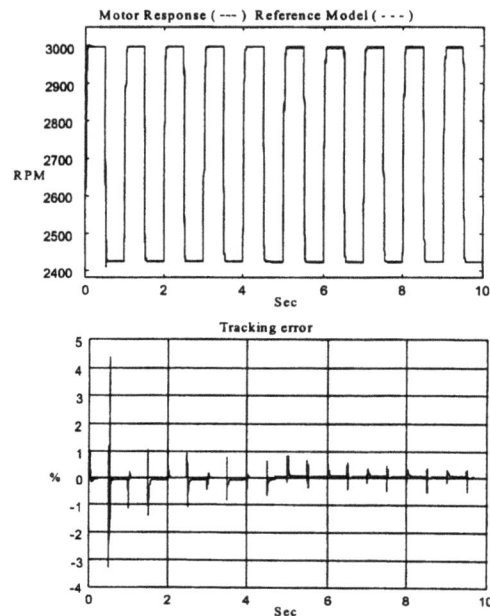

Fig. 7. (Top) Responses of Reference Model and Motor to squared speed reference. (Bottom) Tracking error. Both cases using method II.

It can be seen that in this case the tracking error is significantly smaller than with method I. Also, after many simulations it can be concluded that the method II is more robust and easy to adjust. The convergence is much quicker (one cycle) and the model follows the motor dynamics with high accuracy (smaller than 1%). Also the system reacts very fast to load disturbances.

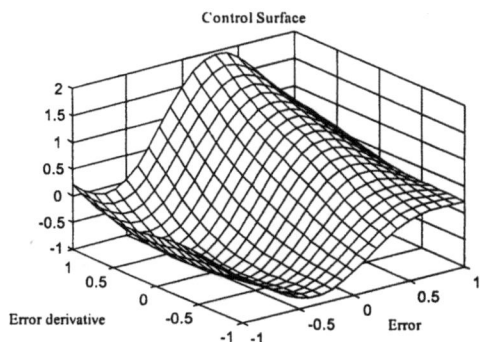

Fig. 8. Control Surface (Method II).

To appreciate in more detail the responses obtained, figure 9 shows simultaneously: the output of the reference model, the Motor speed when the controller is adjusted using both schemes and the response motor-controller without compensation. It should be noted that no appreciable difference could be observed between motor and Reference model when the second method is used. Also, the non-linear open loop response of the motor is plotted since it behaves differently depending if the reference signal is going up or down. Also, it can be seen that controller compensates the slippage.

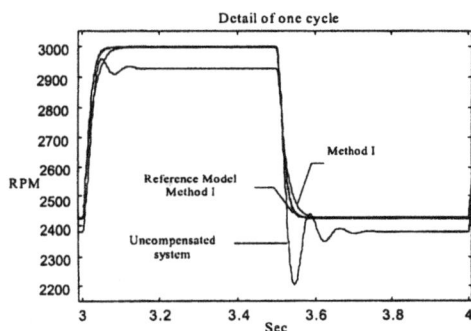

Fig. 9. Response of motor to step changes in speed.

Figure 10 details the responses for the two cases mentioned above against load disturbances. Again the advantages of method II become evident.

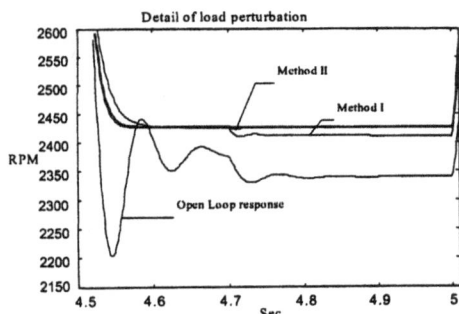

Fig. 10. Motor response to load changes.

The techniques used in this work are an interesting approach to control linear and no-linear systems where the structure is not known very well and they show a variable structure. From the simulations it can be seen that the second method used for identification converges more quickly than using method one, require less real time processing since the identification performed by the ANFIS structure is not necessary.

6. CONCLUSIONS

Different rulebases, with 9 and 25 rules were tried for both the controller and the inverse model and the differences were not significant. A stability analysis if planned as future work. Results obtained using the fuzzy identifier are very sensitive to the selection of the learning parameters and gain coefficients of the controller and identifier. For our case the choice was probably not the optimum set but enough to give adequate results. Adjusting procedure was cumbersome.

Finally, even though this paper presents results from simulations performed with MATLAB, a real time implementation using C running on a Motorola DSP processor is underway and experimental results are expected soon. The method is being used to control a 3-Phase motor of 1.5 CV made by CZERWENY, coupled to a VARIMAK FM140AP brake using a DANFOSS 3002 controller.

REFERENCES

Åstrom, K.J., Wittenmark, B. (1995). *Adaptive Control* Addison Wesley.

Bose, B.K., (1986). *Power Electronics and AC Drives*, Prentice Hall.

Carrera, A.M and Valla, M.I. (1996). Control Vectorial Indirecto de Motores de Inducción, *Proyecto Final Ingeniería Electrónica, Facultad de Ingeniería*, Universidad Nacional de La Plata, Argentina.

Consoli, A., Lorenz, R., Seidl, D.and Schooder D. (1995). Neuro-Fuzzy Drive/Motion Control, *IEEE, IAS'95 Tutorial Course Notes*, Oct 8, 1995.

Jang, J.R. (1992). Neuro Fuzzy Modeling: Architecture, Analysis and Applications. *PhD Thesis*, University of California, Berkeley, CA. July 1992.

Jang, J.R. (1993.) ANFIS: Adaptive-Network-Based Fuzzy Inference System, *IEEE Transactions on System Man and Cybernetics*, **23**, n° 3, May/June 1993.

Layne,J. (1992) Fuzzy Model Reference Adaptive Control. *Master's thesis, Dept. of Electrical Engineering*. Ohio State University. Mar. 3 1992.

Layne, J., Passino, K. and Yurkovich, S., (1992). Fuzzy Learning Control for Antiskid Braking Systems. *IEEE Transactions on Control Systems Technology*, Vol. 1, n°2, June 1993, pp. 122-129.

Narendra, K. and Partasarathy, (1990). A. Identification and Control of Dynamical Systems Using Neural networks, *IEEE Transactions on Neural Networks*, pp. 4-27. Vol. 1, N° 1.

Zadeh, L.A. (1965). *Fuzzy Sets*. Information and Control, Vol. 8, pp. 338-353.

NEURAL NETWORK OPTIMISATION USING GENETIC ALGORITHM: A HIERARCHICAL FUZZY METHOD

S. K. Sharma and M. O. Tokhi

Department of Automatic Control and System Engineering
The University of Sheffield, UK

Abstract: In this paper, fusion of neural networks (NNs), genetic algorithms (GAs) and fuzzy logic (FL) is considered by taking account of the advantages of each. In this process neural networks are used for universal approximation, genetic algorithms are used for optimisation of the structure and weights of the neural network and fuzzy logic is used to give directional and priority approach to genetic evolution. Hierarchical fuzzy approach can simultaneously provide a priority and direction of search to a chromosome so as to achieve an optimal or near optimal set of solutions. The proposed approach dynamically adopts the chromosome and maintains uniformity and diversity in the population to simultaneously provide local and global search. Modelling of flexible manipulator is used to demonstrate the performance of the proposed approach. *Copyright © 2000 IFAC*

Keywords: Evaluation function, genetic algorithm, hierarchical fuzzy approach, neural networks.

1. INTRODUCTION

Genetic algorithms and related techniques offer promising approaches for automatically exploring the design space of neural architectures (Goldberg, 1989). Central to this process of evolutionary design of neural architecture is the choice of fitness function for the GA. Fitness function not only constrains the class of neural architectures that are representable (evolvable) in the system, but also determines the efficiency and the time - space complexity of the evolutionary design procedure as a whole. Since GA is used to find both an optimal architecture and synapse weights of neural network architecture, the evaluation function must include not only a measure of sum-squared-error (MSE), but also a feasibility measure of network structure and its complexity, i.e. number of nodes and their connectivity. Many fitness functions have been used along with LMS or MSE (Korning, 1995; Schmidt and Stidsen, 1995). Korning (1995) compared LMS with a fitness function where the number of output nodes with correct sign were used as a basis for fitness ranking, and found that it outperformed the LMS. Further tests and results

published in (Schmidt and Stidsen, 1995) show that training of a high performance NN classifier with a GA can effectively be done using a fitness function that is strongly based on the number of correctly classified patterns during learning. Whitley *et al.* (1990) proposed an evolution process that rewards feasible networks with fewer nodes and links. Russo (1998) utilised various parameters as a fitness measure to obtain an optimal solution through genetic evolution. Some of the other fitness functions used for genetic evolution can be found in (Takumi and Eiichiro, 1995) and (Schmidt, 1995, 1996).

The evaluation function usually incorporates several constraints. In this work four constraints namely: correctly classified pattern (**Cr**), change in error ($\dot{e}$), error (**e**) and number of hidden nodes (**Th**) are chosen in the evaluation function. The first three factors are responsible for tuning of the neural network whereas the last factor is for optimisation of the neural network structure. In general, the analytical expression of the fitness function of an individual *i* is given by

$$\text{fit}(i) = \frac{k}{k_1\mathbf{Cr} + k_2\dot{\mathbf{e}} + k_3\mathbf{e} + k_4\mathbf{Th}} \qquad (1)$$

Choosing an inappropriate constant (k, k_1, k_2, k_3, k_4) in the evaluation function can lead to a near infeasible structure or non-optimal solution. As the number of constraints increases it increasingly becomes difficult to select a constant. Moreover, there is no consideration given to priority in selection and direction as how a particular chromosome is moving in a problem domain. Such guidance is necessary for dynamic adaptation of a chromosome in the genetic evaluation.

In this paper, a hierarchical fuzzy method is proposed for the selection of an evaluation function. This approach dynamically allocates a priority as well as optimise the direction of movement of a chromosome in the problem domain. Acceptable range of each constraint can be set in simple linguistic terms. Furthermore, to give more resolution and fine local tuning, unequal length and dynamic range of membership function is applied (Sharma and Tokhi, 2000).

2. HIERARCHICAL FUZZY METHOD

Since there are four constraints considered in this work, it is difficult to code the constraints into rules in a simple fuzzy logic. It is also difficult to provide a priority approach in the problem domain. If new constraints are added, it is very difficult to code all the rules with new considerations in a simple fuzzy logic approach. The hierarchical fuzzy method removes all these ambiguities by dividing the constraints into different tiers based on priority. Here two tier structures are considered. Fig. 1 shows the hierarchical fuzzy structure containing two tiers. The first tier contains only one module T1M1 whereas Tier 2 contains three modules, namely T2M1, T2M2 and T2M3.

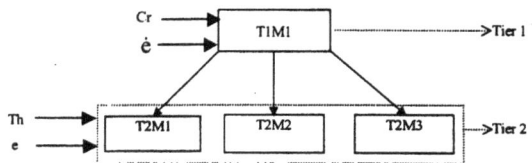

Fig. 1. Hierarchical Tier structure.

In Tier 1, the correctly classified pattern (**Cr**) and change of error ($\dot{\mathbf{e}}$) are used as antecedents and the fitness value as a consequent. Error and change in error at k-th generation are given by

$$e(k) = y_r - y(k) \qquad (2)$$

$$\dot{e}(k) = e(k) - e(k-1) \qquad (3)$$

where, y_r is the desired output and $y(k)$ is the output at k-th generation.

Table 1 shows the inference rules associated with module of Tier 1. It can be seen that a chromosome fires a rule from Table 1 depending upon its correctly classified pattern and change in error.

Table 1: Tier 1 module (T1M1)

Cr $\dot{\mathbf{e}}$	ZR	PS	PM
NM	B	B	T2M1
NS	M	B	T2M2
ZR	S	M	T2M3
PS	Z	S	M
PM	Z	Z	S

Table 2 shows the inference rules associated with modules of Tier 2. Three modules namely T2M1, T2M2, and T2M3 are considered in Tier2. Each module in Tier2 has the error (**e**) and number of hidden nodes (**Th**) as antecedents and the fitness value as a consequent. The range of fitness value of Tier2 is more than Tier1 and in Tier2, range of fitness value for T2M1 is more than T2M3 and so on. The fuzzy sets of consequent are labelled as B, B1, B2, and B3 (Big Positive), M, M1, M2, and M3 (Medium Positive), S, S1, S2, and S3 (Small Positive), and Z, Z1, Z2, and Z3 (Zero) respectively for T1M1, T2M1, T2M2, and T2M3. Since value of B1>B2>B3>B and similarly for M1, S1 and Z1, any chromosome reaching Tier 2 gets more fitness values than Tier 1, and is thus more acceptable. Tier 1 acts as a filter, any chromosome sneaking into Tier 2 will have most acceptable features of Tier 1. Any chromosome stuck in Tier 1 will be refined through genetic operators till it reaches an acceptable level. Thus only those chromosomes having large (PM) number of correctly classified pattern (**Cr**) and of decreasing error ($\dot{\mathbf{e}}$) (see Table 1) are allowed to sneak through Tier 1 for further refinement in Tier 2. In Tier 2, chromosomes are selected on the basis of their mean square error (**e**) and number of hidden nodes (**Th**). Thus, every chromosome in the population gets fitness values depending on the basis of priority and constraints. Since the first priority is to select the chromosome with maximum number of correctly classified pattern (**Cr**) and of decreasing error ($\dot{\mathbf{e}}$), the best chromosome in every generation comes always with feasible solution. As the number of generations increases, the chromosomes become more refined. Since the number of data to be classified and maximum limit of hidden nodes are known at the beginning, the range of their membership functions (Figs. 2 and 3) can be decided. The range for membership functions of the error (**e**) and the change in error ($\dot{\mathbf{e}}$) are not known. These are set dynamically such that a considerable number of chromosomes is always near to the problem domain [9].

Table 2: Tier 2 modules (T2M1, T2M2 and T2M3).

The	ZR	PS	PM
ZR	B 1	B 1	M 1
PS	B 1	M 1	S 1
PM	M 1	S 1	Z 1
	T 2 M 1		
The	ZR	PS	PM
ZR	B 2	B 2	M 2
PS	B 2	M 2	S 2
PM	M 2	S 2	Z 2
	T 2 M 2		
The	ZR	PS	PM
ZR	B 3	B 3	M 3
PS	B 3	M 3	S 3
PM	M 3	S 3	Z 3
	T 2 M 3		

By changing the range of membership function of antecedents, the pressure of local and global search can be changed (Fig. 3). In this case, more stress is given to global search at the beginning of a generation and to local search at a later stage of the generation.

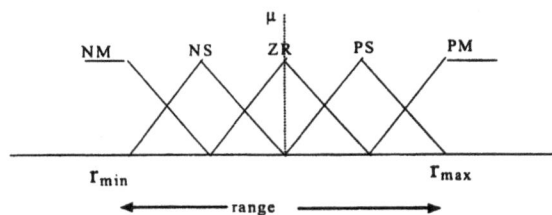

Fig. 2. Membership function of change in error.

Fig. 4 shows the membership function of the fitness value of both tiers. Depending upon the value of F_{min} and F_{max}, fitness value to a chromosome is allotted. Since the value of F_{min} and F_{max} of T2M1 is more than the value for T2M2, T2M3 and T1M1, any

chromosome falling in module T2M1 will be more acceptable and so on.

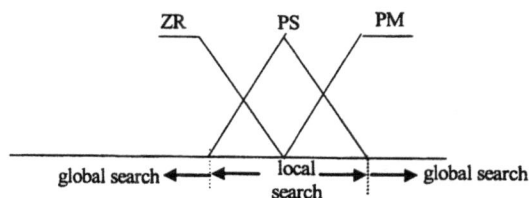

Fig. 3. Membership function of error, correctly classified pattern and number of hidden nodes.

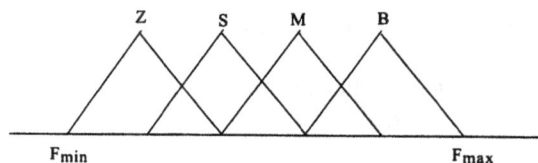

Fig. 4. Membership function of fitness value.

2.1. Genetic neural network learning

Neural network training using genetic algorithms offers several advantages. Since no derivative computations are involved, it is less likely for these algorithms to get trapped in local minima. Another advantage is that these algorithms constitute a parallel search as opposed to point-by-point search. Since the objective in this work is to optimise the structure of neural network, pruning and better generalisation is achieved. Each individual in the population is composed of two segments of strings, which represent the three levels of genomic structure; the higher segment defines the connectivity, the lower encodes the connection weights and biases. The first level denotes the number of hidden units for the neural network with fixed number of input and output units (Fig. 5). The second level denotes the connection, where each bit represents an individual connection, which can take the value '1' or '0'; the value '1' meaning that the connection is established. Thus, $C_{ij}=1$ implies connection between the jth unit of one layer to the ith unit of second layer. Since the weights and biases are both numeric values, the third level is encoded as a fixed length real-number string rather than a bit string. Initially the hidden units and connections are encoded randomly using binary bits '1' or '0' and weights and biases are encoded randomly between ($\pm$ 1.0) as shown in Fig. 5.

With this approach, the optimal architecture and its optimal set of input weight and biases are determined simultaneously. No a priory assumption about the network topology is needed and the only information required is the input and the output characteristic of the task.

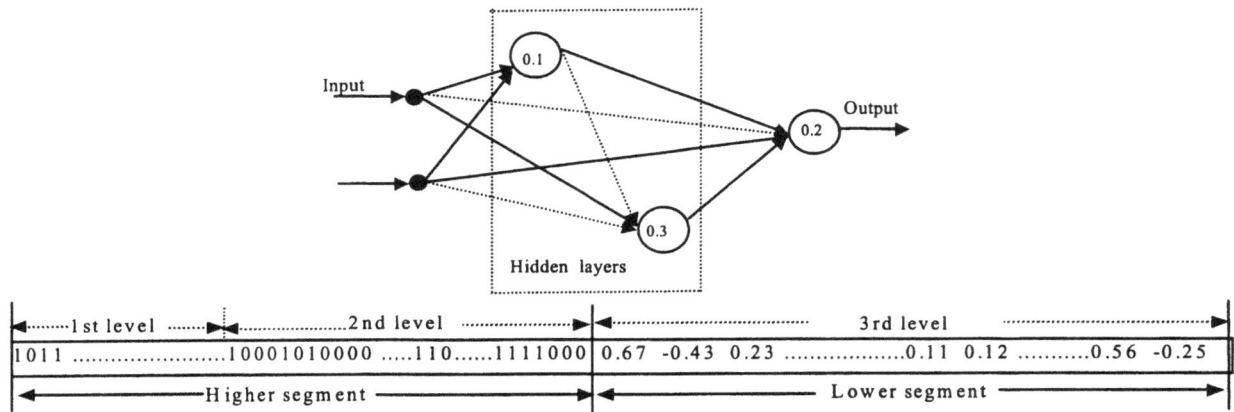

Fig. 5. Neural network architecture encoding.

The following steps outline the training process of the neural network architecture, at time k:

1. A population (P) of neural networks is considered. Steady state GA is used where the children from the best two parents replace least two chromosomes after going through crossover and mutation. Uniform crossover is applied to the bits part and arithmetic crossover is applied to the decimal coded part of a chromosome.

2. Half of the best population is subjected to biased mutation where 1 is replaced by 0 and vice-versa in the case of bits and a small random number is added in case of decimal. The remaining half of the population is subjected to unbiased mutation where selected bits and weights are randomly replaced. This type of technique maintains the diversity as well as uniformity in the population.

3. Selection of a chromosome is done using fuzzy logic. Fuzzy logic with dynamic range and unequal length of membership function increases resolution among the chromosome [9]. Thus, providing more diversity and fine local tuning at the later stage of generations. Hierarchical fuzzy logic is used to give priority in selection of the constraints required for feasible and optimal structure of the neural network. In this case, the first hierarchy of fuzzy logic selects the entire chromosome with more correctly classified patterns and of decreasing error. The second hierarchy of fuzzy logic selects all the prior selected chromosomes with more pruned structure. Thus, the second hierarchy gets chromosomes, which are already refined in the first hierarchy with their constraints. In this manner, priority is exercised in the selection and the approach always selects the chromosomes with feasible solution.

4. The newly formed individuals are assigned fitness values as described above.

5. In spite of selecting random weights, when a particular link is reconnected again, previous trained weights are used to give better convergence.

6. If the maximum number of generations is not exceeded, proceed to step 3; else, the neural-network with the best fitness value in the final generation is chosen as the optimum.

3. EXPERIMENTAL SET UP

Fig. 6 shows the experimental set up of a flexible manipulator. The shaft encoder is used for measuring the hub angle and tachometer is used for measurement of the hub velocity of the manipulator. An accelerometer is located at the end-point of the flexible arm measuring the end-point acceleration. The flexible arm is constructed using a piece of thin aluminium alloy. The parameters of the flexible arm are given in Table 3.

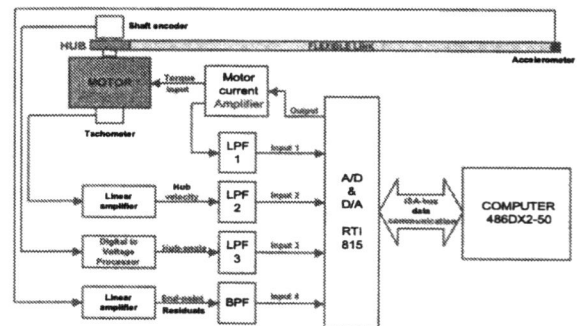

Fig. 6. The flexible manipulator experimental set up.

4. SIMULATION STUDIES

A population of 20 neural networks was considered. Neural network of MLP type with tanh as a sigmoidal function is considered. The initial set of weights was randomly generated between ±1.0. The neural network with 9 inputs in case of hub-angle and end-point acceleration and 11 inputs in case of hub-velocity is considered. It has one output and two hidden layers in each case. Maximum number of hidden units in each hidden layer is 8. The crossover probability (P_C) was set at 0.35 for decimal coded part and uniform crossover for binary coded part. Biased mutation was 0.3 and unbiased mutation was 0.5. In each generation, 70% of the data were selected for training and the remaining 30% for testing. Total runs were for 3000 generations. The goal of the experiments was to model a flexible manipulator for hub-angle, hub-velocity and end-point acceleration. To provide a comparative assessment, two methods were used to select an evaluation function.

Table 3. Physical parameters of the flexible manipulator.

Parameter	Value	
Length	960.0	mm
Width	19.008	mm
Thickness	3.2004	mm
Mass density per unit volume	2710	kgm^{-3}
Second moment of inertia, I	5.192×10^{-11}	m^4
Young modulus, E	71×10^9	Nm^{-2}
Moment of inertia, I_b	0.04862	kgm^2
Hub inertia, I_h	4.86×10^{-4}	kgm^2

Method 1:

In this case, equation (1) is used to evaluate the fitness value of a chromosome. To give more priority to change in error ($\dot{e}$) and correctly classified pattern (**Cr**) as in case of hierarchical fuzzy approach, the constants are chosen accordingly as:

k=1.0; k_1=k_2=0.5; k_3=k_4=1.0;

Two more sets of values were considered as

k=1.0; k_1=k_2=0.1; k_3=k_4=1.0; and

k=1.0; k_1=k_2=0.01; k_3=k_4=1.0;

The best results out of these three are selected for comparison with the hierarchical fuzzy method.

Method 2:

In this case, the hierarchical fuzzy method is used to evaluate the fitness value of a chromosome. The dynamic range of the error (**e**) is 3 times the minimum error in the population at every generation, whereas in case of change in error ($\dot{e}$) it is 4 times the minimum value. Minimum and maximum values for the membership function of hidden nodes and correctly classified patterns are 30% and 70% of the total. Penalty is automatically provided whenever any chromosome goes out of range by the shape of membership function.

Fig. 7 shows the MSE for the hub-angle, hub-velocity and end-point acceleration at every 1000 generations. It is noted that fast convergence and fine local tuning is achieved through hierarchical fuzzy method in every case. This is because of the adaptive and priority nature of the approach in the selection. A trial and error method is always used to select a constant for method 1 and constants are not adaptive as in case of the hierarchical fuzzy method. Thus, selection of chromosomes is not dynamic. Moreover, if there is any new constraint added in the fitness function, it can easily be added in method 2 with adding extra module or tier with priority or relationship to other modules or tiers. In method 1, the chromosomes are selected on the basis of their fitness values, but in case of the hierarchical fuzzy method a chromosome is selected on the basis of its acceptable features, which can easily be changed as required. Furthermore, due to the dynamic range and unequal length of the membership function, more resolution is obtained at the later stage of generations. Thus, this type of approach maintains uniformity and diversity in the population, and provides a simultaneous local and global approach in selection.

5. CONCLUSION

A hierarchical fuzzy approach has been proposed for selection of an evaluation function. The approach effectively shows the dynamic nature of a chromosome in a problem-domain and allows the selection of a chromosome with acceptable features. Experimental results have shown that

1. The method is valuable for simultaneously combining the different constraints within a priority-based approach.

2. The method is easy to design because it requires only the operating range of constraints, which is well known for any problem.

3. The approach does not need human expert knowledge to decide the constants of a constraint in an evaluation function. Simple knowledge of a

problem-domain can effectively decide an evaluation function.

4. The approach is a scaleable integration method that can be applied when the number of constraints increases.

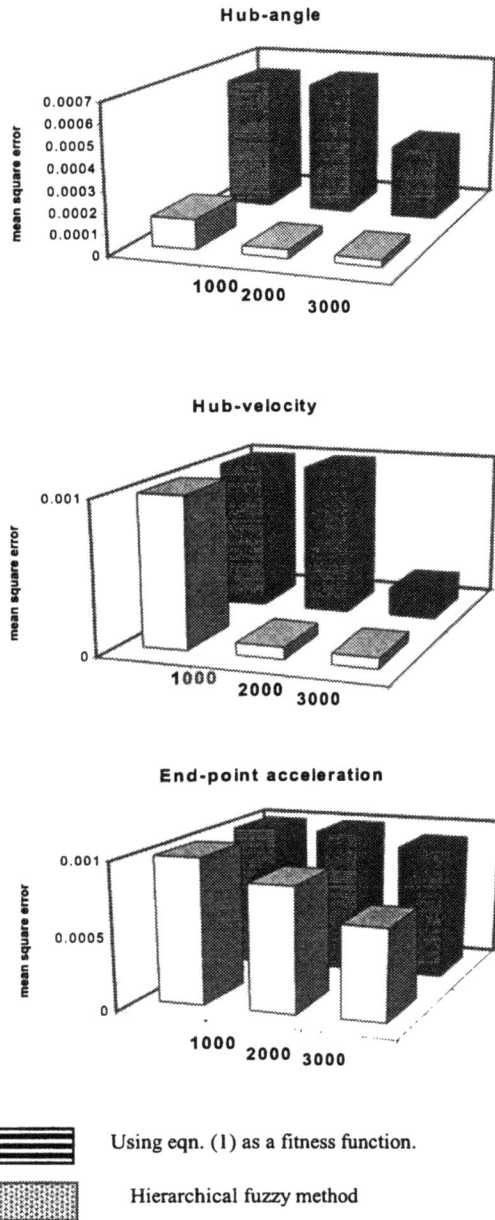

Hub-angle

Hub-velocity

End-point acceleration

Using eqn. (1) as a fitness function.

Hierarchical fuzzy method

Fig. 7: Mean square error vs generations.

5. Since the approach deals with constraints in tier, which, are independent, a parallel approach (algorithm) can be used to save time on execution.

6. The approach automatically assigns penalty for a constraint falling out of range by proper selection of its membership function. Because of a priority approach in the selection, a best gene in every

evaluation always comes with feasible solution. This may not be the case with other fitness functions.

Although the work presented here shows good results, more promising results may be obtained by considering an exponential penalty/reward in membership functions and choosing a membership function with different sets of priority. Future research will concentrate on these issues.

REFERENCES

Goldberg, D. (1989). *Genetic algorithms in search, optimization and machine learning*, Addison-Wesley, Massacheusettes.

Korning, P.G. (1995). Training of neural networks by means of genetic algorithms working on very long chromosome. *International Journal of Neural Systems*, **6**, (3), email:aragorn@daimi.aau.dk.

Russo, M. (1998). FuGeNeSys- A fuzzy genetic neural system for fuzzy modelling. *IEEE Transactions on Fuzzy Systems*, **6**, (3), pp. 373-388.

Schmidt, M. (1995). *A unification of genetic algorithms, neural networks and fuzzy logic: the GANNFL approach*. Research Report, Computer Science Department, Aarhus University, Denmark, 1995.

Schmidt, M. (1996). GA to train fuzzy logic rulebases and neural networks. FLAMOC '96: *Proceedings of Fuzzy Logic and the Management of Complexity*, on ftp.diami.au.dk, anonymous ftp: pub/empl/marsh/FLAMOC96.ps.Z.

Schmidt, M. and Stidsen, T. (1995). GA to train NN's using sharing and pruning: Global GA search combined with local BP search. *Proceedings of the Applied Decision Technologies*, on ftp.diami.au.dk, anonymous ftp: pub/empl/marsh/ADT95.ZIP, also in John Wiley's book "Neural networks and Their application", 1995, chapter 8.

Sharma, S. K. and Tokhi, M.O. (2000). Genetic evolution: A dynamic fuzzy approach. *Proceedings of FUZZ-IEEE2000: The ninth IEEE International Conference on Fuzzy Systems*, San Antonio (Texas), 07-10 May 2000, **2**, pp. 748-752.

Takumi, I. and Eiichiro, T. (1995). Applying adaptive structured genetic algorithm to reasoning and learning method for fuzzy rules using neural networks. *Proceedings of the IEEE International Conference on Neural Networks*, **6**, pp. 3124-3128.

Whitley, D., Starkweather, T. and Bogart, C. (1990). Genetic algorithms and neural networks: optimizating connections and connectivity. *Parallel Computing*, **14**, pp. 347-361.

FUZZY PREDICTIVE ALGORITHMS APPLIED TO FORCE CONTROL OF ROBOTIC MANIPULATORS

J.M. Sousa * Luis Filipe Baptista ** José M.G. Sá da Costa *

* Technical University of Lisbon, Instituto Superior Técnico
Department of Mechanical Engineering/GCAR
Av. Rovisco Pais, 1049-001 Lisboa, Portugal
Phone: +351-21-8418190, E-mail: sadacosta@dem.ist.utl.pt
** Escola Náutica Infante D. Henrique
Department of Marine Engineering
Av. Eng. Bonneville Franco, Paço de Arcos, Portugal
Phone: +351-21-4460010, E-mail: luisbaptista@enautica.pt

Abstract: This paper proposes a combination of classical impedance controller with a fuzzy predictive algorithm. This control strategy allows for the inclusion of a nonlinear environment in the control design in a straightforward way, improving the global force control performance. In order to reduce the oscillations of the optimized reference position, a fuzzy scaling machine is included in the force control strategy. The performance of the force control scheme is illustrated for a two degree-of-freedom robot. Simulation results obtained with the fuzzy control scheme reveal an accurate force tracking performance considering non-rigid environments. *Copyright © 2000 IFAC*

Keywords: Fuzzy systems, predictive control, force control, impedance control, robotic manipulators.

1. INTRODUCTION

A force control strategy for robotic manipulators in the presence of non-rigid environments combining the conventional impedance controller and a model predictive control (MPC) algorithm in the force control scheme was proposed recently (Baptista *et al.* 2000). In this force control methodology, the predictive controller generate the position and velocity references in the constrained direction, in order to obtain a desired force profile acting on the environment. This control strategy improves the global control system performance (Wada *et al.* 1993).

Usually, impedance and environmental models are linear, mainly because the solution of an unconstrained optimization procedure can be analytically obtained with a moderate computational burden. However, the environment has in general a nonlinear behavior, and a nonlinear model for the contact surface

must be considered. Therefore, in this paper the linear spring/damper parallel combination, often used as model of the environment, is replaced by a nonlinear one, where the damping effect depends on the penetration depth (Marhefka and Orin 1996). Unfortunately, when a nonlinear model of the environment is used, the resulting optimization problem to be solved in the MPC scheme is non-convex. This problem can be solved using discrete search techniques, such as the branch-and-bound algorithm (Sousa *et al.* 1997). This discretization, however, introduces a tradeoff between the number of discrete actions and the performance. Moreover, the discrete approximation can introduce oscillations around non-varying references, and slow step responses depending on the selected set of discrete solutions. These effects are highly undesirable, especially in force control applications.

A possible solution to this problem is the use of a fuzzy scaling machine as in this paper, which uses an adaptive set of discrete alternatives based on the fulfillment of fuzzy criteria applied to force control. This approach has been used in predictive control (Sousa and Setnes 1999), and is generalized here for model predictive algorithms. The adaptation is performed by a scaling factor multiplied by a set of alternatives. By using this approach, the number of alternatives is kept low, while performance is increased. Herewith, the problems introduced by the discretization of the control actions are highly diminished.

The paper is organized as follows. Section 2 summarizes the manipulator and the environment models. The impedance controller with force tracking is presented in Section 3. Section 4 present the model predictive algorithm and the branch-and-bound algorithm. The fuzzy scaling machine is presented in Section 5. Simulation results of a two-degree-of-freedom (2-DOF) robot are discussed in Section 6. Finally, some conclusions are drawn in Section 7.

2. MODELING

Consider an $n-$link rigid-joint manipulator constrained by contact with the environment. The complete dynamic model is described by (Sciavicco and Siciliano 1996):

$$M(q)\ddot{q} + C(q,\dot{q})\dot{q} + g(q) + d(\dot{q}) = \tau - \tau_e \quad (1)$$

where $q, \dot{q}, \ddot{q} \in \mathbb{R}^{n \times 1}$ correspond to the joint position, velocity and acceleration vectors, respectively, $M(q) \in \mathbb{R}^{n \times n}$ is the symmetric positive definite inertia matrix, $C(q,\dot{q}) \in \mathbb{R}^{n \times n}$ is the centripetal and Coriolis matrix, $g(q) \in \mathbb{R}^{n \times 1}$ contains the gravitational terms and $d(\dot{q}) \in \mathbb{R}^{n \times 1}$ accounts for the friction terms. The vector $\tau \in \mathbb{R}^{n \times 1}$ is the joint input torque vector and $\tau_e \in \mathbb{R}^{n \times 1}$ denote the generalized vector of joint torques exerted by the environment on the end-effector. From (1) it is possible to derive the robot dynamic model in the Cartesian space:

$$M_x(x)\ddot{x} + C_x(x,\dot{x})\dot{x} + g_x(x) + d_x(\dot{x}) = f - f_e \quad (2)$$

where x is the n-dimensional vector of the position and orientation of the manipulator's end-effector, $f_e = J^{-T}(q)\tau_e \in \mathbb{R}^{n \times 1}$ is the contact force vector and J represents the Jacobian matrix.

In this paper, the contact surface is modeled by a nonlinear spring-damper system. The interaction force on the environment $f_e = [f_n \ f_t]^T$ is given by (Marhefka and Orin 1996):

$$f_n = k_e \, \delta x + \rho_e \, \delta x \, \dot{x} \quad (3)$$

where the terms k_e and ρ_e are the environment stiffness and damping coefficients, respectively, $\delta x = x - x_e$ is the penetration depth, where x_e stands for the distance between the surface and the base Cartesian frame, and f_n is the normal contact force. Note that the damping effect depends non-linearly on the penetration depth δx. The tangential contact force f_t due to surface friction is assumed to be given by (Yao and Tomizuka 1995):

$$f_t = \mu \mid f_n \mid \text{sgn}(\dot{x}_p) \quad (4)$$

where $\dot{x}_p$ is the unconstrained or sliding velocity components and μ is the dry friction coefficient between the end-effector and the contact surface.

3. FORCE CONTROL

The impedance control approach aims at controlling the dynamic relation between the manipulator and the environment (Hogan 1985). The complete form of a second order type impedance control, which is valid for free or constrained motion, is given by:

$$M_d\ddot{x} - B_d(\dot{x}_d - \dot{x}) - K_d(x_d - x) = -f_e \quad (5)$$

where $\dot{x}_d$, x_d are the desired velocity and position defined in the Cartesian space, respectively, and $\dot{x}$, x are the end-effector velocity and position in Cartesian space, respectively. The matrices M_d, B_d and K_d are the desired inertia, damping and stiffness for the manipulator. The reference or target end-effector acceleration u_t is then given by:

$$u_t = M_d^{-1}\left(B_d\dot{e} + K_d e - f_e\right) \quad (6)$$

where $\dot{e} = \dot{x}_d - \dot{x}$, $e = x_d - x$ are the velocity and position errors, respectively. Thus, u_t can be used as the command signal to an inner position control loop in order to drive the robot to the desired trajectory.

From the classical impedance controller presented in (6) it is possible to obtain a force control scheme considering a steady-state contact condition. Virtual position and velocity references, x_v and $\dot{x}_v$ respectively, must be calculated in order to obtain a desired contact force on the environment (Singh and Popa 1995). Therefore, to apply a desired force profile f_d on the environment, the target or reference acceleration in the constrained direction $\ddot{x}_f$ is obtained from (6) as:

$$\ddot{x}_f = m_d^{-1}\left(b_d(\dot{x}_v - \dot{x}) + k_d(x_v - x) - f_n\right) \quad (7)$$

where m_d, b_d and k_d are the appropriate elements of matrices M_d, B_d and K_d of (5) in the constrained direction.

The major drawback of the impedance control approach given in (6), (7) is the precise calculation of x_v, especially for non-rigid environments with unknown stiffness characteristics. In this approach the target velocity $\dot{x}_v$ is assumed to be zero, since in general is a noisy signal. This paper presents a model predictive algorithm based on the impedance and the environment model, given respectively by (6) and (3), in order to obtain an optimized virtual reference. The impedance controller calculates the target acceleration vector $u_t = [\ddot{x}_f \ \ddot{x}_p]^T$, where $\ddot{x}_f$ given by (7) where the unconstrained acceleration $\ddot{x}_p$ is obtained from (6) with a proportional-derivative compensation term, given by:

$$\ddot{x}_p = \ddot{x} + K_P e + K_D \dot{e} \quad (8)$$

where K_P and K_D are proportional and derivative gain matrices, respectively. Notice that $\ddot{x}_p$ is only calculated in (6) for the directions where the position is controlled. The target acceleration vector is then used as the driving signal to the position control loop in order to linearize the robot dynamic model and track the desired force profile. Since the robot controllers are usually implemented in the joint space, it is useful to obtain the correspondent target joint command signal for the inverse dynamics controller. Then, using the appropriate cinematic transformations, $\ddot{q}_t$ will be given by:

$$u_q = J^{-1}(q)(u_t - \dot{J}(q)\dot{q}) \tag{9}$$

which finally leads to the following controller with force tracking, given by:

$$\tau = \hat{M}(q)u_q + \hat{g}(q) + J^T f_e \tag{10}$$

where $\hat{M}(q)$, $\hat{g}(q)$ are estimates of $M(q)$ and $g(q)$ in the robot dynamic model (1).

4. MODEL PREDICTIVE ALGORITHM

Predictive algorithms consist of a broad range of methods, which are used to predict a desired variable in an optimal way. The most common predictive algorithms are model based predictive controllers (Camacho and Bordons 1995), which have one common feature; the controller is based on the prediction of the future system behavior by using a process model. Generalizing, MPA are based on the following basic concepts: 1) Use of an available (nonlinear) model to predict the system outputs at future discrete time instants over a prediction horizon. 2) Computation of a sequence of future inputs using the model of the system by minimizing a certain objective function. 3) Receding horizon principle; at each sampling instant the optimization process is repeated with new measurements, and only the first input obtained is applied to the process.

In general, a MPA minimizes a cost function over a specified *prediction horizon* H_p. The future plant outputs are predicted at each time instant k using a model of the process. The predicted output values $\hat{y}(k + 1), \ldots, \hat{y}(k + H_p)$ depend on the states of process at the current time k and on the future inputs $u(k + 1), \ldots, u(k+H_c)$, where H_c is called the *control horizon* in model predictive control (MPC). For the sake of simplicity, the same nomenclature is used for MPA in this paper. The input signal changes only inside the control horizon, remaining constant afterwards, i.e., $u(k + j) = u(k + H_c - 1)$, for $j = H_c, \ldots, H_p - 1$. The sequence of future inputs is obtained by optimizing an objective (cost) function which describes the control goals. The objective function is usually of the following quadratic form:

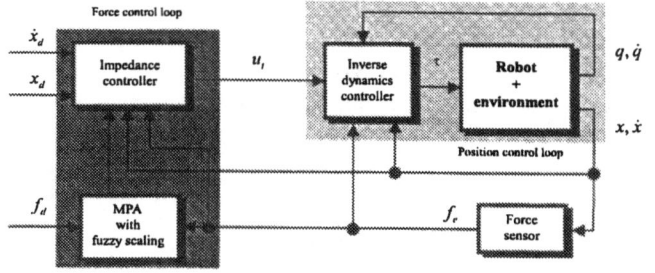

Fig. 1. Block diagram of the proposed force control scheme.

$$J(u) = \sum_{i=1}^{H_p} \left[\alpha_i (w(k + i) - \hat{y}(k + i))^2 + \beta_i (\Delta u(k + i - 1))^2 \right], \tag{11}$$

or some small modifications of it, where $\hat{y}$ are the predicted process outputs, w is the reference trajectory, and Δu is the change in the input weighted by the parameters β_i. The first term of (11) accounts for the minimization of the output errors, and the second term minimizes the energy cost. The parameters α_i and β_i determine the weighting between the two terms in the global criterion. The system inputs and outputs, as well as state variables, can be subjected to constraints, which are then incorporated in the optimization problem.

The following cost function is considered for the application in this paper:

$$J(\tilde{x}_v) = \sum_{i=1}^{H_p} (f_d(k + i) - f_m(k + i))^2 , \tag{12}$$

where f_d is the desired force and f_m is the predicted contact force. Note that a discrete version of the model is required predicting the future values $f_m(k + i)$ based on the measured position $x(k)$ and the measured velocity $\dot{x}(k)$ at time instant k. The sequence of future target position $x_v(k), \ldots, x_v(k + H_p - 1)$ describes the new target acceleration. Thus, the optimal target positions over a specified prediction horizon results in a new target acceleration u_t by the impedance control law (6). Figure 1 presents the block diagram of the proposed force controller.

The MPA performance depends largely on the system model. This must be able to accurately predict the future system outputs, and at the same time be computationally attractive to meet real-time demands. In the presence of a nonlinear model and constraints, the optimization problem to be solved at each time instant is non-convex. In this case, the *Sequential Quadratic Programming* (SQP) method can be considered (Gill *et al.* 1981). However, this method has in general high computational costs and hampers the application of MPA to fast systems. Moreover, convergence to local minima often occurs, resulting in poor performance of the predictive scheme. Alternative optimization methods for non-convex optimization problems can be used when the solution space is discretized. By discretizing the solution space, the problem is transformed into a discrete optimization problem, where techniques such as branch-and-bound (B&B) or genetic algorithms can be applied. In the following, the branch-and-bound algorithm is used to solve the non-convex optimization problem in MPA.

B&B is a widely used technique to solve difficult (usually non-convex) optimization problems (Mitten 1970). In this paper, B&B is used for the optimization problem that must be solved at each time instant k in the predictive algorithm. A B&B algorithm is characterized by the following rules: i) *Branching rule* - defines how to divide a problem into sub-problems, and ii) *Bounding rule* - establishes lower and upper bounds in the optimal solution of a sub-problem, where these bounds allow for the elimination of sub-problems that do not constitute an optimal solution. The model of the system predict its future outputs, $\hat{y}(k+1), \ldots, \hat{y}(k+H_p)$, Let M be the number of discrete possible inputs of the system. Let also these discretized inputs be denoted ω_j. Thus, at each step the inputs $u(k+i-1) \in \Omega$, are given by

$$\Omega = \{\omega_j | j = 1, 2, \ldots, M\}. \quad (13)$$

The problem to be solved in predictive problems is represented by the objective function (11). This optimization problem is successively decomposed by the branching rule into smaller sub-problems. At time instant $k + i$ the cumulative cost of a certain path followed so far, and leading to the output $\hat{y}(k + i)$ is given by $J^{(i)}$. where $i = 1, \ldots, H_p$, denotes the level corresponding to the time step $k+i$. A particular branch j at level i is created when the cumulative cost $J^{(i)}(u)$ plus a *lower bound* on the cost from the level i to the terminal level H_p for the branch j, denoted J_{L_j}, is lower than an *upper bound* of the total cost, denoted J_U:

$$J^{(i)} + J_{L_j} < J_U. \quad (14)$$

Let the total number of branches verifying this rule at level i be given by N. In order to increase the efficiency of the B&B method, this number should be as low as possible, i.e. $N \ll M$. The lower bound is expressed by an estimated lower bound of the cost over the remaining steps $i + 1, \ldots, H_p$, which is generally not known and must be estimated. In order to achieve $N \ll M$, the upper bound should be as low as possible (close to the optimal solution) and the lower bound as large as possible.

Two serious drawbacks of B&B are the exponential increase of the computational time with the prediction horizon and the number of alternatives, and the discretization of the possible inputs. This discretization can cause oscillations of the outputs around the reference trajectory, slow responses and overshoots. A possible solution to this problem is to adapt the inputs in order to better suit the present situation in the system.

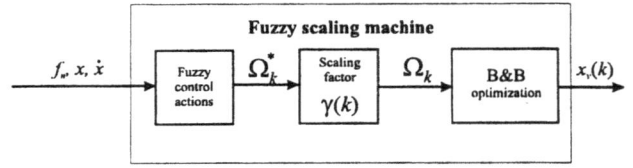

Fig. 2. Branch-and-bound optimization with fuzzy scaling applied to force control.

5. FUZZY SCALING MACHINE

Fuzzy predictive filters, as proposed in (Sousa and Setnes 1999), construct discrete control actions by using an adaptive set of control alternatives multiplied by a gain factor. This approach diminishes the problems introduced by the discretization of control actions in MPC. The fuzzy scaling machine proposed in this paper generalizes the scaling inference for model predictive algorithms. The predictive rules consider an error in order to infer a scaling factor, or gain, $\gamma(k) \in [0, 1]$ for the discrete incremental inputs. For the robotic application presented in section 6 considered in this paper this error is given by $e_m = f_n(k) - f_m(k)$ where $f_m(k)$ is the contact force predicted by a discrete-time version of (3) with constant estimated parameters k_e and b_e. The fuzzy scaling machine reduces thus the problem introduced by the discretization of the inputs, while at the same time the number of necessary input alternatives is kept low, thereby speeding up the optimization. The design consist of two parts: the choice of the discrete inputs, and the construction of the fuzzy rules for the gain filter, as shown in fig. 2.

Let the virtual position $x_v(k - 1) \in X$, which was described in (7), represent the input reference at time instance $k - 1$, where $X = [X^-\ X^+]$ is the domain of this reference position. The upper and lower bounds of the possible change in this reference signal at time k, x_k^+ and x_k^-, respectively, are given by

$$x_k^+ = X^+ - x_v(k-1) , x_k^- = X^- - x_v(k-1) \quad (15)$$

The values x_k^+ and x_k^- are thus the maximum changes allowed for the input reference when it is increased or decreased, respectively. The adaptive set of incremental input alternatives are now defined as

$$\Omega_k^* = \{0, \lambda_l\ x_k^+, \lambda_l\ x_k^- | l = 1, 2, \ldots, N\} , \quad (16)$$

where the designer has to choose the distribution λ_l. The choice of λ_l sets the maximum change allowed at each time instant by scaling the maximum variations x_k^+ and x_k^-, and the l parameter determines the number of possible inputs. From (16) it follows that the cardinality of Ω_k, i.e., the number of discrete alternatives, is given by $M = 2l + 1$ including the zero element.

The fuzzy scaling machine applies a scaling factor or gain, $\gamma(k) \in [0, 1]$, to the adaptive set of inputs Ω_k^* in order to obtain a scaled version Ω_k that is presented to the optimization routine:

$$\Omega_k = \gamma(k) \cdot \Omega_k^*. \quad (17)$$

The scaling factor $\gamma(k)$ must be chosen based on the predicted error between the reference and the system's output, which is defined as

$$e(k + H_p) = f_d(k + H_p) - f_n(k + H_p), \quad (18)$$

where $f_d(k + H_p)$ is the reference to be followed at time $k + H_p$. Further, the change in the error gives an indication on the evolution of the system, and this information should also be considered in the derivation of $\gamma(k)$. The change in error is defined as

$$\Delta e(k) = e(k) - e(k - 1). \quad (19)$$

Considering the predicted error and the change in the error, simple heuristic rules can be constructed for the gain. When both $e(k + H_p)$ and $\Delta e(k)$ are small, the system is close to a steady state situation. The set of input alternatives should then be scaled down to allow finer control based on the given references x_v, i.e., $\gamma(k) \rightarrow 0$, in order to approach zero steady state error without introducing oscillations around the set-point. When the predicted error and the change in error are both high, bigger corrective steps should be taken, i.e., $\gamma(k) \rightarrow 1$. The two fuzzy criteria, "small predicted error" and "small change in error", are defined by the membership functions $\mu_e(e(k + H_p))$ and $\mu_{\Delta e}(\Delta e(k))$, respectively. These two criteria can be aggregated by means of a conjunction that can be represented by the minimum operator (Klir and Yuan 1995). The aggregated fulfillment of these criteria is given by:

$$\mu_\gamma(e(k + H_p), \Delta e(k)) = \min(\mu_e, \mu_{\Delta e}), \quad (20)$$

where min is the minimum operator, and represents a multi-dimensional membership function μ_γ. The scaling factor $\gamma(k)$ can be easily derived by taking the fuzzy complement of μ_γ (Klir and Yuan 1995):

$$\gamma(k) = \overline{\mu_\gamma} = 1 - \mu_\gamma. \quad (21)$$

From the definition of μ_γ in (20) and the definition of $\gamma(k)$ in (21) it follows that when one of the two variables, error or change in error, is not small the gain $\gamma(k)$ is increased. Only when both conditions are fulfilled, i.e., both error and change in error are small, the gain is decreased.

Summarizing, (16) represents an adaptive set Ω_k^* of incremental inputs at time instant k. These are determined by the available reference space at time k, as defined in (15). The inputs are scaled by the gain $\gamma(k) \in [0, 1]$ to create a set of alternatives Ω_k that are passed on to the optimization routine. The value of $\gamma(k)$ is determined by simple fuzzy criteria regarding the error state of the system. Possible constraints on the reference signal are implemented by selecting properly the parameters λ_l.

6. APPLICATION TO ROBOT FORCE CONTROL

The force control scheme presented in Fig. 1 is applied to a 2-degree-of-freedom(DOF) robot through computer simulation for an end-effector force/position task in the presence of robot model uncertainties and geometric uncertainty in the environment and inaccuracy in the correspondent stiffness characteristics (see Fig. 3). The robotic model represents the links 2 and 3 of the PUMA 560 robot. In all simulations a constant time step of 1 ms was used in the controller implementation, and the complete dynamic model of the robot (1) is simulated in MATLAB/SIMULINK. A non-rigid friction contact surface is placed in a vertical plane of the robot workspace where it is assumed that the end-effector always maintain contact with the surface during the complete task execution.

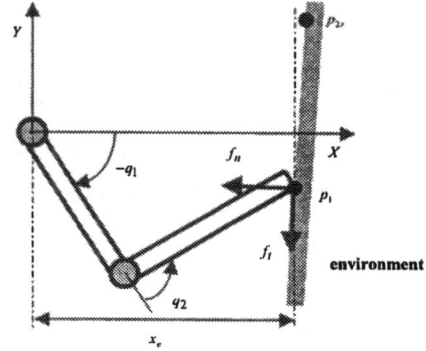

Fig. 3. Two-degree-of-freedom planar robot.

In order to analyze the force control scheme robustness to environment modeling uncertainties, a non rigid time-varying stiffness profile $k_e(t)$ is considered:

$$k_e(t) = 1000 + 100sin(\pi t/2) \quad (22)$$

The damping coefficient and the coefficient of dry friction are settled to $\rho_e = 45$ Ns/m^2 and $\mu = 0.2$, respectively. Note that $k_e = 1000$ N/m is kept constant in the environment model (3) used for predict the contact force f_m. The matrices in the impedance model (6) are defined as M_d=diag[2.5 2.5] and K_d=diag[250 2500] to obtain a good force tracking in the $x-$axis direction and an accurate position tracking performance in the $y-$axis direction. The matrix B_d is computed in order to obtain a critically damped system behavior. The control scheme was tested considering the desired force profile and an end-effector reference position trajectory from $p_1 = [0.5 \quad -0.2]$ m to $p_2 = [0.5 \ 0.6]$ m. Geometric uncertainty is introduced leading to a real final location of the contact surface $p_{2r} = [0.512 \ 0.6]$ m, as shown in fig. 3. The parameters of the predictive controller are $H_p = H_c = 2$. These values revealed to be sufficient to control the system. The prediction horizon can be however increased without increasing significantly the computational demands. The fuzzy scaling machine is applied only during the constant path of the reference force profile. This means that during the reference force transition periods, the fuzzy scaling inference is switched off. The maximum allowed increment Δx_v=0.05 for MPA with fuzzy scaling.

Figure 4 present the optimized virtual trajectory without and with fuzzy scaling and Fig. 5 represent the scaling factor $\gamma(k)$ and the joint torque 1. Note that the scaling machine reduce significantly undesirable oscillations on the virtual trajectory $x_v(k)$ and on the joint torque $\tau_1(k)$. The scaling factor exhibits a fast convergence to values around zero during the constant reference force path, which reduces the chattering present on the target trajectory and in the joint torques.

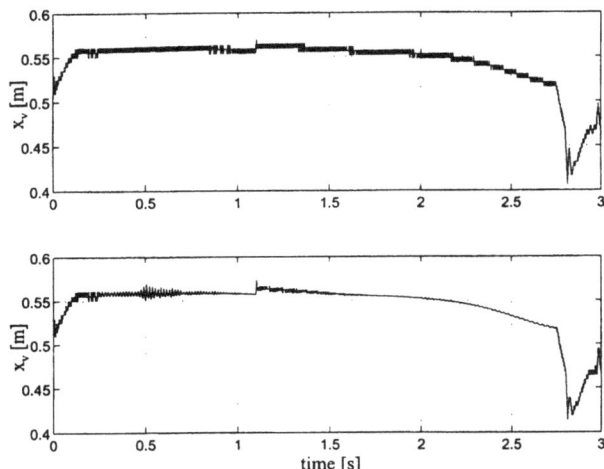

Fig. 4. MPA: virtual trajectory without and with fuzzy scaling machine.

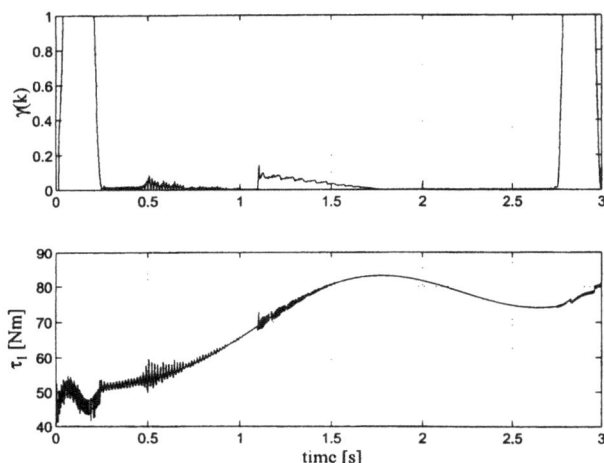

Fig. 5. MPA with fuzzy scaling machine: fuzzy scaling factor $\gamma(k)$ and joint torque $\tau_1(k)$.

7. CONCLUSIONS

In this paper a force control scheme integrating a model predictive algorithm and an impedance controller in the presence of nonlinear environments is proposed. The obtained results reveal that the force control scheme with a fuzzy scaling machine reduces significantly the oscillations of the target trajectory and consequently has a beneficial effect on the joint torques behavior. Simulations results can be easily extended for a robot with more than two joints. This generalization do not involve a large increase in the computational costs, since the optimization routine is applied only in the Cartesian constrained direction. In the near future the algorithm will be applied in real-time to a planar 2-DOF robot in the robotics laboratory.

8. REFERENCES

Baptista, L. F., M. Ayala Botto and J. M. Sá da Costa (2000). A predictive force control approach of robotic manipulators in non-rigid environments. To appear in International Journal of Robotics and Automation.

Camacho, E.F. and C. Bordons (1995). *Model predictive control in the process industry*. Springer Verlag. London, UK.

Gill, P. E., W. Murray and M.H. Wright (1981). *Practical Optimization*. Academic Press. New York and London.

Hogan, Neville (1985). Impedance control: an approach to manipulation: Part I-II-III. *Journal of Dynamic Systems, Measurement, and Control* **107**, 1–24.

Klir, G.J. and B. Yuan (1995). *Fuzzy sets and fuzzy logic; theory and applications*. Prentice Hall.

Marhefka, D. W. and D. E. Orin (1996). Simulation of contact using a nonlinear damping model. In: *Proc. IEEE International Conference on Robotics and Automation*. Vol. 2. Minneapolis, USA. pp. 1662–1668.

Mitten, L.G. (1970). Branch-and-bound methods: General formulation and properties. *Journal of Operations Research* **18**, 24–34.

Sciavicco, L. and B. Siciliano (1996). *Modelling and control of robot manipulators*. Electrical and Computing Engineering. MacGraw-Hill, Inc.. USA.

Singh, S. K. and D. O. Popa (1995). An analysis of some fundamental problems in adaptive control of force and impedance behavior: Theory and experiments. *IEEE Transactions on Robotics and Automation* **11**, 912–921.

Sousa, J. M., R. Babuška and H. B. Verbruggen (1997). Branch-and-bound optimization in fuzzy predictive control: An application to an air conditioning system. *Control Engineering Practice* **5**(10), 1395–1406.

Sousa, J.M. and M. Setnes (1999). Fuzzy predictive filters in model predictive control. *IEEE Transactions on Industrial Electronics* **46**(6), 1225–1232.

Wada, H., T. Fukuda, K. Kosuge, F. Arai and K. Watanabe (1993). Damping control with consideration of dynamics of environment. *Proc. of the IEEE Int. Conference on Intelligent Robots and Systems* **1**, 1516–1520.

Yao, B. and M. Tomizuka (1995). Adaptive control of robot manipulators in constrained motion - controller design. *Journal of Dynamic Systems, Measurement, and Control* **117**, 320–328.

CORBA FOR CONTROL SYSTEMS

Ricardo Sanz

Universidad Politécnica de Madrid

Abstract: The Common Object Request Broker Architecture (CORBA) is a middle-ware specification for the development of interoperable, distributed object systems. Object technology is of extreme importance in complex control system development besides its progressive use in real-time systems. This paper tries to provide a general overview of the topics in distributed object systems, focusing on CORBA aspects that are critical for control systems engineering. Some sample applications are presented. *Copyright ©2000 IFAC.*

Keywords: CORBA, object-oriented systems, distributed systems, real-time systems, development methodologies.

1. INTRODUCTION

1.1 Information Technology in industry

The process of incorporation of information technology (IT) into industrial processes is making profound modifications in production systems. Control and monitorization technology is leaving the *islands-of-automation phase*, entering a new phase of complete systems integration. While enterprise integration architectures (EAI) are hot topics in advanced business engineering, at the production level where controllers live, the incorporation of new technology and designs is confronting difficult problems.

In most cases the problems are mainly due to classical barriers posed to innovation in production systems: lack of predictability, need for non-stop operation, lack of reliability and availability, less than ideal market maturity, exploitation managers resilience, etc.

Two main objectives are being pursued in this effort, namely Complete Horizontal Integration (CHI) and Complete Vertical Integration (CVI). CHI deals with the integration of business units, business-to-business integration or supply chain integration.

In his speech we will address more the topic of CVI. It is time to start thinking in plant-wide integration reaching even the lowest levels in production plants: sensors, actuators and basic controllers.

Distributed object computing (DOC) is gaining an increased audience in information technology and, in new tech sectors, it is the technology of election for new system implementation. From global experience in last years it is pretty clear that -besides other advantages- DOC technology enhances systems integrability, making easier the construction of complex information applications. We will see in this paper how a DOC technology, namely CORBA, can supply us with some tools needed for better development of complex, integrated control systems.

1.2 Control engineering processes

Control systems complexity is increasing at a very fast pace in this days. New needs and new capabilities (nobody knows who come first) are driving control systems development into mainstream systems engineering. Integration capabilities are getting progressively critical as system size increments, because modular development is the only known practical way for complex systems engineering.

From this perspective it is surprising, to some extent, the limited role that software technologies play in control engineering journals and symposia. It looks like this technology does not have relevance enough to be considered a research discipline for control engineers (only small-scoped real-time topics are addressed in control engineering places).

This paper does not contain equations, nor feedback loops or control algorithms ... Can we say it that it is relevant for control engineering? The answer is *Yes, it's very relevant*. Let's take a closer loop at the the *real structure of a control problem*.

The typical development process of a control system can be decomposed, like any other engineering process, in a series of phases that go from the identification of the need to the decommission of the control system.

An example of phasing can be:

(1) Problem identification
(2) Plant Modeling
(3) Control design
(4) Control implementation
(5) Commissioning
(6) Operation
(7) Decommissioning

Research in control systems has been mainly focused in the second and third phases, because the first is considered an *a priori* for control engineering (*i.e.* it is always given) and from fourth to seventh they can be left to implementors (*i.e.* to raw work force). The separation between the control laboratory and the real plant is too wide for real engineering.

The basic technology used today to implement control systems is software technology. But, beyond a classical view of digital implementation of controllers (Åström and Wittenmark, 1997), software technologies are the basis of modern complex control systems, from SCADAs and DCSs to intelligent controllers based on soft computing (Gupta and Singh, 1996).

Software is the main *implementation* tool and this has relegated software technologies from the core *theoretical* control discipline. Any real control engineer can see this problem taking a view on "consolidated" control engineering magazines as *Automatica* or the *IEEE Transactions on Automatic Control*. No paper about a software topic will find a place in any of them; it will be relegated to the fellow *implementations* journal. This is a big mistake. Not only big, but critical for the discipline.

Control engineering is about *systems performance*; this means that the knowledge of the controlled system must involve not only the target system but the controller itself, and when controllers are software-based, giving a guarantee on global performance means a clear analysis and deep understanding of software issues. When controller complexity increases there are no available formal methods to guarantee behavior. Only good development processes can provide an statistically predictable quality. Good processes involve all controller life-cycle; from the problem identification phase to the operation phase and even decommission).

When complexity increases due to software flexibility the probability of failure increases. Systems that were manually operated are now operated by computers and this leads to a critical computer dependence of many artificial systems. The case of the USS Yorktown is paradigmatical. The ship had to go back to the harbor due to a software failure.

While software is becoming a real problem, it is also providing some solutions. For example, advanced research topics on systems fault-tolerance are strongly based in information processing capabilities that are used to detect the fault, isolate it, and devise alternative control strategies that can overcome the fault (Blanke *et al.*, 2000).

1.3 *Complex Software for Control*

Software systems can range from a small shoe shop database to Star Trek's *USSS Enterprise* control software. In a quick effort we can make a quick and dirty classification of software systems based on factors that induce systems complexity:

- Conventional: the shoe shop database.
- Real-time: meeting deadlines.
- Embedded: run within limited resources.
- Fault Tolerant: good behavior under faults.
- Distributed: run on several interacting computers.
- Intelligent: solving ill-posed problems.
- Large: millions of lines of code.
- Integrated: interoperate with alien systems.
- Heterogeneous: run on heterogeneous platforms.

Complexity factors affect negatively the systems development process. Development effort grows with complexity much more than linearly and there are even systems we cannot build; examples are 24x365 systems (total availability), one-shot systems (should work at the first try) [1] or HP-LC (High Performance and Low Cost).

---

[1] Star wars was an example of this problem.

88

Software engineers have always been raiders of the silver bullet (Brooks, 1992) looking to solutions to software development problems. Complex software engineering is just an emerging discipline, that is slightly appearing in frontier areas between those complexity topics mentioned before.

2. INTRODUCTION TO COMPLEX CONTROL SYSTEMS

A typical control system in a modern plant is composed by a heterogeneous collection of hardware and software entities scattered over a collection of heterogeneous platforms (operator stations, remote units, process computers, programmable controllers, intelligent devices) and communication systems (analog cabling, serial lines, fieldbuses, LANs or even satellite communications). This HW/SW heterogeneity is a source of extreme complexity in the control system regarded as a whole.

Apart from the platforms that provide support to the different control system components, the technologies used in control system implementation are quite heterogeneous and provide functionalities that go well beyond the classical sensing-calculating-acting triad.

Examples of this heterogeneity is the use of software systems for controller autotuning, advanced monitorization, filtering and estimation, adaptation and learning, plant-wide optimization, or real-time, in-the-loop simulation. Interception software systems are playing a wide collection of intelligent roles in complex controllers fitting as interfaces between pre-existent systems (pants, controllers and humans). Examples of these roles are data/action filters and monitors.

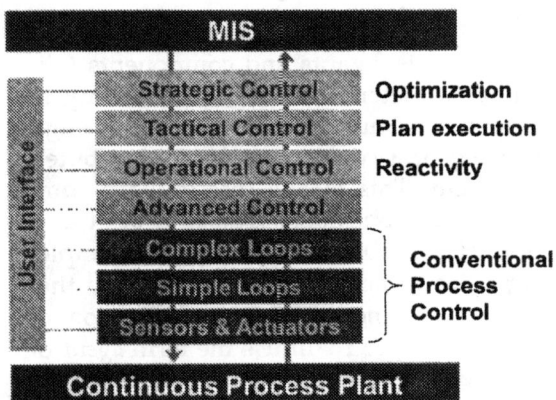

Fig. 1. A classical layering of control entities in a complex continuous process control system. Layer quantity and labeling is somewhat field-dependent, but layer roles can be easily mapped from domain to domain.

Classical hierarchical layering overcomes some of the difficulties of complex systems construction. An example of layering is shown in Figure 1 where some *intelligent* layers are added atop classical control layers in process control systems.

While hierarchies encapsulate low level behavior, simplifying the deployment of higher level controllers, they do not necessarily solve the problem of the *conceptual integrity* of the system. Layers can be difficult to match if they lack a common view of structure and responsibility distribution.

Conceptual integrity –an elusive, difficult to define property– is seen as the core factor affecting systems constructability. Conceptual integrity manifests in several system properties (some of them functional and some non-functional) that are considered extremely important in systems construction. These properties are the basic design principles of systems architecture(Shaw and Garlan, 1996) (See Table 1).

Table 1. Architecture design principles

Conforming	Scalable
Suitable	Simple
Composable	Standard
Modular	Proven
Extensible	Performing
Fast	Efficient

We will see in the next section how object technology can provide us with some ideas and tools to approximate this ideal of system conceptual integrity.

3. THE ROLE OF OBJECT TECHNOLOGY

The very nature of control systems is object-oriented (OO) because a control system couples virtual entities with real ones. A controller correlates control design issues and software implementations –that are very conceptual in its nature– with sensors, actuators and external world entities –that are very physical objects.

Control software makes a continuous mapping between external an internal entities and hence, object-oriented software is a natural way to build these systems. During the last decade OO technology was relegated from mainstream real-time software because OO implementations introduce computational overhead to support some aspects of OO computation (for example, dynamic binding). While this is usually the case, today computational power makes less important this overhead, and OO technology is becoming the technology of election to build complex real-time systems *because it provides better mechanisms for complexity handling*. An example of big importance for us is the case of real-time distributed systems, where OO technology is a clear winner (Shokri and Sheu, 2000).

Industrial plants are *Seas of Objects* and software-intensive controllers for them reflect this nature. The natural plant-modeling mechanisms are object-oriented (Rodríguez and Sanz, 1999) and dealing with preexisting software systems – for example legacy controllers– is best done using object wrapping. Advanced controllers are designed using clear responsibility distribution between control objects (CRC cards are a good approach to distributed controller analysis). This approach enables the development of architectures that exhibit some of the properties of Table 1 (Rushby, 1999).

3.1 *Objects, components and agents*

This discussion about responsibilities lead us to a concept of control systems as collections of interacting agents that match the most classical object-oriented view: objects have inner life and interact by means of message interchange.

Old days' object passivity, like the Smalltalk's approach to OO, do not fit well with our distributed controller model because it employs only one thread of control that leads to a computation model based on the sequentiation of method request and execution. This approach does not fit our needs because activities in the world are *naturally concurrent* and not sequential.

The need of going out of mono-threading was clear very soon and mechanisms for dealing with it were promptly added to OO systems. Exception handling extensions to classical environments and special multi-threading environments were developed for support this concurrent model; but true concurrence is only possible in multiprocessing environments: multiprocessors and distributed systems.

The extension of multi-threading support in operating systems provided a fake but very effective environment for concurrence. The simplified and less resource consuming model of thread interaction has demonstrated a benefit for complex systems development.

This support from the operating systems has brought new life to object systems. Object are no longer passive entities that become active only upon request from other objects. We can distinguish two types of activity:

Re-Activity: Objects are active in response to other objects' requests.

Pro-Activity: Objects are active pursuing object's own goals.

Objects are no longer passive, they can initiate activity by their own *will*. Beyond the big philosophical discussions in artificial intelligence circles about the meaning of the term *autonomy*, this step to object pro-activity has been the first true step to real, practical autonomy. Pro-activity was obvious in past control systems but not from a perspective of inner will of the entity. Only from the fusion of pro-activity and responsibility a true advance to autonomy has been achieved.

We are then reaching fields that go beyond simple, single activity and we get immersed in a process of agentification. Agents' technology provide models of autonomy in limited scopes; where agent interaction is a central issue. There is a lot of wasted words and paper in relation with agency, and, as a result, "agent" has almost lost its meaning (Sanz, 2000). Reading a dictionary, two concepts of agent emerge that fit our purpose: a. Those who act b. Those who act on behalf of others. This last meaning is the common interpretation in the Internet-related agencies (mail filters, web crawlers, etc). The first sense is best suited for our view of complex control systems.

Distributed control systems are agencies, where each agent pursues an objective. The operational cycle of each agent is based on three interrelated activities: Sensing, Reasoning and Acting (*i.e.* a control loop).

Agent-based models of software development are focused on the partial autonomy of agents in relation with an specific task (Sanz *et al.*, 2000). The resulting agency is a community of communicating entities that perform a global, rational decision making process by means of negotiation and collaboration. This involves in many cases policies for resource sharing, creation of markets and contract signing.

Models of objects, agents and components (object and agent materializations) are evolving to a common view. This view fits the control engineering view of a component for a distributed control system. This means that modern complex distributed controllers are being built as a collection of reusable components that implement agencies to achieve a final objective that is shared-by or emergent-from a collection of agents. In this paper I will use the term *agent* or *object* to refer to the same type of *"active object"* entity. *Component* will be the term used to refer to concrete implementations of the agents.

Agent componentization is a good foundation for a research "product line", because it offers a foundation for easy, component-based development of new systems using proven compo-

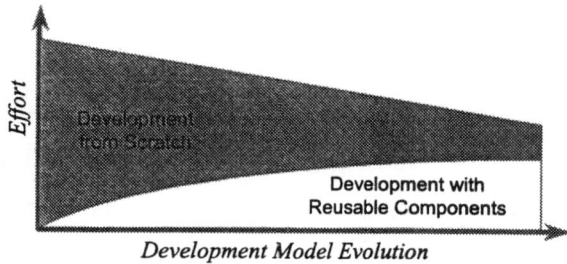

Fig. 2. Resource allocation to custom development and component reuse evolve with the development model.

nents. This let developers concentrate the efforts in new components they are developing and not in well known components that need to be rebuilt to fit in a new schema. Companies moving to a component-based development process have seen that the effort needed for new applications is reduced as is reduced the effort put in custom, application-specific developments (this is due to the product-line focus of most companies).

3.2 Distributed Object Computing

Distributed Object Computing (DOC) or Object-Oriented Distributed Processing (OODP) is a software model based in the use of services provided by objects that are running in different hosts. Distribution means true concurrence even when most distributed applications serialize behavior of the application using some form of centralized controller.

DOC can be considered a generalization of the client/server model. In DOC, client and server roles are relative to a specific request and not to the whole life-cycle of the object (an object can be the client in a request and the server in the next one).

DOC is a "natural" way of modeling distributed systems because it hides implementation details (OS, protocols, languages) behind "interfaces". Encapsulation, abstraction and inheritance are valid and very useful concepts to model distributed control systems.

There are many benefits of using DOC for control systems engineering. In many cases they are the same as for any other type of system, but in most situations they are of critical importance for control software due to the special requirements posed to control systems. Examples of these benefits are:

- Object collaboration through connectivity and interworking;
- performance through parallel processing;
- reliability and availability through replication;

- scalability and portability through modularity;
- extensibility through dynamic configuration and reconfiguration;
- cost effectiveness through resource sharing and open systems;
- maintainability through hot swapping and
- design flexibility through transparency.

DOC is an extremely valuable model for control software development.

3.3 Integration

DOC technology addresses particularly well one of the main problems of complex systems construction: *integration*.

If we consider the interaction between two pieces of code (let's call them the *client* and the *server*) we can identify four relative positions (*i.e.* four coarse types of integration mechanisms:

In-Thread: Client and server are parts of the same thread. Interaction is done by method call. This means serialization (no concurrence) and a simple integration vehicle (programming language routine invocations). This is easy to use, extremely fast and reliable. It is strange to have client and server in a different state –from a reliability perspective– due to external factors.

In-Process: Client and server are parts of the same process but in different threads. We have inter-thread requests usually based on ITC [2] mechanisms provided by the operating system. This is relatively complex but is very fast and reliable.

In-Host: This situation is similar to the previous, but in this case client and server are in different processes. Inter-process requests are based on operating systems IPC [3]. This is also a fast and reliable mechanism.

In-Net: Client and server are in different hosts. The basic integration mechanism is some form of remote procedure call (RPC) [4] Inter hosts requests rank lower in speed and reliability because it is easier to have different host states in client, server or even communication channel. Distribution means in many cases unpredictability and unreliablility.

Middleware is a generic name used to refer to a class of software whose sole purpose is to serve as glue between separately built systems.

---

[2] Inter-Thread Communication.
[3] Inter-Process Communication.
[4] Lower level mechanisms can also be used but in most cases it is not worth the effort.

Object-oriented middleware is used to simplify the development and use of ubiquitous objects. Middleware tries to simplify the implementation of clients and servers for different relative locations; for example making possible the implementation of clients that are unaware of server locations.

A big simplification is achieved using the same interface to be used by client and servers independently of the base integration mechanism; *i.e.*the same interface is used to wrap an IPC and an RPC (see Figure 3).

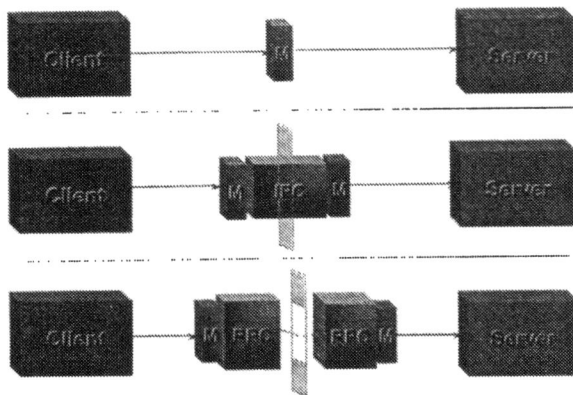

Fig. 3. A great simplification is achieved using the same interface to be used by client and servers to use/provide the service.

But the real big step is when this interface is independent of the relative location of the opposite object (see Figure 4).

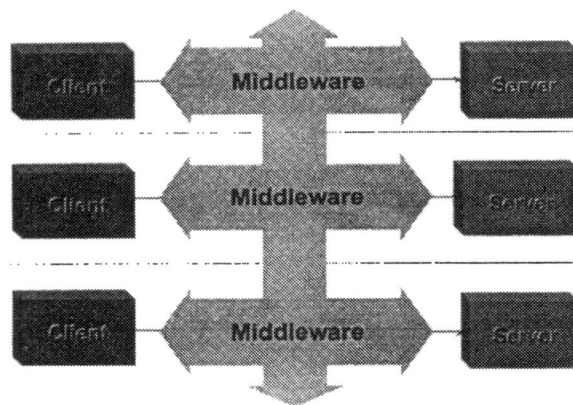

Fig. 4. Middleware hides underlying mechanisms to provide an homogeneous platform for ubiquitous computing.

Brokering middleware is based on the use of an intermediary entity between the client and the server: the broker (See Figure 5). The process of remote invocation is decomposed in eight steps:

1. The client makes a call to the client stub (the client plug to the broker).
2. The client stub packs the call parameters into a request message and invokes a wire protocol.

3. The wire protocol delivers the message to the server side stub(the server plug to the broker).
4. The server side stub then unpacks the message and calls the actual method on the object.
5-8. The response -if any- uses the same process to reach the client.

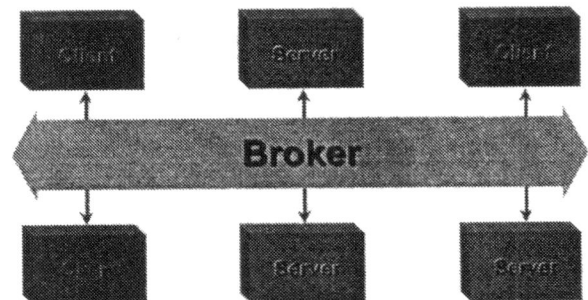

Fig. 5. Brokering middleware is based on the use of an "intelligent" intermediary between clients and servers.

3.4 *Middleware and muddleware*

There are many contenders in the object-oriented middleware arena. The three main technologies are Microsoft's COM+, Sun Microsystems' Java RMI and Object Management Group CORBA.

There are big discussions about what "is the best technology" but –as is the usual case– there is no clear winner. If we try to understand all the arguments we easily get into the muddle (of terminology, of arguments, of policies, of money and lost business opportunities, etc.).

From my point of view there are some clear – although partial– criteria to follow:

- Homogeneous-platform applications on MS desktop machines: COM+
- Internet-wide heterogeneous platforms: Java
- Fully heterogeneous and special requirements: CORBA

But what are those "special requirements"? The answer to this question is pretty long but extremely interesting for control engineers: Real-time behavior, fault tolerance, small memory footprint, pervasive heterogeneity (hardware, operating system and programming language), platform resource control, vendor independence, open specification process, modularity, embedability, etc.

Does it mean that you cannot use COM or Java for control applications? Not so, it only means that you cannot use them if you have some of these requirements unless you want to put a huge effort to fulfill them. It is simpler to build applications with tough requirements using CORBA [5]. There are many good books on CORBA (many of them published by OMG Press/Wiley) but the book of Jon Siegel is particularly important (Siegel, 2000).

Things are somewhat changing for Java. While it was originally developed for the embedded market it reaches the public recognition in website and Internet programming. Now, after the approval of the Real-time Java specification, perhaps it can regain the embedded and real-time markets.

4. OMG, UML AND CORBA TECHNOLOGY

OMG stands for Object Management Group [6], an organization created to foster object technology by means of the creation of a software marketplace for object technology. Using OMG's object technology any organization can leverage previous efforts in building control systems. Two of the main components of this technology spectrum are CORBA and UML.

The OMG is an *standardization organization* with an open, vendor-neutral, international, widely recognized and rapid standardization process based on demonstrated technology. It is composed by more than 800 members (for profit and not for profit organizations), with tens of concurrent technology processes, ranging from networking infrastructure to air traffic control or human genome data management. It maintains a strong liaison with other organizations as ISO, ITU-T, W3C, TINA-C, Meta Data Coalition, etc. OMG's object technology is the object technology of reference: IDL, UML, MOF, XMI [7], etc.

The main contribution of OMG to the OO world is the Object Management Architecture (OMA). This is an specification for the construction of open distributed object systems based on brokering and a collection of predefined services (OMG, 1998c).

The technology provided by the OMA can be grouped in:

Object Request Broker (ORB): is the run-time integration vehicle that forwards requests and responses.
Interface Definition Language (IDL): is the interface definition mechanism for implementation independence.

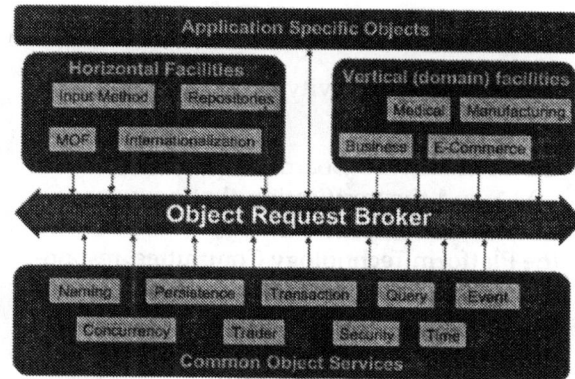

Fig. 6. OMA Overview. The main parts are shown but only some of the services are detailed.

Language Mappings: are the mappings from IDL to several programming languages for the implementation of client an servers.
Repositories: are stores that provide run-time information about interfaces and implementations.
Interoperability: between CORBA systems and with external entities (like Microsoft COM).

Most of these specifications are contained in the main CORBA document: *Common Object Request Broker Architecture and Specification* (OMG, 1998a) [8].

The OMG provides extensions and profiles over the base specifications as separate documents that specify the points of departure from the main specification. Of special importance for control systems engineering are the Minimum-CORBA specification; the Real-time CORBA specification and the Fault-tolerant CORBA specification (see Section 5).

4.1 *CORBA IDL*

OMG IDL (Interface Definition Language) is an implementation independent language used to specify CORBA object interfaces. It is now an ISO standard and has several interesting characteristics: it supports multiple-inheritance (not so common in OO technology); it is –obviously– strongly typed; it is independent of any particular language and/or compiler and can be mapped to many programming languages (some mappings are specified by the OMG and others are contributed specifications); it enables interoperability because it isolates interface from implementation.

---

[5] Even while being knowledgeable in CORBA is a daunting task.
[6] Find it at http://www.omg.org/.
[7] Interface Definition Language, Unified Modeling Language, MetaObject Facility, XML-Based Metadata Interchange.

[8] In release 2.3, the language mapppings were taken out to constitute separate documents with own evolution.

4.2 *OMG Structure and Activity*

The OMG technical activity is organized around three bodies:

- the Architecture Board, responsible of the OMA and the verification that new specifications are compliant with it;
- the Platform Technology Committee, responsible of CORBA core technology, and
- the Domain Technology Committee, responsible of specifications in vertical domains.

The work is performed by a collection of working groups in the different areas; from core technology like the interoperability protocols to domain specific activities like data acquisition or financial security.

The OMG specification process is based on the submission of specifications from private organizations in accordance with Request For Proposals (RFPs) done by the OMG. This means that the specification elaboration process is not done by an standardization committee (ISO C++ took more than eight years) but by an –usually– small group of OMG members *based on their own criteria and previous developments.*

This means that if a company possess a technology that fits an RFP, the company can send the specification of that technology as a proposal to the OMG and it has a good chance of getting it approved as an OMG specification. This has been the case of UML proposed by Rational or the Fault-tolerance specification proposed by Sun.

If there are several proposals, the different submitters try to find a consensus and deliver a single, consolidated version, supported by all them. This is usually called a Joint Revised Submission.

5. CORBA ASPECTS FOR REAL-TIME CONTROL

Apart from the importance of having a platform for integration and development of modularized controllers, there are some new issues in CORBA that are specially relevant for distributed control systems engineering. These issues are: predictable behavior, fault tolerance and embeddability.

The Real-time PSIG [9] (Platform Special Interest Group) is addressing all these topics because they have focused their activities on real-time systems, and most real-time systems are also embedded and have some fault tolerance requirements.

The Real-time PSIG goal is the recommendation of adoption of technologies that can ensure that OMG specifications enable the development of real-time ORBs and applications.

To achieve this goal, the Real-time PSIG gathers real-time requirements from industry, organize workshops and other activities and involve real-time technology manufactures to elaborate Requests For Information and Requests For Proposals for these technologies.

The main results of this work an be organized in the three categories:

Real-time CORBA: The Real-Time CORBA specification in addition to the Messaging specification provides mechanisms for controlling resource usage to enhance application predictability (OMG, 1999c; OMG, 1998b).

Fault-tolerant CORBA: The specification provides mechanisms for fault tolerance based on entity redundancy (OMG, 1999b).

Minimum CORBA: Addresses the construction of CORBA applications on systems with little resources like embedded computers (OMG, 1999c; OMG, 1999a; OMG, 1998b). This specification eliminates most dynamical interfaces that are not necessary in frozen applications (most embedded applications are ROMmed applications).

5.1 *Real-time CORBA*

RT-CORBA standardizes the the mechanisms for resource control (memory, processes, priorities, threads, protocols, bandwidth, etc.) and handling of priorities in a distributed sense (for example forwarding client priorities to the server).

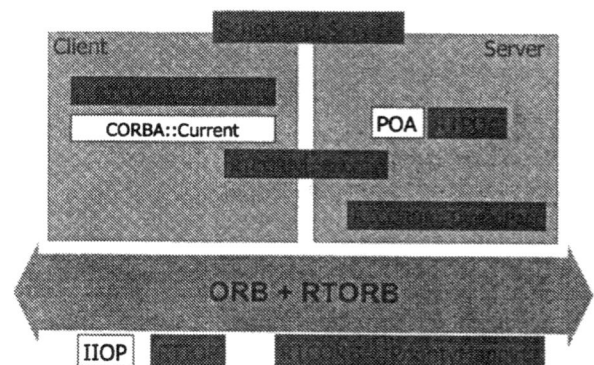

Fig. 7. Real-time CORBA extensions to provide strong control of resources to both clients and servers.

Using these mechanisms, the Real-time CORBA developer can control:

- Request time-outs
- Resource allocation and sharing
- Priority control and propagation

---

[9] They can be found at http://www.omg.org/realtime/.

- Priority inversion
- Method invocation blocking
- Routing
- Transport selection

5.2 *Fault-tolerant CORBA*

Fault-tolerant CORBA tries to enhance application fault tolerance reducing to a minimum the impact to the application (computing overheads and increase of complexity). Fault tolerance is increased by means of entity replication: cold passive replication, warm passive replication, active replication or active replication and majority voting.

5.3 *Minimum CORBA*

Embedded CORBA applications reduce memory footprint by means of elimination of some features (dynamic interfaces and repositories), the use of standardized operating system services or special transports. The elimination of a specific service from the specification does not mean that the application cannot use it, only that it will not be necessarily provided by a compliant CORBA implementation.

5.4 *Bridging Domains*

While the Minimum CORBA specification reduces the requirements posed to the ORB, the *Real-time CORBA* and *Fault Tolerant CORBA* specifications can increase the size an complexity of the application.

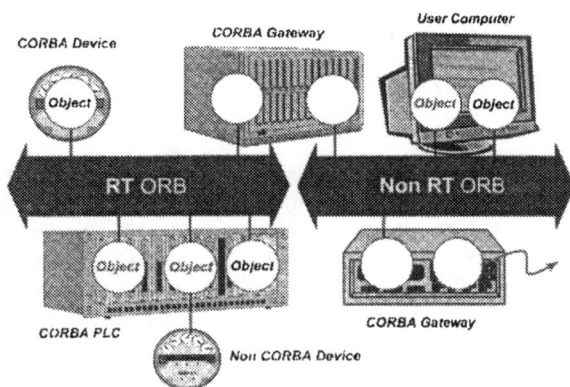

Fig. 8. A CORBA Gateway can bridge between two worlds of different protocols or ORBs.

Thanks to interoperability, it is not necessary at all to have all the application running atop he same ORB. It is possible to have the critical part of an application running over a Real-time ORB and the rest over a more conventional one. Figure 8 shows an example of using a CORBA gateway to bridge between two different worlds in a control application.

6. APPLICATION EXAMPLES

In this section I will present succinctly some examples on the use of CORBA technology in control systems. The broker used is all of them is the ICa Broker(Sanz *et al.*, 1999b), a broker specially tailored for control applications that was developed by us during the ESPRIT DIXIT project [10].

6.1 *Robot Teleoperation*

As a laboratory experiment we have used CORBA to build a robot teleoperation application (see Figure 9. The application contains three CORBA objects: a six DoF [11] full force feedback master, a seven DoF robot slave and a coordinate space mapper (transforms master axis space into robot axis space).

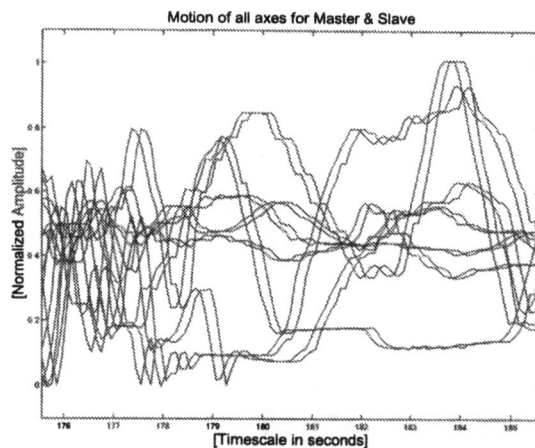

Fig. 9. Axis position evolution in master and robot during a test.

A big delay is appreciated ($\approx$ 250 ms) but with a small jitter. The test was done using the common laboratory 10baseT network in normal state (about 20% load).

6.2 *Risk Management*

Another application of interest is RiskMan. This is a system for emergency management in a chemical complex with nine plants (see Figure 10). The system supports the whole life-cycle of emergencies: prevention, detection, firing, diagnosis, handling, follow-up & cancellation.

---

[10]Now it is a commercial product distributed by an UPM spin-off company called SCILabs.
[11]Degrees of Freedom.

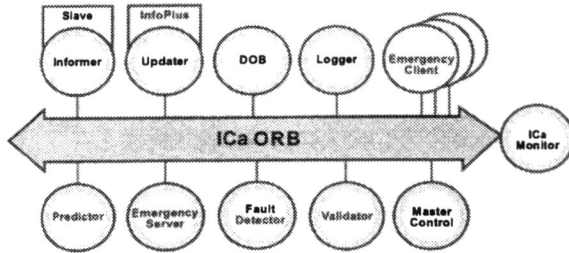

Fig. 10. Some of the CORBA objects that compose the RiskMan application. Informer and Updater are wrappers of external systems.

The application is composed by a collection of CORBA objects running on heterogeneous platforms (VAX/VMS, Alpha/UNIX, x86/Windows NT) performing an heterogeneous collection of functions: expert systems, user interfaces, wrappers of real-time plant databases, data filters based on fuzzy rules, predictors based on neural networks, etc.

6.3 HydraVision

HydraVision is a real-time video system for the support of remote operation of hydraulic power plants. It uses a country-wide fiber optics WAN network of a electric company to integrate a collection of objects that wrap physical entities in the system (see Figure 11).

Fig. 11. The HydraVision main user interface and one of the wrapped cameras.

The physical systems that are wrapped as CORBA objects are: cameras, MPEG compressors, image/audio multiplexers, microphones, loudspeakers, video monitors, video stores, still image printers, etc.

The system is supports multicasting and bidirectional streaming. It used by human operators to: get a visual confirmation of the status of the remote plant, video-conference, faking human presence, remote diagnosis, etc..

6.4 PICMAK

PIKMAC is an operator support system designed to address plant-wide strategic decision making in a cement plant. The system is used by operators specially in night and weekend shifts when there is only one one person in the plant(see Figure 12).

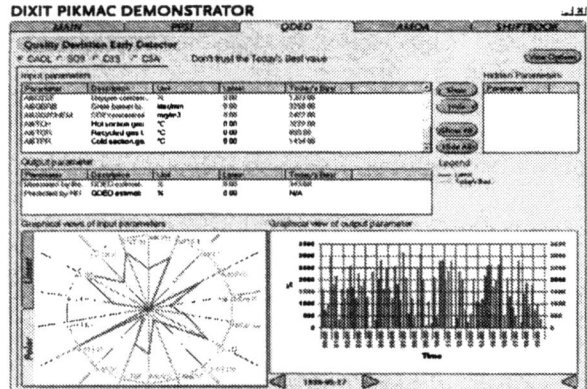

Fig. 12. Part of the user interface that shows the results of the on-line quality estimator QDED. It uses neural networks to estimate present clinker quality because it is not possible to have an direct real-time measure of it.

The system is composed by a collection of interacting CORBA objects that provide four top level functionalities:

- Clinker quality estimation using neural network technology.
- Instantaneous cost estimator using deep models.
- Alarm management using expert systems.
- Inter-shift communication using multimedia technology.

6.5 Present work

Our present work is focused in lowering the requirements posed to the platforms to run CORBA objects. We are working in the development of embedded modular systems based on this technology.

As examples, we are using CORBA in mobile robotics and also in the development of standardized implementations of electric substation automation systems (SAS).

In this last case, we try to build CORBA automation objects embedded in field devices based on the Electric Utilities IEC 61850 Draft standard. To do this work, we have funding from the European Comission through the IST DOTS project.

7. STATE AND FUTURE OF THE TECHNOLOGY

CORBA technology is impressive but perhaps too impressive for normal control systems developers. It suffers what is called a second system effect, trying to address all possible functionality or requirement. We must identify our own needs and determine if the CORBA way fits our needs. If not, we are still in time to modify it.

Perhaps the main question is *Why we need integration ?*. Beyond many obvious answers (to build TotalPlants, to achieve total safety, to be the first in the market, to spend less money, etc.) I would like to stress one door that this approach opens for us: The modular approach fostered by CORBA will let us develop true modular control systems, and this will eventually lead to reach human-like complexity levels in artificial minds. For sure CORBA will not be the integration technology for future C3POs, but it will open that way. If you remember the movie 2010, HAL 9000 goes back to life (or conscience) when Dr. Chandra re-connect the modules that encapsulate high-level mental functions using an integrational backbone.

The second point I want to mention is *design freedom*. Design freedom is necessary in the complex control systems domain to explore alternative controller designs. Excessively restrictive technologies will collapse –unnecessarily– dimensions of the controller design space (Shaw and Garlan, 1996). This is, for example, the case of some fieldbus technologies that support several slaves but only *one* master. While design restrictions (in the form of prerequisite design decisions) simplify development they sacrifice flexibility.

Can we get both, simple development and flexibility ? The key are no-compromises frameworks, *i.e.* frameworks where design dimensions are still open even when pre-built designs are available. To continue the example of the fieldbus, the one-master/several-slaves approach is one type of pre-built, directly usable, design; but the underlying field bus mechanism should allow for alternative, multi-master designs. This can be done by means of the development of agent libraries that provide predefined partial designs in the form of design patterns (Sanz *et al.*, 1999c), and a transparent object-oriented real-time middleware.

This approach will let developers construct their own agencies to support their own designs because it is impossible to fight the *not-invented-here* syndrome. Let the people do what they think they need. Do not define *ultimate* solutions. Provide reusable assets that can be adapted to any problem in a domain progressively focused (Sanz *et al.*, 1999a).

8. CONCLUSIONS

Software technology is of extreme relevance in any area of engineering activity. In the case of automatic control systems, we can say that it is not only relevant but a mandatory technology in a wide variety of control system implementations. Control engineers must know more about software because it is as basic as differential equations for the proper construction of control systems.

While there are many research developments in DOC, three are the main contenders in the DOC wars: COM, CORBA and Java. But it is pretty clear that the only widely available technology that is addressing -more or less- the full range of topics in our automatic control business is CORBA. The three main objectives we are searching in a software technology are embedded, real-time and robust. All they are being addressed by CORBA specifications: Minimum CORBA, Real-time CORBA, CORBA Messaging and Fault-Tolerant CORBA. It is worth note however that most commercial products are ignoring the real-time market because they think it is a very small market.

It should be clear however, that selection of any one of these technologies does not hamper the application of the others. In fact, in relation with Java and CORBA, it is worthy note that both distributed object models are pretty the same, and evolution of distribution for Java applications is being fostered over CORBA compliant brokers. On the other side, CORBA interoperability specifies mappings to COM and OLE Automation, making possible the integration of COM based applications in broad-class CORBA systems.

While CORBA has been developed for distributed applications, the transparent integration mechanism it offers serves also for non-distributed applications. Some broker implementations provide local transports that do not use network protocols and hence do not induce a big overhead. There are even broker implementations that can reduce overhead for local invocations to zero (Sanz *et al.*, 1999b). This means that the programmer can transparently decide where to put the objects and CORBA can do it with a minimal impact in performance.

Real-time CORBA is very new (we have only release 1.0 and some errata corrections) and it has been developed as a compromise usable in many fields and hence it has only a fixed priority model (instead of other dynamic priority models better suited for control applications). Dynamic scheduling service will appear eventually but it is suffering a painful specification process.

We should mention the strong bias in RT-CORBA to telecoms, that make it sacrifice strong predictability. Real-time ORBs are being deprecated by mainstream ORB vendors and hard real-time ORBs are far in the future. Next major issues for our business will be pluggable transports (that are not politically correct because they are not IIOP) and real-time services; like the mentioned scheduling service, real-time events or transactions. Some of them have been demonstrated by OMG contributors. Other relevant issues are combinations of specifications; for example Realtime + Fault-tolerant or Real-time + Minimum. Stay tuned.

Remember: Ignoring software topics in control systems research is a big mistake. Not big, but critical for the discipline. Control engineering is not only a discipline of mathematical modeling and differential equation solving. Control engineering is the discipline of *artificial behavior* and software is what makes the behavior. CORBA is a good tool to support our designs, but we must work hard to make CORBA more oriented towards control systems engineering.

9. REFERENCES

Åström, Karl Johan and Björn Wittenmark (1997). *Computer Controlled Systems*. third ed.. Prentice-Hall. New York, NJ.

Blanke, Mogens, Christian Frei, Franta Kraus, Ron J. Patton and Marcel Staroswiecki (2000). Fault tolerant control systems. In: *Control of Complex Systems* (Karl Åström, Alberto Isidori, Pedro Albertos, Mogens Blanke, Walter Schaufelberger and Ricardo Sanz, Eds.). Springer. In Press.

Brooks, Fred (1992). No silver bullet. *Computer*.

Gupta, M.M. and N.K. Singh (1996). *Intelligent Control Systems*. IEEE Press. Piscataway, NJ.

OMG (1998a). Common object request broker architecture and specification. Technical Report 2.3. Object Management Group.

OMG (1998b). Corba messaging. Technical Report orbos/98-05-05. Object Management Group.

OMG (1998c). A discussion of the object management architecture. Technical report. Object Management Group.

OMG (1999a). Errata for real-time corba joint revised submission. Technical Report orbos/99-03-29. Object Management Group.

OMG (1999b). Fault tolerant corba. Technical Report orbos/99-10-05. Object Management Group.

OMG (1999c). Real-time corba. Technical Report orbos/99-02-12. Object Management Group.

Rodríguez, Manuel and Ricardo Sanz (1999). HOMME: A modeling environment to handle complexity. In: *IASTED Modeling and Simulation Conference*.

Rushby, John (1999). Partitioning in avionics architectures: Requirements, mechanisms, and assurance. Technical Report NASA/CR-1999-209347. NASA.

Samad, Tariq (1998). Complexity management: Multidisciplinary perspectives on automation and control. Technical Report CON-R98-001. Honeywell Technology Center. Minneapolis, MI.

Samad, Tariq and Weyrauch, John, Eds.) (2000). *Automation, Control, and Complexity: New Developments and Directions*. John Wiley and Sons.

Sanz, Ricardo (2000). *Agents for Complex Control Systems*. in Samad and Weyrauch (2000).

Sanz, Ricardo, Fernando Matía and Santos Galán (2000). Fridges, elephants and the meaning of autonomy and intelligence. In: *Proceedings of ISIC'2000*. Patras, Greece.

Sanz, Ricardo, Idoia Alarcón, Miguel J. Segarra, Angel de Antonio and José A. Clavijo (1999a). Progressive domain focalization in intelligent control systems. *Control Engineering Practice* 7(5), 665–671.

Sanz, Ricardo, Miguel J. Segarra, Angel de Antonio and José A. Clavijo (1999b). ICa: Middleware for intelligent process control. In: *International Symposium on Intelligent Control*. Cambridge, MA.

Sanz, Ricardo, Miguel J. Segarra, Angel de Antonio, Fernando Matía, Agustín Jiménez and Ramón Galán (1999c). Patterns in intelligent control systems. In: *Proceedings of IFAC 14th World Congress*. Beijing, China.

Shaw, Mary and David Garlan (1996). *Software Architecture. An Emerging Discipline*. Prentice-Hall. Upper Saddle River, NJ.

Shokri, Eltefaat and Phillip Sheu (2000). Real-time distributed object computing: An emerging field. *IEEE Computer* pp. 45–46.

Siegel, Jon (2000). *CORBA 3: Fundamentals and Programming*. second ed.. OMG Press/Wiley. New York.

REAL-TIME CONTROL OF AN AIR MOTOR INCORPORATING A PNEUMATIC H-BRIDGE

M. O. Tokhi*, M. Al-Miskiry* and M. Brisland**

* Department of Automatic Control and Systems Engineering,
The University of Sheffield, UK.
** Aquadraulics Ltd, Barnsley, UK.

Abstract: This paper presents a feasibility study of real-time control of an air motor using the concept of a pneumatic equivalent of the electric H-bridge. A radial piston air motor is considered. A pneumatic H-bridge is developed allowing speed and direction control of the motor. A PID control strategy is adopted to control the motor speed. The control strategy is tested in controlling the motor at high and low speeds. Experimental results show that a good level of speed control of the motor is achieved, and that both speed and direction control of the motor is feasible in real time with the pneumatic H-bridge concept. *Copyright © 2000 IFAC*

Keywords: Air motor, PID control, pneumatic H-bridge, real-time control, motor control.

1. INTRODUCTION

Air motors are compact, lightweight sources of smooth vibrationless power. They start and stop almost instantly, and are not affected by continuous stalling or overload. They play a very significant part as prime movers because they are relatively cheap, easy to maintain, and have the versatility of variable speed, high starting torque, are intrinsically safe in hazardous areas, and will operate in exceptionally harsh environments (Hitchcox, 1995; Morgan, 1984; Simnett and Anderson, 1983).

Common designs of air motors include axial piston, radial piston, rotary vane, gerotor, turbine, V-type and diaphragm (Hitchcox, 1995; Mahanay, 1986. Among these, piston air motors are used in applications requiring high power, high starting torque, and accurate speed control at low speeds. They have two to six cylinders arranged either axially or radially within a housing. Pressure acting on pistons that reciprocate within the cylinders develops output toque. Motors with four or more cylinders produce relatively smooth torque at given operating speeds, due to overlapping power pulses as two or

more pistons are undergoing a power stroke at any time within a revolution.

The power developed by a piston motor depends on the inlet pressure, the number of pistons, and the piston area, stroke, and speed. At any given inlet pressure, more power can be obtained from a motor that runs at a higher speed, has a larger piston diameter, more pistons, or longer stroke. Speed limiting factors are the inertia of the moving parts and design of the valves that control the inlet and exhaust to the pistons.

The motor utilised in this work is of the radial piston type (Automation Airmotors, 1998). This type of motor constitutes a robust, oil-lubricated construction and is well suited to continuous operation. The motor has the highest starting torque of any air motor and is particularly beneficial for applications involving high starting loads. Overlapping power impulses provide smooth torque in both forward and reverse directions.

The basic characteristics of pneumatic motors are well understood and have been published extensively. However, studies concerned with its control have been so few that a universally accepted method is still a research topic. A feedback speed control mechanism with an electro-pneumatic proportional valve has been studied and reported by Noritsugu (1987). In this study the I-PD control scheme has been adopted. With this approach, the controller is simply implemented by electronic proportional, integral and derivative elements. The design of the control system is based on the partial model matching method, which is easily applicable to a system including a dead-time component. An electro-pneumatic servo type speed control method with a pulse width modulated (PWM) technique was proposed by Noritsugu (1988), which requires an inexpensive on-off solenoid valve and uses the I-PD controller in a similar manner as indicated above.

Speed control of air motors is commonly achieved through regulating the flow into the motor by using electro-pneumatic proportional valves. For forward and reverse direction, 3-way or 4-way spool valves are employed. In this study, airflow to the motor is controlled by means of a single spool valve, both ends of which are acted upon by proportional solenoids. This approach has some fundamental problems that emerge with the age; e.g. as wear between the spool and the sleeve affects the sealing performance.

The air motor is required to have bi-directional shaft rotation and position control. As the motor decelerates to a stop position before changing direction, air flow is reduced and suspended before flow is reversed. During this suspended period, or dead-band, the motor shaft is not under control. This situation can affect the target position. Thus, inherent dead band in the characteristics of the spool valve is not desired.

In this paper a pneumatic H-bridge method is proposed to reduce or eliminate the above problem. In the H-bridge, the flow direction is no longer mechanically linked, but is software controlled. The magnitude of the dead band is variable and may be tuned to suit the mechanical variations in the motor. The dead band offset may also be software tuned to suit the environment in which the motor is installed.

2. THE PNEUMATIC H-BRIDGE

The radial piston motor under consideration is motivated by the need for a positioning device that could produce a relatively high torque output at a much lower cost than electrical devices, the prices of which tend to become excessive for applications above 8.1349 Nm (Noritsugu, 1988). There have been some applications of this equipment. These include an XY table in glass works where the requirements for a motor, are to accelerate heavy loads under dirty and wet conditions, and in food machinery where the motor offers a clean oil-free system that can be washed down without fear of harm to the operator. The motor uses four pistons in a radial configuration with a central tri-lobed cam (Billany, 1990). This design gives 12 power strokes per revolution, providing virtually cog-free shaft output (above minimum speed). By employing computer modelling techniques during the design of the motor, the torque ripple amplitude is harmonised to achieve constant torque output and torque efficiency in excess of 83% (Automation Airmotors, 1998).

Previous control of the motor has incorporated a flow control valve (spool valve), linking the motor ports. The function of the valve is to control the direction and the rate of airflow through the motor, hence the speed or position. Actuation was provided by a proportional solenoid. With this arrangement, the dead-band effect during change of direction of shaft rotation as indicated earlier, exists, see Fig. 1. This affects the shaft resolution; the motor shaft may oscillate or suddenly "jump" if the encoder detects that the shaft moves outside the position window (Automation Airmotors, 1998).

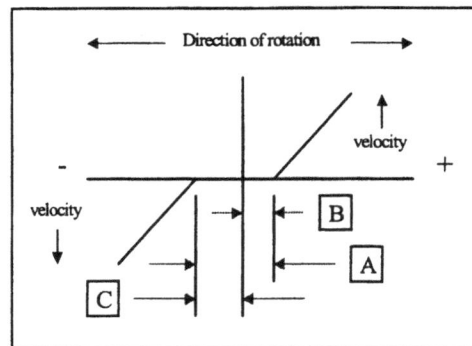

Fig. 1. Spool valve flow control characteristics.

It is proposed that the flow characteristics shown in Fig. 1 can be produced using the pneumatic equivalent of an H-bridge. Since the change of flow direction is no longer mechanically linked but software controlled, the magnitude of the dead-band (A) can be variable and may be tuned to suit the mechanical variation in the motor. The dead band offsets (B and C) may also be tuned to suit the environment in which the motor is installed. Unlike an electrical H-bridge, here the effect of short circuit, in which the dead band A is absent, is not critical. This feature allows flexibility in software tuning.

The functional set-up of the proposed H-bridge system is shown in Fig. 2. Four piezo poppet type valves are arranged to form an H-bridge around the two motor ports. Poppet valves A1 and A2 control a variable rate of airflow through the motor, and poppet valves B1 and B2 control the motor direction. Poppet valve A1 in association with poppet valve B2 will allow air flow through the motor in a particular direction, whilst A2 in association with B1 will allow air flow in the opposite direction. For operation in any one direction, only one of this set will be active whilst the other pair will be closed.

Fig. 3 shows a poppet needle valve (A1 and A2 in Fig. 2) in which a tapered "needle" attached to the piston is able to vary the annular area of the poppet seat according to the position of the piston in its enclosing cylinder. Supply pressure entering the valve is able to act beneath the piston and lift the needle from its seat. A thin line is also tapped from this pressure to act on the top of the piston. A compression spring also acts on the topside of the piston. Thus, a combination of air pressure and spring force pushes the piston down to close the valve. The piston is lifted if the pressure on the top is bleed off;

this is achieved by venting some air into the atmosphere in a controllable fashion. The cross-sectional area of the port into the chamber is smaller than the cross-sectional area of the vent port; thus permitting decay of the pressure in the chamber. Mark space signal is applied to a piezo membrane that operates the flapper 'on' and 'off' for a duration, which is determined by the duty cycle of the PWM signal. Thus, by controlling the piston position, which in turn throttles the air into the motor, the speed of the motor can be regulated.

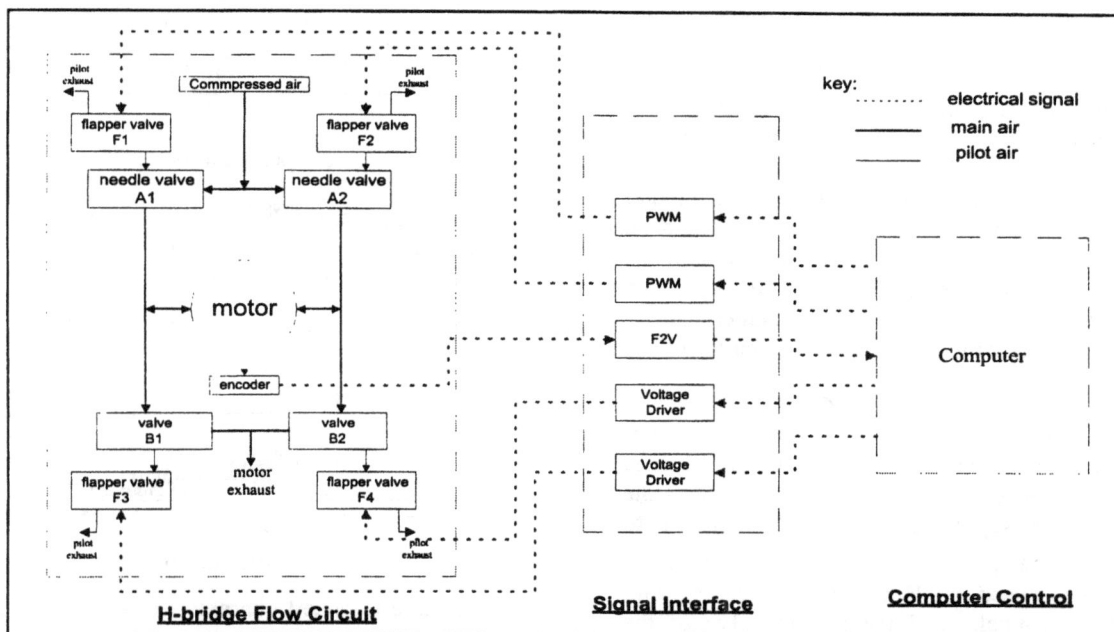

Fig. 2: The proposed motor control system.

Fig. 4 shows the poppet valve (B1 and B2 in Fig. 2) which is used for controlling the direction of flow through the motor. The operation is similar to that of the needle valve above, but operates in a simple open or closed mode. Two flappers X and Z are associated with controlling the pressure in the piston chamber. Flapper X opens and closes the air into the chamber, and flapper Z opens and closes the vent to the atmosphere.

During experimental tests it was found that the piezo valve (A1 and A2 in Fig. 2) orifice was of insufficient size to vent the pressure force acting on the poppet needle. As a result the needle would not lift from its seat to allow airflow into the motor. It was decided to replace the piezo valve stages with Horbiger pressure control valves. Similarly, the piezo devices at the two outlet sides of the H-bridge were replaced with two conventional solenoid operated valves (ON/OFF), each controlled by a 24V dc. The airfit tecno pressure controlling Hoerbiger valve (PCV) is electronically controlled proportional pressure regulating valve with Piezo 2000 (Hoerbiger Pneumatic, 1999). This device is used for regulating the air pressure applied on the topside of the piston of the needle valve, which in

turn controls the air flow entering the motor. The PCV regulates pressure at the rate of 10^5 v/Nm^{-2}. Since the supply pressure is maintained at 4.8×10^5 Nm^{-2} in this work, the operating range is restricted to 0 - 4.8 volts corresponding to 0 - 4.8×10^5 Nm^{-2} output. The Horbiger valve has the facility of controlling the pressure force acting on the needle relative to a 0 to 10v input (or 4 to 20mA current).

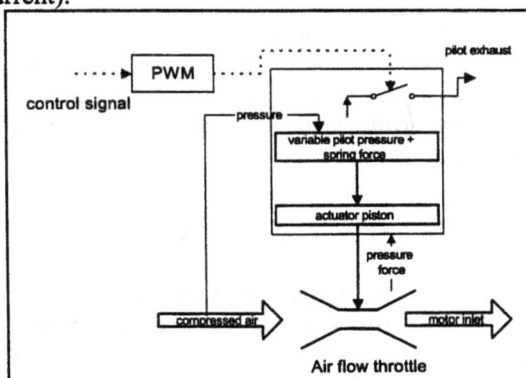

Fig. 3: Poppet needle valve.

101

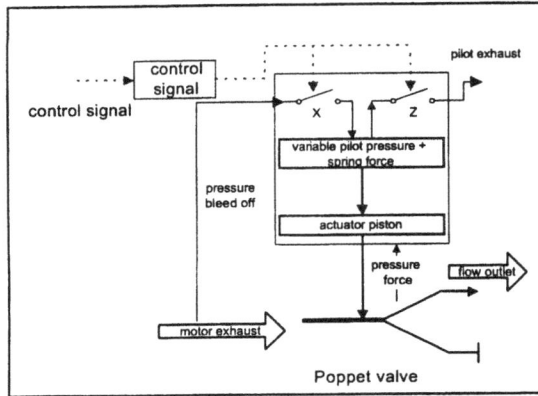

Fig. 4: Poppet valve.

A PC with necessary hardware is used to source and read all plant devices. Where signal compatibility is a problem, an appropriate preconditioning is employed to facilitate proper interfacing between the computer and the plant. An RTI815 I/O board is used to provide real-time interface between the PC and the airmotor (Analogue Devices, 1994). An Applocon Liveline PC14AT digital I/O timer counter card is also incorporated within the interface circuitry of the system.

A set of experiments, to obtain a general characterisation of the performance of the H-bridge mechanism, was conducted. In these experiments the compressed air supply pressure was adjusted to 4.8×10^5 Nm^{-2}, a spring rate (for the needle valve) of 0.11 N/mm was used, the H-bridge was supplied with air on both channels, and the needle valve of the opposite direction fully closed while the system operated in one direction. The pressure to speed characteristics of the H-bridge were recorded for the direct and reverse modes of operation of the motor. These are shown in Fig. 5. It is noted that the H-bridge exhibits a linear characteristic over a certain range of operating points in each direction, where an increase in the pressure applied to the PCV results in a decrease in the motor speed. It is evident from these results that the pneumatic H-bridge concept is feasible in practice, that is, by regulating the applied voltage to the PCV, the input pressure and hence the speed as well as direction of rotation of the motor can be controlled. The linear operating region of the system spans over a narrow pressure range. This, however, can be increased with suitable adjustments to the system parameters surrounding the needle valve and the spring type.

3. PID CONTROL OF THE MOTOR

For purposes of this study, a PID control strategy is adopted. PID control is one of the most widely used approaches in the design of continuous-data control systems (Kuo, 1992). Many computer control implementations have simply taken over the well-established analogue PID. Algorithms developed using various digital control design techniques can be

equally effective and a lot more flexible than the three-term controller. However, the art of tuning PID controllers is well established and the technique gives a well-behaved controller.

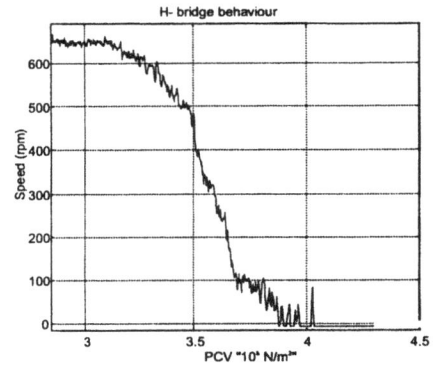

(a) Direct operation mode.

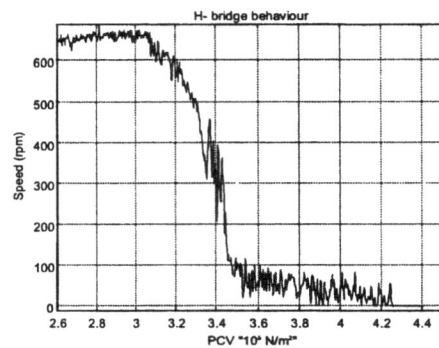

(b) Reverse operation mode.

Fig. 5: Pressure to speed characteristics with the H-bridge.

Fig. 6 shows a block diagram of the control structure used, where the speed measurement can either be done through frequency to voltage converter (F2V) or directly through the PC14AT. The controller in Fig. 6 represents a digital PID, obtained through a discretisation of the corresponding continuous PID, using first-order finite difference methods (Bennett, 1994);

$$s(n) = s(n-1) + e(n)$$
$$u(n) = K_p e(n) + K_i s(n) + K_d [e(n) - e(n-1)]$$

where, $K_i = K_p (T_s / T_i)$ and $K_d = K_p (T_d / T_s)$ with K_p representing the proportional gain, T_i the integral action time, T_d the derivative action time and T_s the sample interval.

Calculation of the control action within the PID depends on the sampling interval T_s, which in most cases is lower than the computer's execution speed. Thus, for correct operation, some means of fixing the sampling interval is required. Synchronisation can be achieved through, for example, using a real-time clock signal or an external interrupt. The use of real-time clock signal is a general solution to timing a control loop. It is provided by utilising the existing real-time clock (dos clock). Use of external interrupt for synchronisation can be very effective. The control loop is written as an interrupt, which is associated with a particular interrupt line. The interrupt line is activated by the Am9513 on-board timer chip of the RTI815 card. Both these synchronisation techniques were used in this work, although results with timing using the dos clock are presented only.

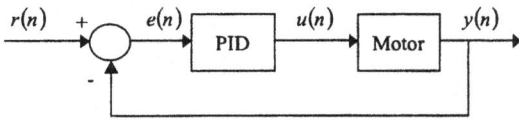

Fig. 6: Block diagram of the control system.

4. EXPERIMENTAL RESULTS

In these experiments an air motor with power rating of approximately 350 watts at $7 \times 10^5 \, \text{Nm}^{-2}$ was utilised. The experiments were carried out under a "no load" condition, which as there is no load dumping, this can be considered a "worst condition". the pre-conditioned shaft encoder signal was used directly through counter/timer of the PC14AT to infer the rotational speed of the motor. Both channels of the H-bridge were tested for high and low speed performance. A PID algorithm was devised to facilitate multiple set point changes at predefined intervals. Initial tests revealed that the H-bridge behaviour at high speed was different from the low speed. For this reason the PID tuning parameters and hence the system response was expected to be different in these regions.

For high-speed control of the motor, the PID controller was realised with an air pressure of $4.8 \times 10^5 \, \text{Nm}^{-2}$, $K_p = 0.005$, $T_i = 0.6 \, \text{sec}$, $T_d = 3.5 \, \text{repeat/sec}$ and $T_s = 50 \, \text{ms}$. Figure 7 shows the response (speed) of the motor with set points at 300, 400 and 450 rpm in both direct and reverse operating modes. It is noted that the system follows the desired speed reasonably well. The speed of response is slightly lower for a rising than for a lowering speed. To evaluate the set point tracking ability of the system at high speeds, the control system was subjected to a linearly changing set point. Fig. 8 shows the performance of the system for both direct and reverse modes of operation of the motor. It is noted that the system tracks the set point well.

For low-speed control of the motor, the PID controller was realised with an air pressure of $4.8 \times 10^5 \, \text{Nm}^{-2}$, $K_p = 0.02$, $T_i = 1.1 \, \text{sec}$, $T_d = 1.5 \, \text{repeat/sec}$ and $T_s = 50 \, \text{ms}$. Fig. 9 shows the system response with set points at 130, 160 and 200 rpm. It is noted that the system achieved the desired set point reasonably well. For the H-bridge in the reverse operation mode a large spike (although still

within safe operating region of the motor) appeared when the set point changed from 130 rpm to 200 rpm. The spike was observed to come from the actual system and not the signal processing circuitry. In comparison with conventional air motors and electric motors, the operating characteristics of the motor are unusual in low speed shaft output. This renders response time as "relative".

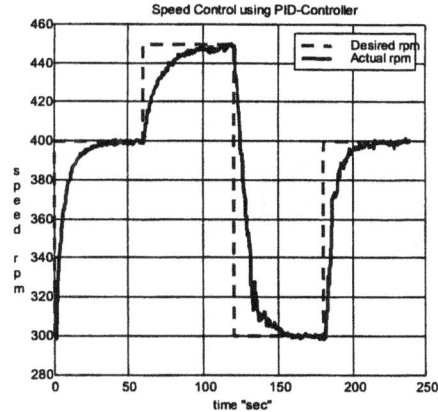

(a) Direct operation mode.

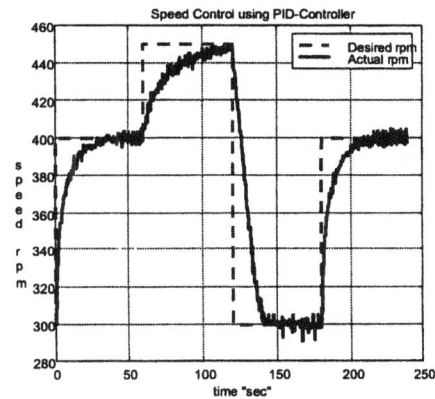

(b) Reverse operation mode.

Fig. 7: High-speed control of the motor.

It is noted in the results above that, good control can be achieved at high speeds. Low-speed control, on the other hand, is achieved well with set point changes from high to low. However, with set point changes from low to high a spike in the system response may appear. A possible cause to the spike appearing in the response of the system when the set point changes from low to high is related to the mechanical design of the system. At low speeds the needle valve is just crack opened. This means that the cross-sectional area of the needle valve, exposed to the main pressure, is less than the topside of the piston. When a higher speed is demanded, the pressure on the top of the piston (needle valve) is reduced by the PCV, and only when enough pressure is released the piston starts to move back to open the port. It is during this movement that the cross-sectional areas become equal, but the balance is lost and an overshoot is experienced.

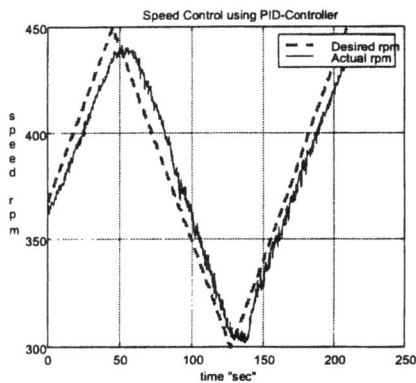

(a) Direct operation mode.

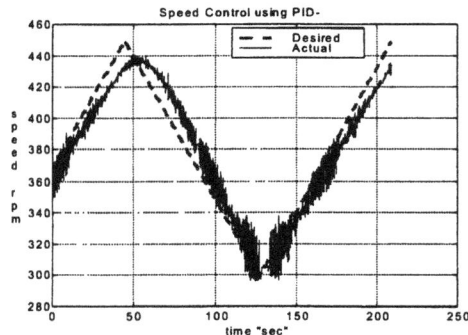

(b) Reverse operation mode.

Fig. 8: High-speed set point tracking of the system.

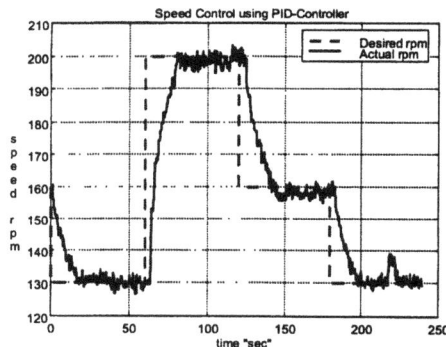

(a) Direct operation mode.

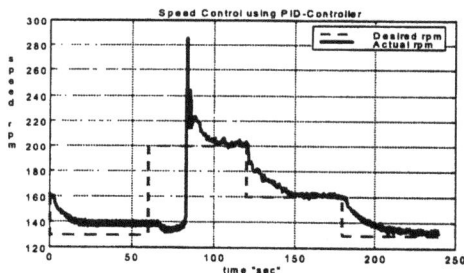

(b) Reverse operation mode.

Fig. 9: Low-speed control of the system.

The experimental results above show that speed control of the motor can be achieved through the H-bridge mechanism. In these experiments the PID parameters were tuned for achieving a response without overshoot. The system response, however, was slow. The limiting factors associated with this were, avoiding response overshoot and the use of

slow sample rate. Using a more powerful processor, allowing faster sample rates, can enhance the performance of the system.

5. CONCLUSION

A feasibility study of real-time control of an air motor using a pneumatic equivalent of the H-bridge concept has been carried out. It has been demonstrated that, speed as well as direction control of the motor can be achieved with the proposed concept. Good speed regulation of the motor is achieved with a PID controller at high speeds, where the characteristics of the H-bridge are fairly linear. At low speeds, where the characteristics of the H-bridge change, as good level of control as with high speeds is not achieved. The performance of the system can be enhanced within this speed range, with a more powerful computing platform that allows higher sampling rates. Moreover, to achieve overall control of the system at both high and low speeds, an adaptive control strategy can be adopted.

REFERENCES

Analogue Devices (1994). *RTI-800/815 user's manual*, Analogue Devices.

Automation Airmotors Ltd. (1998), *Air motor flow control system - development project.* Automation Airmotors Ltd., Barnsley, UK.

Bennett S, (1994). *Real time computer control- an introduction,* 2nd edn, Prentice-Hall.

Billany_M. (1990). Microprocessor-controlled air servo motor provides high speed, allows prolonged stall. *Power Conversion & Intelligent Motion,* **16,** (10), pp.35-36, 45-46.

Hitchcox_A.L. (1995). Performance insurance for air motors. *Hydraulics & Pneumatics,* **48,** (10), p.63-68.

Hoerbiger Pneumatic. (1999). *Automation technology – piezo – technology,* Hoerbiger/Origa Pneumatic, Filderstadt.

Kuo, B.C. (1992). *Digital control systems,* 2nd edn, Saunders.

Mahanay_J. (1986). Gerotor air motor: new otion for low-speed output. *Machine Design,* **58,** (3), pp.75-77.

Morgan, G. (1984). Motors that run on air. *Power Transmission Design,* **26,** (9), pp.35-38.

Noritsugu, T. (1987). Electro-pneumatic feedback speed control of a pneumatic motor. I. With an electro-pneumatic proportional valve. *Journal-of-Fluid-Control,* **17,** (3), pp.17-37.

Noritsugu, T. (1988). Electro-pneumatic feedback speed control of a pneumatic motor. II. With a `PWM operated on-off valve. *Journal of Fluid Control,* **18,** (2), pp.7-21.

Simnett_R.W, Anderson_E. (1983). Air motor drives for small pumps. *Chemical Engineering,* **90,** (25), pp.73-75.

DESIGN OF DISTRIBUTED MANUFACTURING SYSTEMS USING UML AND PETRI NETS

B. Bordbar, L. Giacomini and D.J. Holding

*Department of Electronic Engineering, School of Engineering, Aston University,
Aston Triangle, Birmingham B4 7ET, UK
Tel: Tel: +44 (0)121 359 3611 Fax: +44 (0)121 359 0156
e-mail: {B.Bordbar,L.Giacomini,D.J.Holding}@aston.ac.uk*

Abstract: This paper describes the design of a supervisory control system for a distributed
manufacturing process, which forms part of a wider manufacturing system. The focus of the paper
is on the design of a verifiable discrete event controller using a UML based method. The approach
adopted involves (i) using Petri net models instead of conventional Statecharts to provide analytic
Dynamic Models; and (ii) using compositional Petri net techniques to synthesise the
Interconnection Model. The model of the complete controller can be then analysed and verified
using Petri net theory. The approach is demonstrated by application to a prototype packaging
machine. *Copyright© 2000 IFAC*

Keywords: Discrete-event dynamic systems, Petri-nets, Object modelling techniques,
Manufacturing systems

1. INTRODUCTION

Recent advances in computer technology have
resulted in a widespread use of Discrete-Event
Dynamic Systems or DEDSs in manufacturing,
robotics, traffic management, logistics, and computer
and communication networks (Cassandras, 1999).
DEDSs require complex control systems (Ramadge
and Wonham, 1987) to ensure correct and optimal
operation. To model complex DEDSs, researchers
have developed bottom up, top down and hybrid
synthesis techniques. However, these approaches
concentrate on functional abstraction, and have
produced incomplete specifications and designs
(Firesmith, 1993). In order to facilitate the design of
complex systems, produce more understandable
designs and specifications, facilitate the transition
between design and implementation and to enable
software re-use, several researchers including Booch
et al. (1999), Douglass (1999), have advocated a
paradigm shift towards object oriented (OO)
techniques.

The Unified Modelling Language (UML), originally
a methodology for software designers, is the most
recent product generated by the aggregation of
previous generation Object Oriented methodologies
(Booch et al., 1999). UML takes the designer
through the design life cycle, starting from the
description provided by users or experts down to the
final software product. It preserves convergence and
clarity in design by prescribing a set of steps that
generate an evolving model of the system, and
facilitate the rigorous examination of this model.
Thus, the application of UML by different people
with different skills results in comparable and highly
portable final designs.

UML consists in a set of nine main graphs or charts
with explanatory comments that can be expressed in
a formal way or in plainspoken language. These fall
into two categories, static aspect diagrams and
dynamic aspect diagrams, and the designer can
choose quite freely to use a subset of them. In UML

the dynamics of objects are described using a form of state diagram known as a Statechart (Harel, 1987). Concurrent or distributed systems are formed by creating parallel State Charts that are inter-connected and synchronised using *interaction diagrams*.

Although Statecharts are very popular and are well supported by implementation tools, they currently lack analytic capabilities and thus software tools cannot ensure the functional consistency of the overall design. Conventionally the behaviour of such designs is investigated using process considerations, such as completeness arguments (Levenson, 1995) and are demonstrated by simulation. However, to facilitate analysis, any system described by a State Chart can be replaced by an analytic representation such as Process Algebras, Automata, or Petri nets. Among the alternatives, Petri nets have a graphical approach that is easy to understand (Murata, 1989) and they are more effective in describing concurrent and asynchronous systems. Petri net theory can be used to analyse DEDS characteristics such as synchronisation, concurrency, conflicts, resource sharing, precedence relations, event sequences, non-determinism and system deadlocks (Desrochers and Al-Jaar, 1995). Also Petri nets, unlike state diagrams, are modular, and larger nets can be formed by simply merging places or transitions.

In this paper, to enhance analytic capabilities we shall improve our model by substituting the State Chart representation of dynamic models with a Petri-net. The paper also presents a method of synthesising coordination and synchronisation logic for distributed or large scale designs using Use Case information and compositional Petri net techniques. The approach is demonstrated by application to a manufacturing system comprising a prototype packaging machine.

2. UML BASED DESIGN

2.1   *Use Case and Class diagrams.*

The UML design procedure (Booch et al., 1999) starts with the study of the Use Cases which are detailed written descriptions of 'what the objectives are' and 'how the job is carried out'. Studying the use cases enables the designer to recognise different '*key agents*' of the system (*Objects* in UML terminology).

Considering common features and operations of key agents, objects are extrapolated into collections called *Classes*. Classes can be organised in a graph (or a collection of graphs), to build a '*class diagram*', that describes the *static relationship* between the

classes. The classes are represented graphically by rectangular boxes accomodating lists of attributes and operations and are connected together by lines or links that can be either of association type or of generalisation type. An *association* is a structural relationship that specifies the connection between one or more members of the classes. A *generalisation* is a relationship between a general class and a derived class, i.e. one defined from another class by means of inheritance. The operations defined in the class diagram include all the services that can be requested from an object to effect the behaviour. For the manufacturing system applications we have in mind, the system can be arranged in such a way that all the synchronisation issues can be expressed in terms of Boolean attributes of the involved classes.

2.2   *Petri Net Dynamic Model.*

The dynamic model describes behavioural aspects of the object classes, in the sense that they describe the sequence of operations that occur without regard for what the operations do, what they operate on, or how they are implemented. To improve the representation and facilitate analysis of the UML dynamic model, in this paper the dynamic model is represented by a Petri net. For general information regarding Petri nets, we refer to Murata (1989). Generally, the Petri Net of a class is formed by using a place to represent each Boolean attribute and a transition for each operation that changes the attributes values. A token in a place means that the attribute value is set to true (false otherwise).

2.3   *Graph of Desirable States (GDS) and Compositional Petri Net*

The process of compositional synthesis is not an ad-hoc procedure. Simply decomposing the Use Case diagrams into a bag of rules that are imposed on the objects ignores the important sequence information and will over constrain the model.

To maintain the precedence relationships and attain the synchronisation objectives specified in the Use Case we construct a directed graph, which we shall refer to as the *Graph of Desirable States* (GDS), which enumerates all desirable states and their relationships. The GDS maps the Use Case information into the Petri net domain, i.e. the sentences in the Use Case are translated in sets of rules in terms of places and transitions. The word "desirable" reflects the facts that the graph embraces all we expect the system to do, and any unwanted or undesirable behaviour is prohibited by identifying (for the design of constraint or inhibition logic) all enabled transitions that lead to undesirable behaviour.

2.4   *The Graph of Desirable States*

Let us assume that our system is made of m objects. For each object a Petri net is instantiated. Assume that Γ denotes the part of the Use Case dealing with the synchronisation of n of the above components into an overall system (typically, the objects are synchronised two at a time, until the compositional approach encompasses the whole system).

Assume that $(N_1, \mathbf{m}^1_0), \ldots, (N_n, \mathbf{m}^n_0)$, where $\mathbf{m}^i_0$, i=1, $\ldots$, n, denote the initial markings, are bounded and live Petri Nets, representing object instances of these n components of the system. A proportion of the information provided by Γ has already been captured in the body of the dynamics of the Petri nets $(N_1, \mathbf{m}^1_0), \ldots, (N_n, \mathbf{m}^n_0)$. Let $R_\infty(N_i, \mathbf{m}^i_0)$ denotes the set of all reachable markings of the Petri Net $(N_i, \mathbf{m}^i_0)$. For each $\mathbf{m}^i \in R_\infty(N_i, \mathbf{m}^i_0)$, let *enabled*($\mathbf{m}^i$) denote the set of all enabled transitions of N_i under the marking $\mathbf{m}^i$. Each node of GDS is labelled by a (n + 1)-tuple of the form $a = (\mathbf{m}^1, \ldots, \mathbf{m}^n, U)$ where $\mathbf{m}^1, \ldots, \mathbf{m}^n$ are reachable markings of the components $N_1, \ldots, N_n$ and U is the set (possibly empty) of undesirable enabled transitions under $\mathbf{m}^1, \ldots, \mathbf{m}^n$, as derived from the use case Γ. Thus U is a subset of of *enabled*($\mathbf{m}^1$) $\cup \ldots \cup$ *enabled*($\mathbf{m}^n$). For the node labelled with $a = (\mathbf{m}^1, \ldots, \mathbf{m}^k, U)$ we shall write $m(a) = (\mathbf{m}^1, \ldots, \mathbf{m}^k)$ and $U(a) = U$.

The GDS can be generated as follows. Consider the set E_0 of all transitions enabled under initial marking $\mathbf{m}^1_0, \ldots, \mathbf{m}^n_0$. From the above, a possibly empty subset U_0, of the transitions (or more properly their associated actions) are undesirable. Create the first node, which shall be referred to as the *initial node*, and label it with $a_0 = (\mathbf{m}^1_0, \ldots, \mathbf{m}^n_0, U_0)$. From this node start firing each of the desirable transitions $E_0 \backslash U_0$, to obtain another set of nodes each with their marking and a set of undesirable transitions. Put arcs connecting node a_0 and the newly created ones labelling them with t, where t is the name of the corresponding firing transition. The procedure is repeated for each of the new nodes created.

The GDS captures the behaviour expected from the composite net. For example, starting and ending in the same node of GDS represents a cyclic phases of the system. The GDS will also reveal problems with a design: for example, if there is a node a of GDS with no output then our design of the system expects a deadlock, which is anomalous.

Remark: The algorithm creates at most $\alpha^n 2^{n\beta}$ nodes, where α is the maximum number of reachable states of components and β is the maximum number of enabled transitions under different markings. Notice that each subset of the set of enabled transitions can be potentially a non-desirable set of transitions.

2.5   *Connecting the components Petri nets*

Consider the task of interconnecting together the Petri net dynamic models $(N_1, \mathbf{m}^1_0), \ldots, (N_k, \mathbf{m}^k_0)$ to create a composite Petri net $(N, \mathbf{m}_0)$ with the desired coordination and synchronisation as described in the Use Case. The composition is performed using standard Petri net techniques (Juan et al., 1998) and the information in the GDS concerning desirable and undesirable transitions. For example, an almost general rule applicable when we want to prevent a transition t_k in Petri Net N_j from firing under a certain marking $\mathbf{m}^i_k$ in Petri Net N_i, a new place is added and connected as input/output to the transition t_k. The place is also connected to transitions in Petri net N_i in such a way that, when the transitions give rise to the marking $\mathbf{m}^i_k$, the token is removed. The firing of the transitions moving out of the marking $\mathbf{m}^i_k$ will put the token back in the place (see figure 1).

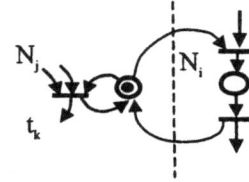

Fig. 1

Although, the composition process is straightforward for a GDS in which for $\forall$ a, b nodes, $m(a) = m(b) \Rightarrow U(a) = U(b)$, in other cases, significantly, the sequence information must be used to distinguish the states in the compositional process. When the composite Petri net is complete, it can be analysed using to Petri net theory to ensure that it is deadlock-free, live and bounded. The method is illustrated in the following example.

3   APPLICATION TO A PRODUCTION LINE PROCESS

The approach is demonstrated by considering the design of a controller for a simplified production line comprising loosely-coupled independently-driven mechanisms as shown in figure 2. The major components of the system are controlled individually and independently and perform motion profiles corresponding to different tasks. Supervisory (discrete event) control is to be used to synchronise the components.

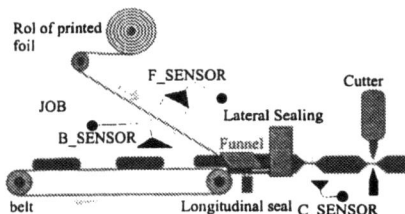

Fig. 2. Production Line.

The wrapping system of figure 2 is made of 4 objects: the belt, the foil roll unwinding device (film), the welder, the cutter. The product (JOB) and the foil that carries a printed tag (TAG) are identified with their supports, i.e. the belt and the film, respectively. JOB and TAG are displaced with respect to the belt and film. Let us examine the Belt Use Cases. When the JOB arrives (*new* JOB) in the proximity of a decision point sensor (**dp**), the state of the TAG is evaluated. If the TAG is at decision point the wrapping can take place (*go*). However, if the TAG is still **out**side the wrapping area, the JOB will stop, **wa**iting for the TAG to arrive at its decision point (*ab*ort operation). When the TAG arrives (*new* TAG), the JOB is restarted (*st*art leading to the **wrapping** state). When JOB and TAG are both in the **wrapping** state, the packaging foil is formed into a tube via a funnel, and a longitudinal sealing roller welds the two edges of the film together. The tube is sealed between packs by a lateral sealer (welder) and the wrapped product exits from the **wrapping** area (*exit* leading to the state **out**). The sealed products are then separated by a cutting machine (cutter) to produce individually packaged products, and the whole cycle restarts.

Similar dynamic models have been derived for the Film (TAG), Welder and Cutter. The welder and film, and film and cutter, are synchronised by applying a heuristic similar to the one between the belt and film.

3.1 *The class diagram*

The description in Section 3 plays the role of the Use Case for the production line of figure 2. The underlined terms represent the classes: Film, Belt, the Welder, and Cutter. The product to be wrapped, JOB, is identified with the belt. The printed film and the motor driving the unwinding are also identified with the Film object. The terms in bold typeface are the attributes of the classes (for the Belt, B_dp, B_wait, B_wrap, B_out; F_dp and similarly for the

class Film). The terms in italic typeface are the operations of the class. As an example, for the Belt:

B_new() {B_out=False; B_dp=True;}
B_ab() {B_dp=False; B_wait=True;}
B_go() {B_dp=False; B_wrap=True;}
B_go() {B_wait=False; B_wrap=True;}
B_exit() {B_wrap=False; B_out=True;}

The class diagram for the production line of figure 2 is shown in figure 3, without the attributes/operations lists.

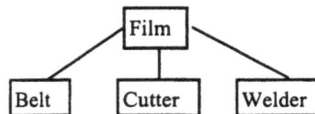

Fig. 3. Class diagram.

3.2 *Synthesising Petri net model*

For conciseness, we will focus on the interaction and synchronisation of the Belt and Film. First Petri nets are derived for each of the classes by assigning one place to each attribute and one transition to each operation. The Use Case description of the dynamics of each object is then used to construct the Petri net dynamic model. Specifically, places associated with attributes that an operation sets to False (or True) form inputs (or outputs) of the associated transition. In this particular application, the dynamic models of the classes are all structured in the same way as shown in figure 4 (a)-(b).

The initial marking for each instantiated object is obtained by considering the initial state of the corresponding components of the production line. The system starts with B_out and F_out.

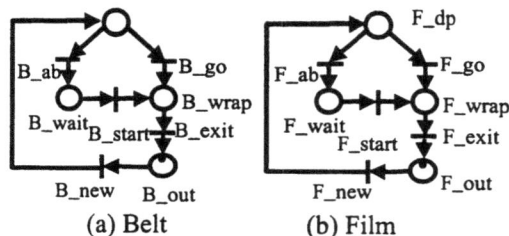

(a) Belt   (b) Film
Fig. 4. Petri Net for the classes Sensor and Belt.

The GDS for the mutual synchronisation of Belt and Film. The next stage in the compositional process is designing the synchronisation logic that enforces the mutual synchronisation heuristic for the Belt and Film, as defined in Section 3. To do this, we make use of the discursive Use Case provided in Section 3, and the Petri Nets of the components, to generate the GDS, using the procedure described in Section 2.4.

For example, starting with (B_out, F_out) we will have a set of enabled transitions enabled: B_new and F_new. From the Use Case, since no JOB or TAG are yet present, none of the two transitions is undesired, therefore U = ∅. Let us put two arrows labelled with B_new and F_new coming out of the current state. We then examine the Use Case and each arc in turn. Let us suppose, that JOB arrives first, i.e. B_new fires. This generates a new marking (B_dp, F_out) and set of enabled transitions {B_ab, B_go, F_new}. From the Use Case, if the JOB is at decision point but the TAG is still out of scope, then we want to decelerate the Belt, until complete rest if needed. Thus the transition B_go is undesirable: U = {B_go}. Proceeding in this way the GDS of figure 5 is built, and contains all the information about the dynamics of the two co-operating subsystems.

Co-ordination and synchronisation. For the GDS in figure 5, let us examine the set of undesirable transitions one by one. For example, B_new is not allowed to fire if and only if the marking is (B_out, F_wrap). This is achieved adding the place SP1, which is always marked except when F_wrap is marked (in fact its token is removed from the firing of F_go or F_start and F_exit replaces it). The same applies to F_new.

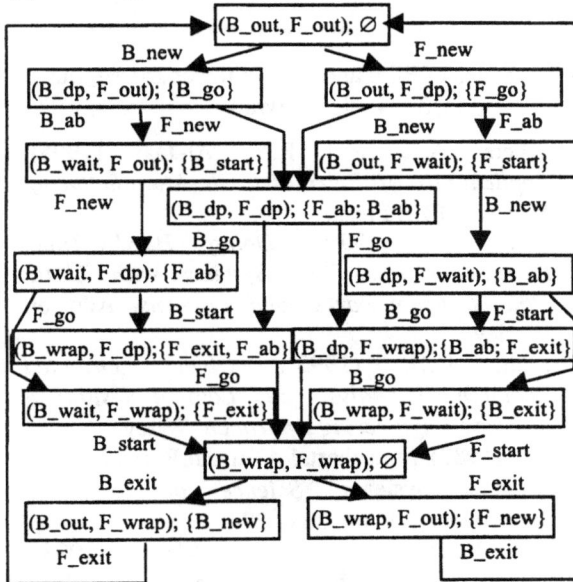

Fig. 5. GDS.

Similarly, B_ab is always an undesirable transition except when F_wrap is marked, therefore a double-sided arc between F_wrap and B_ab is added (and the same applies to F_ab). Also, B_exit is not desired before F_wrap gets marked (for the wrapping to take place the places F_wrap and B_wrap should be both marked), therefore place SP6 is added with an arc to B_exit; it is marked by the firing of F_go or F_start. Similarly SP4 is added for F_exit. Finally, B_go and B_start should fire as soon as F_dp is marked, therefore SP5 is added. SP2 is added to enable F_go and F_start as soon as B_dp is marked.

All this results in the Petri of figure 6; this graph is live and bounded and has the reachability graph shown in figure 7. The reader can notice the strong similarity with the graph of desirable case, figure 5.

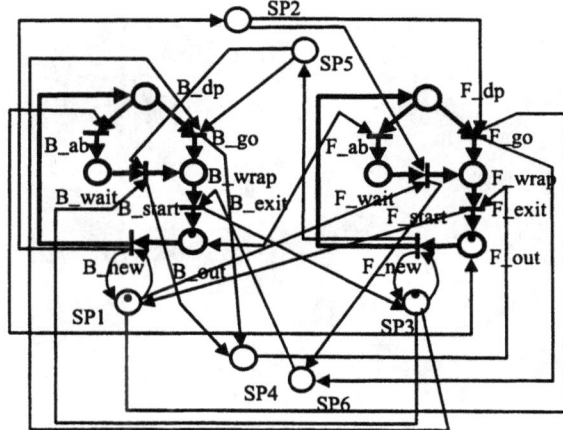

Fig. 6. Petri Net for the discrete part of the production line.

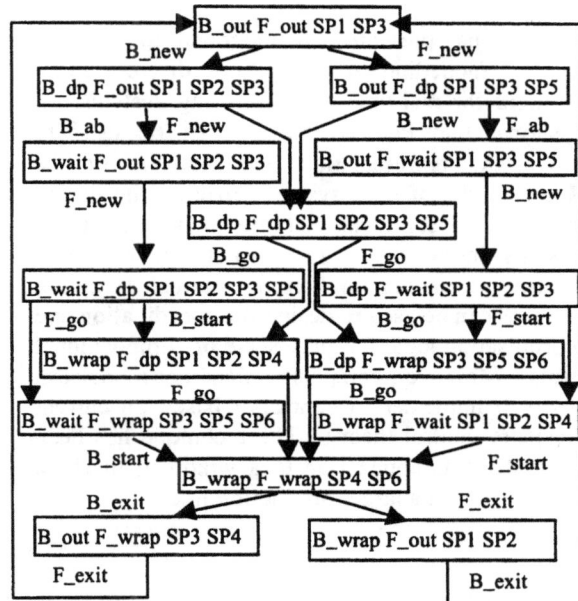

Fig. 7. Reachability graph for the petri net in figure.

4. IMPLEMENTATION

To demonstrate the design a continuous system simulation of the Film-Belt subsystem has been implemented using Matlab (vers. 5.3). To demonstrate the discrete event system involved using Stateflow: the reachability graphs of the component Petri nets were used as specifications for the design of Statechart components and interconnection of the Petri nets was modelled in using global variables which are updated when a transition takes place. The Simulink model is shown in figure 8 and the full state-chart is shown in figure 9. The belt and film systems are in two parallel sections (indicated by the dashed smoothed box). The StateFlow states of the Film and Belt during a typical synchronisation operation are shown in figure 10.

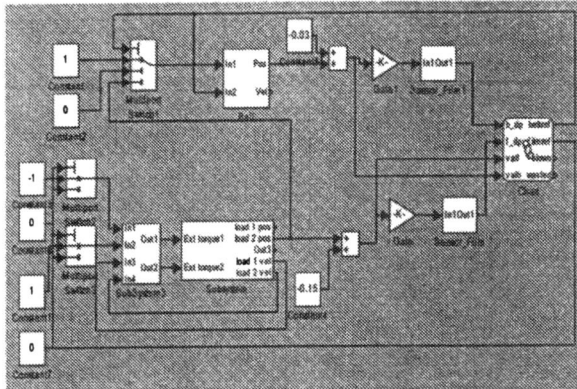

Fig. 8. Simulink scheme.

5   CONCLUSIONS

This paper has presented an integrated approach to UML for modelling and analysing discrete event controllers for real-time manufacturing systems. It has shown that Petri-net theory can be used to improve the representation and analysis of the dynamic model of such systems, making the design engineer more confident that the model accurately represents the system. It has also shown that UML use case information and compositional Petri net techniques can be used to design the coordination and synchronisation logic for large scale or compositional systems. Moreover, composite Petri-net model can be used to implement a controller based on current supervisory control theory. The technique has been illustrated by its application to a wrapping machine that forms part of a larger production line.

ACKNOWLEDGEMENTS

This work was supported by EPSRC (UK) Grant GR/L31234.

REFERENCES

Booch, G., J. Rumbaugh and I. Jacobson (1999). *The Unified Modeling Language User Guide*. Addison Wesley.

Cassandras, C.C. and S. Lafortune (1999). *Introduction to Discrete Event Systems*. Kluwer Academic Publishers.

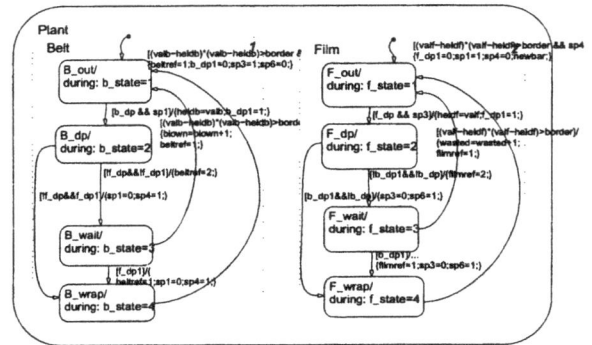

Fig. 9. Stateflow chart.

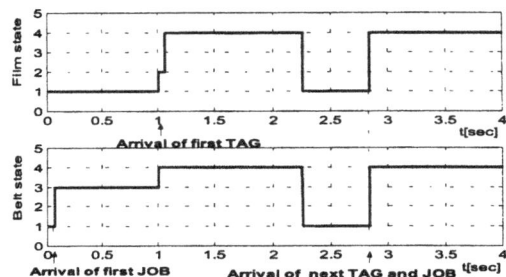

Fig. 10. States of Film and Belt in the chart: 1=_out, 2=_dp, 3=_wait, 4=_wrap.

Desrochers, A.A. and R.Y. Al-Jaar (1995). *Applications of Petri Nets in Manufacturing Systems*. IEEE Press.

Douglass, B.P. (1999). *Doing Hard Time. Developing Real-Time Systems with UML, Objects, Frameworks, and Patterns*. Addison Wesley.

Firesmith, D.G. (1993). *Object Oriented Requirement Analysis and Logical design: A Software Engineering Approach*. Wiley.

Harel, D. (1987). Statecharts: A Visual Formalism for Complex Systems. *Science of Computer Programming*, **8**, pp. 231--274.

Juan, E.Y.T., J.J.P. Tsai and T. Murata (1998). Compositional verification of concurrent systems using Petri-net-based condensation rules. *ACM Trans. on Programming Languages and Systems*, **20(5)**, pp. 917--979.

Levenson, N.G. (1995). *Safeware, Systems safety and Computers*. Addison Wesley.

Murata, T. (1989). Petri Nets: properties, analysis and applications, *Proceedings of the IEEE*, **77(4)**, pp. 541-580.

Ramadge, P.J. and W.M. Wonham (1987). Supervisory control of a class of discrete event processes. *SIAM Journal on Control & Optimization*, **25(1)**, pp. 206-230.

110

PROBABILITY ESTIMATION ALGORITHMS FOR SELF-VALIDATING SENSORS.

A. W. Moran, Dr. P. O'Reilly and Prof. G. W. Irwin

Intelligent Systems and Control Group,
School of Electrical and Electronic Engineering,
The Queen's University of Belfast,
Ashby Building, Stranmillis Road,
Belfast. UK BT9 5AH

Abstract: Three alternative approaches are investigated for probability estimation for use in a self-validating sensor. The three methods are Stochastic Approximation (SA), a reduced bias estimate of this same approach by Naim and Kam (NK) and a method based on the Bayesian Self-Organising Map (BSOM). Simulation studies show that the BSOM based method gives superior results when compared to the NK algorithm. It has also been demonstrated that the BSOM method is more computationally efficient and requires storage space for fewer variables. Copyright© 2000 IFAC

Keywords: Probability, Sensors, Stochastic approximation, Gaussian distributions.

1. INTRODUCTION

The use of smart instruments in industry has increased over recent years as engineers have taken advantage of the added features that they offer over more conventional "dumb" alternatives. These devices are able to apply corrections to the raw measurement in order to provide such features as outputs in engineering units and compensation for thermal drift (de Sá, 1988).

This has generated research interest in instruments that can not only compensate for some undesired physical property, such as non-linearity, but that can also detect and, more importantly, counteract internal faults. Such types of instruments have been termed "Self-Validating" (Henry and Clarke, 1993), in that they are able to provide an indication as to the validity, or confidence, in the measured value and also allow an indication of the health of the instrument to be generated. Such an instrument must therefore be able to extract more than just the process measurement from the sensor output. Various self-validating instruments have been developed including a ther-

mocouple (Yang and Clarke, 1997), a DOx sensor (Clarke and Fraher, 1996), and a coriolis massflow meter (Henry, 1996) and (Henry, 2000). One self-validating approach was described by (Yung and Clarke, 1989) and also implemented on a thermocouple system (Moran et al., 2000). This involves developing a parametric model of the sensor output during fault-free operation. This model is then used as an inverse filter to generate an innovations sequence which should be white noise (Ljung, 1987) if the model constitutes an accurate description of the sensor output. Any change to the statistics of the innovations sequence can be related to the occurrence of sensor faults (Upadhyaya, 1985).

Previous work by the present authors has demonstrated how a change to the variance of such an innovations sequence can be detected, both in simulation (O'Reilly, 1998) and using data from a practical temperature system (Moran et al., 2000). The detection method involved is a Likelihood Ratio test for two hypotheses, the null hypothesis, H_0, where there has been no change and the test hypothesis, H_1, where a change has occurred. The probabilities of H_1 and H_0 were found using on-line stochastic approximation (SA). (Naim and Kam, 1994) have shown that probability estimates, produced by stochastic approximation, will only be unbiased if the decision on the alternative hypotheses is error free, i.e. the probabilities of false alarm and missed detection are zero. This situation is unlikely to occur in practice and the fact that errors occur in the decision process may then lead to biased probability estimates.

The trend towards instruments with higher levels of sophistication means that sensors are constructed with an increasing amount of local processing power. However, the computing power of any one instrument will still be limited and so the best use must be made of the available resources. It is therefore desirable that the on-board algorithms required for self-validation should be efficient both computationally and in their storage requirements.

The aim of this work is to devise an improved probability estimation algorithm suitable for on-line implementation in the self-validating sensor test-bed described in (Moran et al., 2000).

Naim and Kam's (1994) method for generating reduced bias estimates relies on producing an estimation of the bias and then using this to correct the probability estimates. This paper will confirm that the technique described does indeed lead to estimates with reduced bias but that similar, if not superior, performance can be achieved by means of a very simple and computationally efficient method based on the Bayesian Self Organising Map (BSOM).

The decision theory background is outlined in Section 2 which describes the requirement for accurate probability estimates. Section 3 presents the reduced bias stochastic approximation approach of Naim and Kam (referred to here as 'NK') together with an approach based on a Bayesian Self Organising Map (referred to as 'BSOM').

A simple illustrative example is described in section 4, which also contains the main comparative results for the three approaches SA, NK and BSOM. The implementation aspects for a real-time self-validation sensor are discussed in Section 5 while section 6 contains conclusions and future work.

2. DEFINING THE PROBLEM

The decision process described for detecting a change to the innovations variance used a Likelihood Ratio test together with dual hypothesis testing (Moran et al., 2000), (O'Reilly, 1998). The null hypothesis, H_0, assumed that the variance

had not changed, the test hypothesis, H_1, that a change had occurred due to a sensor fault.

The probabilities that a given data set, $\mathbf{d}$, asserts, or refutes, each of these two hypotheses can be calculated. The decision, as to which of the two hypotheses is more likely, can be expressed as the Likelihood Ratio, L:

$$L = \frac{p(\mathbf{d}|H_1)}{p(\mathbf{d}|H_0)} \qquad (1)$$

The decision is now one of determining whether this Likelihood Ratio is above or below a threshold value, τ.

$$\begin{aligned} &\text{if } L > \tau \text{ then } H_1 \text{ is accepted;} \\ &\text{if } L \leq \tau \text{ then } H_0 \text{ is accepted.} \end{aligned} \qquad (2)$$

If the threshold, τ, in (2) is calculated with reference to P_{H0} and P_{H1} (the probabilities of H_0 and H_1), then the Bayes risk can be minimised:

$$\tau = \frac{(C_{10} - C_{00})P_{H0}}{(C_{01} - C_{11})P_{H1}} \qquad (3)$$

If the costs of making a correct decision (C_{00} and C_{11}) are zero and the costs of making an incorrect decision (C_{10} and C_{01}) are equal, the Likelihood Ratio test can be simplified to:

$$\begin{aligned} &\text{If } L > \frac{P_{H0}}{P_{H1}} \text{ , then } H_1 \text{ is accepted;} \\ &\text{If } L \leq \frac{P_{H0}}{P_{H1}} \text{ , then } H_0 \text{ is accepted.} \end{aligned} \qquad (4)$$

3. THE ALGORITHMS

To use (4) it is necessary to know the probabilities of the two hypotheses. Although the exact values may not be known in practice, they can be estimated on-line by a recursive stochastic approximation (SA) method (Naim and Kam, 1994). Thus:

$$\begin{aligned} \widehat{P}_{H1}^{(k)} &= \widehat{P}_{H1}^{(k-1)} + \frac{1}{k}\left(u^{(k)} - \widehat{P}_{H1}^{(k-1)}\right) \\ \widehat{P}_{H0}^{(k)} &= 1 - \widehat{P}_{H1}^{(k)} \end{aligned} \qquad (5)$$

Here: $\widehat{P}_{H1}^{(k)}$ is the estimate of P_{H1} at time-step k and $u^{(k)}$ is the binary value of the decision at time-step k (i.e. $u^{(k)} = 1$ if H_1 is accepted and $u^{(k)} = 0$ if H_0 is accepted).

As can be appreciated, there is a need for accurate estimation of P_{H1} and P_{H0} and (5) will only give unbiased estimates if the decision process is error free such that the probabilities of false alarm and missed detection are zero. Since this is unlikely to occur in practice, other methods will be required to provide the probability estimates. The two methods for generating these probability estimates will now be described.

3.1 *Reduced Bias Estimates*

Naim and Kam (1994) were concerned with distributed detection whereby the decisions of a number of local sensors were transmitted to a central data fusion centre. The decision process for each sensor used an on-line stochastic approximation method for determining the probabilities of two hypotheses. This method of probability estimation is employed here.

The Naim and Kam (NK) algorithm for reduced bias estimation is as follows :-

Step 1: Set initial conditions, i.e. randomly select a value for $P(H_1)$ within the range 0 to 1.

Step 2: Calculate the decision threshold,

$$\tau = \left(\frac{1 - \widehat{P}(H_1)^{(k)}}{\widehat{P}(H_1)^{(k)}}\right) \qquad (6)$$

Step 3: Calculate the Likelihood Ratio of the two pdfs for the scalar input value d:

$$L = \frac{p(d|H_1)}{p(d|H_0)} \qquad (7)$$

Step 4: Determine the binary decision

$$u^{(k)} = \begin{cases} 1, & L \geq \tau \\ 0, & L < \tau \end{cases} \qquad (8)$$

Step 5: Update the probability estimate according to,

$$\widehat{P}(H_1)^{(k)} = \widehat{P}(H_1)^{(k-1)}$$
$$+ \frac{1}{k}\left(u^{(k)} - \widehat{B}(H_1)^{(k-1)} - \widehat{P}(H_1)^{(k-1)}\right) \qquad (9)$$

Step 6: Approximate the probabilities of false alarm and missed detection, $\widehat{P}_F$ and $\widehat{P}_M$, respectively.

Step 7: Approximate the bias in $P(H_1)$ by,

$$\widehat{B}(H_1)^{(k)} = \left(1 - \widehat{P}(H_1)^{(k)}\right) \widehat{P}_F^{(k)}$$
$$- \widehat{P}(H_1)^{(k)} \widehat{P}_M^{(k)} \qquad (10)$$

Step 8: $k = k + 1$, Go to step 2

3.2 *Bayesian Self Organising Map*

Yin and Allinson (1997), described how an arbitrary probability density function can be approximated using a mixture of Gaussian kernels. These kernels each have three parameters that can be adjusted during training, the *mean, variance* and *priors*. These are all adjusted at each time-step by way of an adaptive gain, α, that is set to decrease with time.

In the application described here, the means and variances of the source data are known. It is then only necessary to adjust the *priors* of each of the two kernels. Note that in this application the *priors* are equivalent to the probabilities. Since the problem to be solved here is that of estimating two probabilities, the method described by (Yin and Allinson, 1997) simplifies even further as only two kernels are involved.

The BSOM based method is as follows:

Step 1: Set the initial conditions. Since the probabilities are unknown, randomly select a value for $P(H_1)$ ($\triangleq w_i$) within the range 0 to 1.

Step 2: Calculate the weighted output for each of the two kernels.

$$P_i(x) = w_i \cdot p_i(x|\mu_i, \sigma_i), i = 1, 2 \qquad (11)$$

Step 3: Calculate the output over the sum of the weighted kernel outputs:

$$P(x) = \sum_1^2 w_i \cdot p_i(x|\mu_i, \sigma_i) \qquad (12)$$

Step 4: Update the weights:

$$w_i^{(k)} = w_i^{(k-1)}$$
$$+ \alpha^{(k-1)}\left[\frac{w_i^{(k-1)} \cdot P_i^{(k-1)}}{P(x)^{(k-1)}} - w_i^{(k-1)}\right] \qquad (13)$$

Step 5: Adjust adaptive gain, α, and go to step 2.

$$\alpha^{(k)} = \left(\frac{10}{1000 + k}\right) \qquad (14)$$

4. ALGORITHM PERFORMANCE

For the three cases described below the innovations variance data was simulated as the mixed output from two Gaussian noise sources with equal variances. The means were chosen to give a reasonable overlap of the two probability density functions and for each case different values for $P(H_1)$ and $P(H_0)$ were chosen.

A change to the variance of the innovations sequence could be indicative of a change to the time constant (Yung and Clarke, 1989) and (Moran *et al.*, 2000).

As an illustration of the performance of these three algorithms the Figures (1), (2) and (3), below, show the development of the estimate of $P(H_1)$, for a true value of 0.7. For all cases the means and standard deviations of the noise sources are identical, $\mu_0 = 0.0$, $\mu_1 = 4.0$, $\sigma_0 = \sigma_1$ = 1.58.

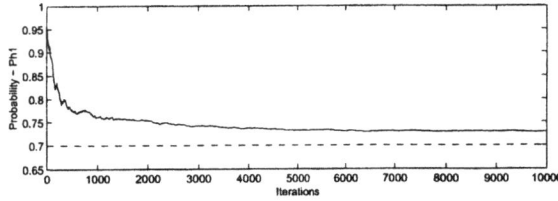

Fig. 1. Estimate of $P(H_1)$ using the SA algorithm.

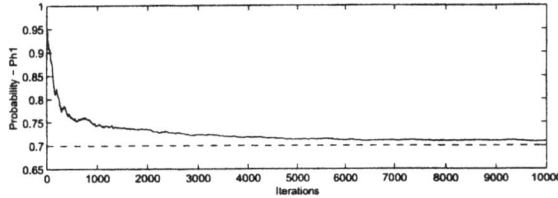

Fig. 2. Estimate of $P(H_1)$ using the NK algorithm for reduced bias estimation.

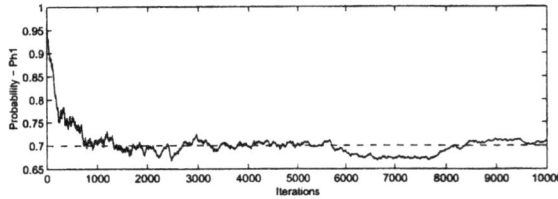

Fig. 3. Estimate of $P(H_1)$ using the BSOM algorithm.

At first glance the performances appear to be very similar. The SA algorithm definitely shows a biased estimate of the probability, the final value being approximately 0.73. The NK algorithm does give a more accurate estimation, however a bias is still evident after 10,000 data points. The BSOM probability estimates actually are "noisy" as the probability estimates, in fact the *priors*, are updated directly from the data, which is in effect a noise source. What is evident, on closer inspection, is that the estimate from the NK algorithm converges to the true value, 0.7, fairly slowly. By comparison the BSOM estimate converges within 1,000 data points.

These figures are for only one run and a clearer indication of the performance of the algorithms can be seen from the results of Monte Carlo simulations of 200 runs, each of 10,000 data points.

In tables 1, 2 and 3:

$\widehat{P}(H_1)$: mean - this is the average of the final values of $P(H_1)$ over the 200 runs.

$\widehat{P}(H_1)$: SD - this is the standard deviation of the final values of $P(H_1)$ over the 200 runs.

Bias (estimate): - these are the estimated biases for the SA and NK algorithms as given by equations 11 and 16 in (Naim and Kam, 1994). Note that there is no equivalent estimation of the bias for the BSOM algorithm.

Bias (actual): $P(H_1) - \widehat{P}(H_1)$ - this is the mean of the final values of the bias, $B(H_1)$, over 200 runs.

Case One

$P(H_1) = 0.3$, $\mu_0 = 0.0$, $\mu_1 = 4.0$, $\sigma_0 = \sigma_1 = 1.58$

Table 1. True $P(H_1) = 0.3$

	SA	NK	BSOM
$\widehat{P}(H_1):mean$	0.2570	0.3109	0.2987
$\widehat{P}(H_1):SD$	0.0411	0.0199	0.0101
Bias (est):	0.0352	01271	—
Bias (actual):	0.0430	-0.0109	0.0013

Here the SA approach did indeed give a biased estimate for $P(H_1)$, while the NK method reduced it. However, the bias from the BSOM algorithm was appreciably lower again and the standard deviation of the probability estimate was approximately halved.

Case Two

$P(H_1) = 0.5$, $\mu_0 = 0.0$, $\mu_1 = 4.0$, $\sigma_0 = \sigma_1 = 1.58$

Table 2. True $P(H_1) = 0.5$

	SA	NK	BSOM
$\widehat{P}(H_1):mean$	0.5005	0.5004	0.4991
$\widehat{P}(H_1):SD$	0.0134	0.0103	0.0116
Bias (est):	0.0000	-0.0009	—
Bias (actual):	-5.37e-4	-3.79e-4	9.48e-4

In this example the SA and NK algorithms both produced very good results. However, this is a special case because with equal probabilities for the distributions and the means being at 0 and 4, the probabilities of false alarm and missed detection were equal which results in the bias actually being zero. The BSOM results were not quite so good, but the standard deviation of the probability estimate was comparable to the previous case.

Case Three

$P(H_1) = 0.7$, $\mu_0 = 0.0$, $\mu_1 = 3.0$, $\sigma_0 = \sigma_1 = 1.58$

Table 3. True $P(H_1) = 0.7$

	SA	NK	BSOM
$\widehat{P}(H_1):mean$	0.7129	0.6953	0.6989
$\widehat{P}(H_1):SD$	0.0133	0.0104	0.0101
Bias (est):	-0.0145	-0.0667	—
Bias (actual):	-0.0129	0.0047	0.0011

Once again the performance of the SA algorithm was bettered by the NK one as would be expected. The results for the BSOM algorithm were better still.

In general it was found that the BSOM approach was the most consistent of the three methods, with very similar standard deviations for the estimate of $P(H_1)$ in all three cases.

5. DISCUSSION

The NK algorithm in (9), (10) is fairly complex. The recursive part comprises seven distinct stages. To calculate the Likelihood Ratio, L, the outputs of two distributions, assumed to be Gaussian, must be calculated at each iteration. In order to find the false alarm and missed detection probabilities, P_F and P_M respectively, it is necessary to carry out integration in order to find the area under the probability density function curve. Since this is a non analytical process, it is necessary to use some other method such as the error function or a look-up table. This obviously adds computational complexity to the algorithm, if the error function is to be coded, or an additional storage requirement should a look-up table be used.

For the BSOM algorithm, however, the recursive part comprises of only four steps. This is much simpler than the NK method and the computational load is, consequently, much lighter. In fact, the only thing that the two algorithms have in common is the need to calculate the outputs of two Gaussian distributions at each iteration. However, in the BSOM case this is a major part of the algorithm and not just the calculation of a subsidiary value as for the NK algorithm.

It is worth pointing out that the increased complexity of the NK algorithm also gives rise to an increased number of program variables for which storage space would be required. This is especially true if the error function were not coded but replaced by a look-up table. For an embedded system, the efficient use of storage space is necessary as it is often limited.

As an indication of the computational complexity of the algorithms used a simple test was carried out. The test comprised of timing the execution speed of each algorithm for a Monte Carlo simulation of 200 runs, with each run comprising of 10,000 iterations. The algorithms were each encoded as 'C' programs and run as DOS programs under windows 95^{TM}.

Table 4. Algorithm timings

	SA	NK	BSOM
Execution time (mS)	9.79	49.75	15.39

The actual execution times are not particularly important, it is the relative difference between the results that is of interest. Although these results are only a rough indication of algorithm performance, they do show that the BSOM algorithm is approximately 3.2 times faster than the NK

algorithm. The result for the SA algorithm is included as an indication of its simplicity.

6. CONCLUSIONS

It has been shown that accurate probability estimates can be found by a method based on the Bayesian Self-Organising Map. The accuracy of the method has been compared to that of the Reduced Bias Estimate algorithm described by (Naim and Kam, 1994), and has been shown to be superior. It has also been demonstrated that the method is more computationally efficient and requires storage space for fewer variables.

Further work will be to extend the work in simulation, shown here, in an effort to implement the method as part of the test-bed system described in (Moran et al., 2000).

7. ACKNOWLEDGEMENT

The first author would like to thank the Department of Education for Northern Ireland (DENI) for financial support.

8. REFERENCES

Clarke, D. W. and P. M. A. Fraher (1996). Model-based validation of a dox sensor. *Control Engineering Practice* 4(9), 1313–1320.

de Sá, D. (1988). The evolution of the intelligent measurement. *Measurement and Control* 21, 142–144.

Henry, M. P. (1996). Programmable hardware architectures for sensor validation. *Control Engineering Practice* 4(10), 1339–1354.

Henry, M. P. (2000). A self-validating digital coriolis mass-flow meter (1): Overview. *Control Engineering Practice*.

Henry, M. P. and D. W. Clarke (1993). The self-validating sensor: Rational, definitions and examples. *Control Engineering Practice* 1(4), 585–610.

Ljung, L. (1987). *System Identification : Theory for the User*. Prentice-Hall. London.

Moran, A. W., P. G. O'Reilly and G. W. Irwin (2000). A case study in on-line intelligent sensing. In: *Proc. of the American Control Conference*.

Naim, A. and M. Kam (1994). On-line estimation of probabilities for distributed bayesian detection. *Automatica* 30(4), 633–642.

O'Reilly, P. G. (1998). Local sensor fault detection using bayesian decision theory. In: *UKACC International Conference on Control*. pp. 247–251.

Upadhyaya, B. R. (1985). Sensor failure detection and estimation. *Nuclear Safety* **26**(1), 32–43.

Yang, J. C.-Y. and D. W. Clarke (1997). A self-validating thermocouple. *IEEE Transactions on Control Systems Technology* **5**(2), 239–253.

Yin, H. and N. M. Allinson (1997). Bayesian learning for self-organising maps. *Electronics Letters* **33**(4), 304–305.

Yung, S. K. and D. W. Clarke (1989). Local sensor validation. *Measurement and Control* **22**, 132–141.

A MLP PREDICTION MODEL FOR POWER PLANT NOX EMISSION

S. Thompson, K. Li

School of Mechanical & Manufacturing Engineering
The Queen's University of Belfast
Ashby Building, Stranmillis Rd., Belfast BT9 5AH, UK

Abstract: In this paper, a MLP (Multi-Layer Perceptron) model is developed for long
time period prediction of NOx emission in a coal-fired power generation plant. In order
to achieve this purpose, a novel training algorithm is used to improve the generalisation
capacity of the model. The application results show the merits of this MLP model.
Copyright © 2000 IFAC

Keywords: Power generation, Neural networks, Modelling, Training, Prediction

1. INTRODUCTION

Due to its environmental impact, power station NOx
emission has been the subject of much research over
the last ten years (ETSU, 1997; Muzio and Quartucy,
1997; Visona and Stanmore, 1996). For an existing
plant, to develop a real-time advanced control system
to control the NOx emission is of great importance.
Advanced control systems for real plant could be
considered as consisting of two development stages.
A modelling stage, in which some form of plant
model using both historical and current plant data,
attempts to capture the relationship between the
plants operational inputs and the NOx output. In the
second stage some form of constrained optimisation
is used to manipulate the inputs of the model in order
to minimise the NOx output. These values are then
presented to the operator (open-loop mode) or in
some cases used to automatically adjust the inputs
(closed-loop mode). It is claimed that existing
techniques (and those being developed) will produce
NOx reductions between 15% and 25% (ETSU,
1997).

Clearly, an essential step in reducing NOx using the
operational aspects of a power plant is to develop an
appropriate NOx emission model. It therefore seems

reasonable to look at existing models (and modelling
techniques) particularly for coal-fired plant.

Most research effort has been devoted to CFD
modelling based on combustion chemistry and
physical principles (Visona and Stanmore, 1996).
However, such models tend to be complex to build,
require significant computing facilities and are
mainly used for new plant design. Alternatively, as
universal approximators, Artificial Neural Networks
(ANN) such as the Multi-Layer Perceptron network
(MLP) and Radial Basis Function Neural Network
(RBF) have found wide applications in non-linear
system modelling and control (Fausett, 1994).
However, some researchers and control engineering
companies[1] argued that Neural Network technology
requires copious amounts of historical data before
producing useful models. And conclude that the
approach is slow to implement, not adaptable to new
circumstances and severely restricted when
optimising in a region outside past "experience".

---

[1] for example Ultramax Corporation who produced the
'ULTRAMAX METHOD' for boiler optimisation and
NOx emission control (http://www.umaxcorp.com
/home.htm)

It is true that like other black box modelling technologies, neural network models are not able to nest the true structure of the studied system, and therefore their generalisation capacity is restricted. However, through appropriate selection of training data, ANN configuration and training algorithm, it should be possible to ease these restrictions and develop an ANN model suitable for on-line implementation.

In this paper, a MLP (Multi-Layer Perceptron) model is developed for prediction of NOx emission in a coal-fired power generation boiler. In particular, emphasis is placed on the ANN training for generalisation. In order that the MLP model will have long-term prediction capacity, a new type of ANN training algorithm is used to improve the generalisation capacity of MLP models.

2. A MLP TRAINING ALGORITHM FOR IMPROVING GENERALISATION CAPACITY

Generalisation refers to the ability of a model to predict the outcome of a system for unseen data. It is well known that if a network is trained until it minimises the error on the training set, generalisation performance could be very poor (McLoone and Irwin 1999). In Neural Network training, there are two solutions leading to generalisation. One is the early stopping technique, that is, two sets of data are involved, one for training and the other for validation, iteration stops when the generalisation error begins to increase. The other technique is regularisation, that is, a penalty term (ΔE) is added to the standard training cost function E which restricts the variance of the model (Fausett, 1994). In this section, a MLP training algorithm for improving ANN generalisation capacity is proposed, and it is based on the early stopping technique.

Suppose that a neural network is used to model a nonlinear MISO dynamics system:

$$\hat{y}(t) = ANN(y^{t-1}, u_1^{t-d_1}, ..., u_m^{t-d_m}; \omega) + \varepsilon(t)$$
$$(1)$$

where t is the time index, ω is the modifiable vector parameter, $\hat{y}(t)$, $y(t)$, $u_1(t),...,u_m(t)$ are ANN prediction, measured output and input sequences respectively, $\varepsilon(t)$ the modelling error. $ANN(\bullet; \omega)$ is some non-linear function determined by the corresponding neural network, $d_1, ..., d_m$ are time delays and

$$\varepsilon(t) = y(t) - \hat{y}(t) \qquad (2)$$

With the early stopping technique, two sets of data are used, which are denoted as Ω_1 and Ω_2

respectively, and it is assumed that Ω_1 has N_1 samples and Ω_2 has N_2 samples. Using the sum-squared error:

$$E(\omega) = \sum_{j=1}^{N} (y_i^j - \hat{y}_i^j)^2 = \|\varepsilon\|^2 \qquad (3)$$

where N is the number of training samples and ω is the adjustable vector.

Let the sum-squared-error functions for the two sets of data Ω_1 and Ω_2 be denoted as $E(\omega)|\Omega_1$ and $E(\omega)|\Omega_2$ respectively, the adjustable vector ω (which is searched within the error space $E(\omega)|\Omega_1$) is denoted as $\omega\Omega_1$, and the adjustable vector ω based on the calculation of $E(\omega)|\Omega_2$ is denoted by $\omega\Omega_2$. For these two sets of data, one is used for training and the other is used for cross-validation, and the early stopping technique is formulated as:

$$\omega^* = \min_{\omega\Omega_1} E(\omega)|\Omega_2 \qquad (4)$$

In this paper, the performance index for ANN training takes into account both the training precision on data set Ω_1 and the generalisation on data set Ω_2, denoted as $SSE(\omega^*)$:

$$SSE(\omega^*) = E(\omega^*)|\Omega_1 + E(\omega^*)|\Omega_2 \qquad (5)$$

For the early stopping technique described in (4), only Ω_1 is used to search the optimal adjustable vectors, while Ω_2 leads a passive role in the ANN training, used for validation. In the technique suggested in this paper, the two data sets are actively involved in training.

The algorithm for this improved technique comprises of several *training cycles*. The key issues in the training algorithm are as follows:

1). In each training cycle, one data set (e.g. Ω_1) is used for training, and the other (e.g. Ω_2) for validation (early stopping technique). To perform the training in each training cycle, any training algorithms for BP networks may be used to minimise $E(\omega)|\Omega_1$ and to update the adjustable vector $\omega\Omega_1$ e.g. the Levenberg-Marquardt (LM) or gradient methods are both suitable for this purpose. In one training cycle, the iteration of training stops when the cost function for the validation set (e.g. Ω_2) can not decrease further, that is:

$$E(\omega_{\Omega_1}^{(i+1)})\big|_{\Omega_2} > E(\omega_{\Omega_1}^{(i)})\big|_{\Omega_2} \qquad (6)$$

where $\omega_{\Omega_1}^{(i)}$ is the adjustable vector in the *ith* iteration on minimising $E(\omega)\big|_{\Omega_1}$. However, if the generalisation error has multiple local minima, then in the process of minimising $E(\omega)\big|_{\Omega_1}$

$$E(\omega_{\Omega_1}^{(i+1)})\big|_{\Omega_2} > \delta E(\omega_{\Omega_1}^{(i)})\big|_{\Omega_2} \qquad (7)$$

could be used as the stopping criteria, where δ is some positive number with range (1,10] in practice.

2). In the latest training cycle, if one data set, say Ω_1, was used for training and Ω_2 was used for validation, then, in the following training cycle, Ω_2 will be used for training, and Ω_1 becomes the validation set.

3). The initial value for each cycle of training is based on the results acquired in the previous training cycle. Suppose $\omega^{(old)}$ is the solution to the previous training cycle and $\omega^{(present)}$ is the solution to the latest training cycle. Then the initial value of the adjustable vector ω_0 for next training cycle is:

$$\omega_0 = [\alpha(\omega^{(present)} + \omega^{(old)}) + \beta \aleph] / \gamma \qquad (8)$$

where $\alpha, \beta \in (0,1)$, $\gamma > 1$ are scalars, $\aleph$ is a normally distributed random matrices with the same dimensions as ω. For the first and second training cycle, the initial value ω_0 is selected randomly.

In conclusion, the overall training process can be described as follows.

Algorithm 1: Neural network training for improving the generalisation capacity

Step 1. Initialisation phase. Select two appropriate sets of training data, and if there is only one data set, It can be always split into two appropriate data sets. Randomly select the initial value of the weights for the neural network, give an initial value (randomly designed) to $\omega^{(old)}$, $\omega^{(present)}$ and $\omega^{(best)}$, where $\omega^{(old)}$ is defined to be the solution of the previous training cycle, $\omega^{(present)}$ is the solution of the latest training cycle, and $\omega^{(best)}$ is the best solution from all previous training cycles. Let

$E(\omega^{(best)})$ be a very large number. Define one data set as the training set Ω_1 and the other as the validation set Ω_2. Select the number of training cycles N_{tr}, and select the maximum epochs N_{ep} for each training cycle. Select an appropriate BP training method (eg Levenberg-Marquardt or, gradient descent method).

Step 2. Training phase. Start to train the network using training set Ω_1. Stop the iteration for this training cycle when (7) is satisfied, or the training epoch in this training cycle exceeds the maximal number N_{ep}.

Step 3. Update phase. Update $\omega^{(present)}$ and $\omega^{(best)}$. Record the final adjustable weight satisfying (7) as $\omega^{(present)}$, and the sum squared error on both data sets as defined in (5). And if $E(\omega^{(present)}) \le E(\omega^{(best)})$, then

$$E(\omega^{(best)}) = E(\omega^{(present)})$$
$$\omega^{(best)} = \omega^{(present)} \qquad (9)$$

Step 4. Preparation phase. Prepare for the next training cycle.
1). Exchange the validation set and training set, that is, the previous training set becomes the validation set, and the previous validation set becomes the training set.
2). Calculate the initial value ω_0 according to (8).

3). Finally, update $\omega^{(old)}$, that is let
$$\omega^{(old)} = \omega^{(present)} \qquad (10)$$

Step 5. Check phase. Let $N_{tr} = N_{tr} - 1$, if $N_{tr} \ne 0$ go to *Step 2*, otherwise, go on.

Step 6. Stop, $\omega^{(best)}$ is the final solution in the neural network training.

Remark 1: The number of cycles for training N_{tr} and N_{ep} depend on the specific recursive training algorithm used, for example for the LM method, $N_{tr} \le 10$, $N_{ep} \le 60$ are generally enough. As for $\alpha, \beta \in (0,1)$, $\gamma > 1$ in (9), $\alpha = 0.67$, $\beta = 0.1$, $\gamma = 5$ are chose for most simulation examples.

Remark 2: It is common that ANN models some times differ significantly in different training. The reason is that for most training algorithm, the training data and the initial values for the adjustable vectors are selected randomly. Furthermore, the artificial systems are generally highly non-linear and have multiple local minima. However, intensive simulation study shows that for this algorithm, there is no significant difference between different training results.

3. NOX EMISSION MODELLING

3.1 *General description of the power plant*

In this paper, NOx emission in a power station located in Northern Ireland is studied. This is a cyclic power plant with two generating units. Each unit has a dual fired (oil or coal) drum boiler and produces full load 300 MWe with oil firing or 200 MWe with coal firing. The boiler was designed to supply its turbine with steam at a temperature of 540°C and up to a pressure of 162 bar. There is one burner box on each corner. All of the sections in each burner box tilt in unison through ±25°, relative to the horizontal. Coal is the major source of fuel. Coal is taken from the bottom of a coal bunker. Pulverised coal exiting the grinding zone is entrained in the hot primary air stream and carried to the static separator where coarser particles are separated and returned to the grinding zone for further processing. Primary air then dries and carries the PF to the burners. Each corner of the furnace houses a burner box and a separated overfire air (SOFA) box, which admit fuel and secondary air streams into the furnace. These streams are directed at tangents to imaginary firing circles in the centre of the furnace. The tangential firing creates turbulence in the combustion area that ensures the thorough mixing of fuel and air streams necessary for complete combustion. Low momentum burners are employed to achieve longer flame paths, leading to reduced flame temperatures. The nozzles in the burner box tilt in unison in the vertical plane to control the position of the fireball, and thus the temperatures in the superheat and reheat regions of the boiler.

The major part of NOx emission in power generation plant has been found to be NO. According to De Soete (1975), there are three main sources of NO in combustion, namely thermal NO, Prompt NO and Fuel NO. Thermal NO results from the high temperature reaction of atmospheric nitrogen and oxygen, prompt NO is formed by the reaction of nitrogen with hydrogen derived radicals in the fuel-rich zone of combustion, while fuel NO results when nitrogen compounds present in the fuel are released and react with oxygen. In coal-fired power plant, fuel NO is the major contribution to NOx emission. Some of the fuel NO can be released from the devolitisation of the fuel and some from the oxidation of the char. According to the NOx formation mechanism, the following operational variables are identified to be of contribution to the overall NOx emission in this power generation plant:

(1). The overall mass flow of fuel, denoted as m_f with unit (Kg/s):

(2). The overall mass flow of air, denoted as m_a (unit Kg/s):

(3). Specifically, mass flow of primary air and mass flow of secondary air, denoted as m_{pa}, m_{sa} respectively with unit of Kg/s, are considered.

(4). The burner tilt position, denoted as θ (degree).

Considering that m_{pa}, m_{sa}, m_a have only two independent variables, there are actually four independent manipulated variables.

3.2 *NOx emission modelling*

In the studied power station, operational conditions and NOx emission are measured and recorded every minute. From the collected data, two sets of data Ω_1 and Ω_2 with 200 samples each are used in MLP model training. The ranges of values for data set 1 and set 2 are as follows:

NOx: $set1 : [254.1, 431.1]$, $set2 : [289.5, 353.0]$
Fuel: $set1 : [9.1, 20.2]$; $set2 : [15.29, 19.19]$
Primary air: $set1 : [31.9, 51.1]$; $set2 : [29.5, 32.7]$
Secondary air:
$set1 : [143.3, 182.9]$; $set2 : [146.1, 169.1]$
Burner: $set1 : [40.1, 70.2]$; $set2 : [61.8, 84.7]$

Since these variables differ significantly in value, all variables are pre-processed by applying a linear transformation, such that the transformed variables have zero mean and unit standard deviation over the transformed training set.

In the NOx emission modelling, a two-layer MLP with 6 nodes in the hidden layer is used. The proposed algorithm is used to train the neural network. The parameters in the training algorithm are selected as follows: $N_{tr} = 4$, $N_{ep} = 5$, $\alpha = 0.67$, $\beta = 0.1$, $\gamma = 5$.

After the training, the selected model is used to predict the NOx emission over a period of a fortnight. Predicted emission is based on the following model:

$$\hat{y}(t) = ANN(\hat{y}^{t-1}, u_1^{t-d_1}, ..., u_m^{t-d_m} ; \omega) + \varepsilon(t)$$

(11)

Remark 3: The difference between (1) and (11) is the past measured outputs are replaced with the past-predicted outputs. It is a well known that in non-linear dynamic system modelling, many ANN models are not able to achieve long term prediction.

Typically, a one step ahead prediction capacity rather than the 18,000 to 36,000 steps demonstrated in this paper.

Figs. 1, 2, 3, 4 show the predicted NOx emission over different time periods. If the error to target ratio is defined as in (12), then the error to target ratio for fig. 1 is 3.5%, fig. 2 is 4.1%, for fig. 3 is 9.1 %, and for fig. 4 is 4.5%. The overall error to target ratio can be kept around 7%.

$$Ratio = \sqrt{\sum_{i=1}^{N} ((y_i\,(j) - t_i\,(j))^2\,) / \sum_{i=1}^{N} (t_i\,(j))^2\,)}$$

(12)

For comparison reasons, the early stopping technique was also used for MLP training, using the same training data and the same neural network configuration with the same number of hidden nodes. This model is also used to predict the NOx emission based on (14) and again the predicted outputs are used as the inputs. However, in order to get a comparable MLP model the early stopping technique is performed many times and only the best are selected. Fi.gs. 5, 6, 7 and 8 show its prediction capacity.

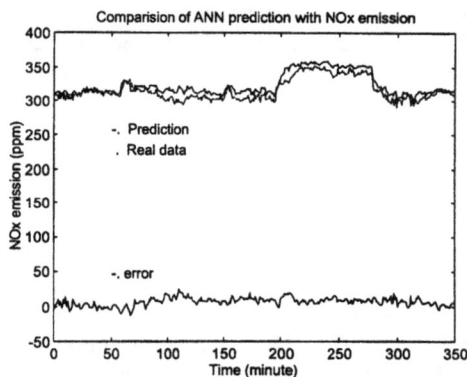

Fig .1 MLP prediction over a period of 350 minutes

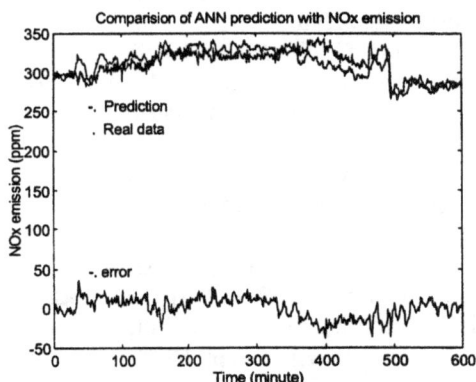

Fig .2 MLP prediction over a period of 600 minutes

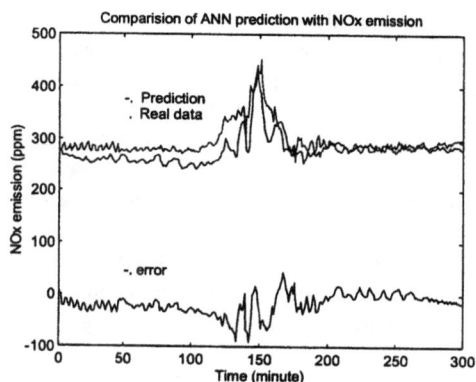

Fig .3 MLP prediction over a period of 300 minutes

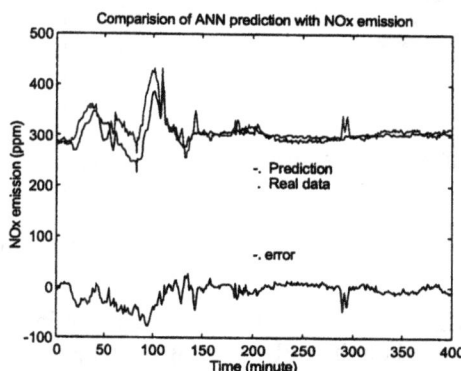

Fig .4 MLP prediction over a period of 400 minutes

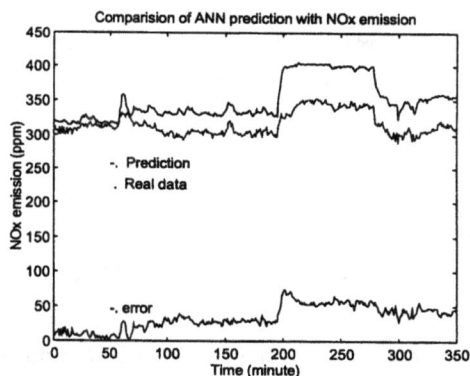

Fig .5 MLP prediction over a period of 350 minutes (based on the early sopping technique)

The error to target ratio defined in (14) for fig. 5 is 12.2%, fig. 6 is 14.0%, for fig. 7 is 14.1 %, fig. 8 is 22.9%. The overall error to target ratio is around 14%.

Remark 4: In this paper, the LM method was adopted as the basic training algorithm in both algorithm 1 and the early stopping technique. When the early stopping technique was used for training, the computation takes about 1.2×10^7 flops compared with 1.4×10^8 flops for the proposed algorithm. This is based on the MATLAB 'flops' function. Using a Pentium II of 233 MHz PC, training takes 11.5

seconds for the new algorithm, and about 1.7 seconds for the early stopping technique. However, the sampling rate for NOx emission within the power plant is every minute and therefore this new algorithm is adequate for on-line application. Further in this study, regardless of training, the early stopping MLP model was always worse than that produced by the new algorithm. Finally, this new algorithm produced a stable result, in that the resulting model did not significantly alter with different initial starting points during training.

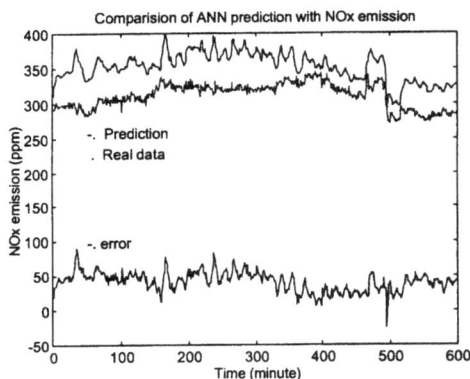

Fig .6 MLP prediction over a period of 600 minutes (based on the early stopping technique)

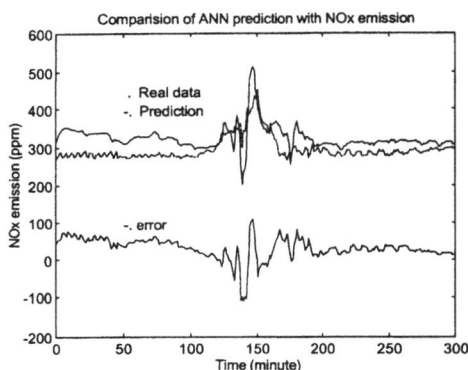

Fig .7 MLP prediction over a period of 300 minutes (based on the early stopping technique)

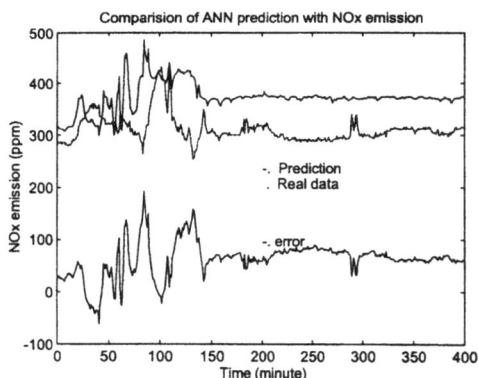

Fig .8 MLP prediction over a period of 400 minutes (based on the early stopping technique)

4. CONCLUSION

In this paper, a MLP (Multi-Layer Perceptron) model is developed for predicting the NOx emission in a coal-fired power generation boiler. In order that the MLP model would have long-term prediction capacity, a novel training method is proposed. The MLP model is used to predict the NOx emission over a period of about a fortnight, and in the model prediction, no measured outputs are used as the model inputs.

Since identification methods based on ANN models are not able to nest the true system structure, generalisation capacity is not unlimited. However, for NOx emission modelling, the methodology developed in this paper is able to produce an ANN model with better generalisation capacity and less model complexity than an equivalent early stopping technique.

ACKNOWLEDGEMENTS

Acknowledgement is made to the British Coal Utilisation Research Association and the UK Department of Trade and Industry for a grant in aid of this research but the views expressed are those of the authors, and not necessarily those of BCURA or the Department of Trade and Industry. The authors also would like to thank Nigen-Kilroot power station, Northern Ireland, for their assistance in providing the operation data.

REFERENCES

Coal R&D programme (1997). *Technology status report: NOx control for puliverised coal-fired power plant*, ETSU, Harwell.

De Soete, G.G., (1975). Overall reaction rates of NO and N$_2$ formation from fuel nitrogen. *15th Symposium (international) on Combustion*, The Combustion Institute, 1093-1102.

Fausett, L. (1994*). Fundamentals of Neural Networks*. Englewood Cliffs, New Jersey, USA: Prentice Hall.

McLoone, S. and G. W. Irwin (1999). Improving neural network training solutions using regularization, submitted to *Neural Computing* (in revision).

Muzio, L.J., Quartucy G. C. (1997). Implementing NOx control: Research to application. *Progress in Energy and Combustion Science*. **23 (3)**, 233-266.

Visona, S.P., Stanmore, B.R. (1996). 3-D modelling of NO$_x$ formation in a 275 MW utility boiler. *J of Institute of Energy*, **69**, pp 68-79.

IDENTIFICATION OF AIRCRAFT GAS-TURBINE DYNAMICS USING B-SPLINES NEURAL NETWORKS

J. C. Lopes[1], A. E. Ruano[1,2] , P. J. Fleming[3]

1. *Unidade de Ciências Exactas e Humanas, Universidade do Algarve*
Campus de Gambelas, 8000 Faro, Portugal
2. *Institute of System & Robotics, Portugal*
3. *Department of Automatic Control and Systems Engineering, University of Sheffield*
Sheffield, UK
emails: aruano@ualg.pt, P.Fleming@shef.ac.uk

Abstract: This paper describes preliminary results regarding the development of a B-splines neural network model of the fuel feed to shaft-speed dynamics of a twin-shaft gas turbine engine. Data recorded from practical testing of the turbine to a multisine input were employed, and models were identified at different points along the turbine operating curve. B-splines neural networks have been found to be good models of the system, delivering good results especially for long-range predictions. *Copyright © 2000 IFAC*

Keywords: Gas Turbines, Systems Identification, Neural Networks

1. INTRODUCTION

A gas turbine is made up of three basic components: a compressor, a combustion chamber and a turbine. Air is drawn into the engine by the compressor, which compresses it and delivers it to the combustion chamber. There, the air is mixed with the fuel and the mixture ignited, producing a rise of temperature and therefore an expansion of the gases. These are expelled through the engine nozzle, but first pass through the turbine, designed to extract energy to keep the compressor rotating (Evans, 1998).

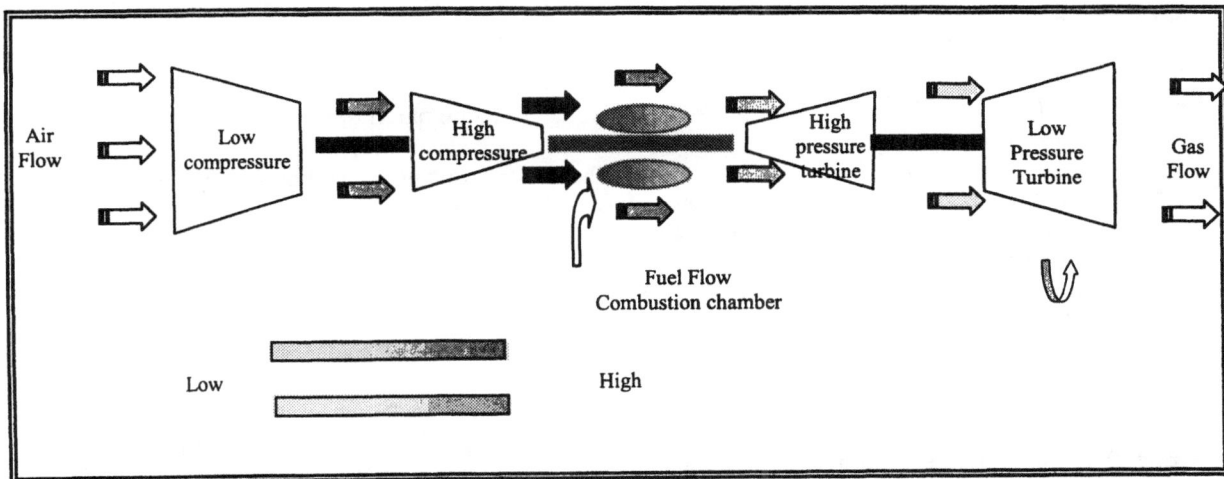

Fig. 1. Simplified diagram of a gas turbine

The work described here uses data recorded from a Rolls Royce Spey MK 202 turbine, whose simplified diagram can be seen in Fig. 1. Both the compressor and the turbine are split into low pressure (LP) and high pressure (HP) stages. The HP turbine drives the HP compressor and the LP turbine drives the LP compressor. They are connected by concentric shafts that rotate at different speeds, denoted as NH and NL.

In order to obtain data needed to identify the turbine dynamics the engine speed was operated in open loop, and perturbed fuel demand signals were applied to the fuel feed system. As the fuel feed system exhibits both linear and nonlinear dynamics, the actual fuel flow was measured using a turbine flow meter. This value, normalised as a percentage of the fuel flow that gives the corresponding percentage of the maximum high-pressure-shaft speed, constitutes the model input. Its outputs are the normalised HP and LP shaft speeds.

Several different perturbations were applied to the fuel feed system of the gas turbine, and the fuel flow and shaft speeds recorded, to be used afterwards for identification purposes.

2. PREVIOUS AND PRESENT IDENTIFICATION APPROACHES

Different identification approaches (Arkov, et al., 1998; Borell, et al., 1998; Norton et al., 1998; Vasques and Fleming, 1998) were applied to the data sets described earlier.

Their main findings were:

- The HP shaft dynamics can be considered first-order over almost the whole of the operating range;
- The LP shaft dynamics can be considered second order, with a common pole with the HP shaft;
- Slow dynamics, probably due to thermal effects, have been also identified, preventing large-transient responses;
- Both HP and LP dynamics vary with the operating point considered.

Different perturbation signals were employed in the approaches referenced above. Of particular importance for the present work, are multisine tests. These are a sum of harmonically related cosines:

$$u(t) = \sum_{i=1}^{F} a(k) \cos \left(i(k) 2\pi f_0 t + \phi(k) \right), \quad (1)$$

where a is the vector of their amplitudes, i is the vector of their harmonic numbers, ϕ is the vector of their phases, and f_0 is the fundamental frequency. All the signals employed had odd harmonics, so that

nonlinearities introduced by even-order nonlinearities would produce only even harmonics, easily detected by inspection. As employed in previous works (Borell, et al., 1998; Vasques and Fleming, 1998), signals with mean shaft-speed values of 53.8%, 75.3%, 75.45% and 88.7% were employed here.

It was previously identified that the engine dynamics vary with the operating point considered, and also that slow dynamics were identified, their influence varying also with the operating point. For this reason, it was decided to assess the performance of neural network models, specifically B-splines neural networks, which have been proved to constitute a valid alternative to classical nonlinear systems identification.

3. B-SPLINES NEURAL NETWORKS

B-splines networks are a kind of Associative Memory Networks which differ from other neural networks because they store the information locally ("learning" about one part of the input space minimally affects the rest of it).

The basic structure of a B-splines neural network has three different levels, as illustrated in Fig. 2. In the first layer, a lattice is defined in order to normalize the original input space. The second layer is composed of basis functions defined over the lattice and in the third layer a weight vector (**w**) is defined, with each weight associated with the output of each basis function.

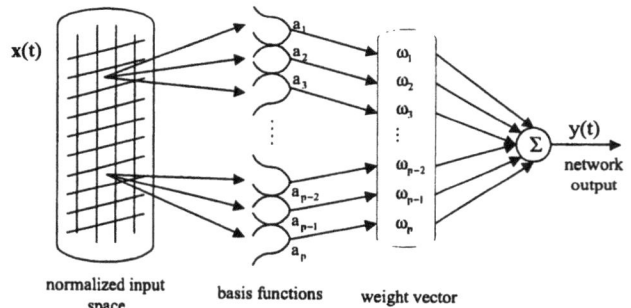

Fig. 2. Structure of a B-splines neural network

The network output at iteration k, ($y[k]$), is a linear combination of the p outputs of the basis functions using the adjustable weights as coefficients:

$$y[k] = \sum_{i=1}^{p} a_i[k] * \omega_i[k] = \mathbf{a}^T[k]\mathbf{w}[k] \quad (2)$$

In order to define a lattice on the input space, a set of n knot vectors must be given, one knot vector for each input axis. As usually there are a different number of knots on each axis, placed in different positions, the designer is able to incorporate a priori knowledge in the network design.

The order, shape and overlap of the basis functions determine the modeling capabilities of the networks. B-splines basis functions are piecewise polynomials, as is shown in Fig. 3.

The B-splines neural networks allow the order of the splines in each axis to be different although that order cannot be changed during training.

Multivariate basis functions are formed from the combination of n univariate basis functions, each one defined on each input axis. The number of multivariate basis functions of order k_i defined on an axis with r_i knots is $k_i + r_i$, therefore the total memory requirements are:

$$p = \prod_{i=1}^{n} (k_i + r_i), \qquad (3)$$

which depend exponentially on the dimension of the input space. This problem is the major drawback in the B-splines neural networks and it is known as *the curse of dimensionality*.

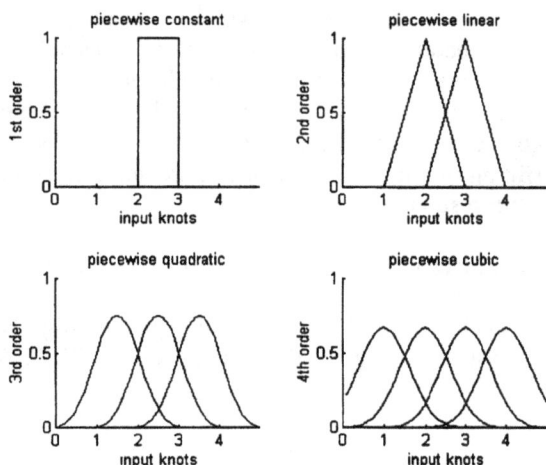

Fig. 3. Univariate *B-splines* basis functions, of 1st to 4th orders, activated by an input within the interval $[\lambda_2, \lambda_3[$

To overcome this problem and in order to obtain, during off-line training, a good network structure for later on-line learning, we employ the ASMOD construction algorithm (Weyer and Kavli, 1995).

The ASMOD algorithm is based on the idea of evolving a network by refining a very simple initial model. It starts from an empty model or one consisting of a small number of relevant sub-networks and gradually new inputs are included and dependencies are identified, achieving this way a better representation for each input. At each iteration, the possible ways of network evolution are identified

and the performance of each is measured, the best refining step being included in the current model. The resultant model is more complex but exactly reproduces the previous one.

During the refinement process, growing steps alternate with pruning ones, in a way to prevent possible redundancies.

Performance measures are used to determine the new model to employ or to stop the refining processes. These measures make a balance between the network complexity, the quantity of training data and the reduction of the mean square error (MSE). Among some known formulas, the *Bayesian Information Criterion*, which is given by:

$$K = m \ln(J) + p \ln(m) \qquad (4)$$

has been employed. In (4), K is the performance measure employed, m is the number of training patterns, J is the MSE of the model and p is the size of the model.

Usually, the new model taken is the one that presents the lowest K value. The refining step is stopped if in consecutive steps of growing and pruning, it is impossible to reduce the value of K.

Considering now the problem of determining the linear output weights, there is a variety of training rules that may be used to train B-splines neural networks. The *Normalised Least Mean Squares (NLMS)* (Brown and Harris, 1994) (also referred as "Error Correction Rule") is the most appropriate to on-line adaptation. As we are interested in off-line learning, this training rule was replaced by the solution in least squares sense, for the whole training set, this is:

$$\mathbf{w} = \mathbf{A}^{+} \hat{\mathbf{y}} \qquad (5)$$

where the symbol '+' denotes a pseudo-inverse, $\mathbf{A}$ is the matrix of the basis functions outputs, and $\hat{\mathbf{y}}$ is the desired output vector. As the determination of the weights is a linear problem, this approach proved to be much faster, and to produce much better results.

4. RESULTS OBTAINED

Some pre-processing was performed before the data was used to derive the model. The data was firstly filtered by a Chebyshev filter of order 8, with cut-off frequency of 1.6Hz, and was normalised in a range [-1,1].

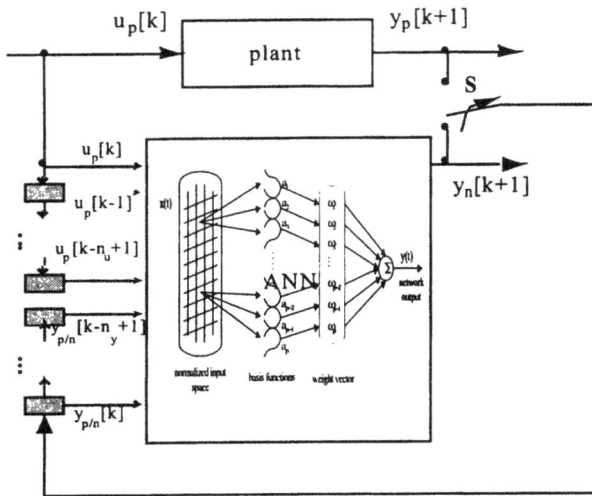

Fig. 4. Off-line training scheme

For the training process, 4000 training patterns, corresponding to 1 signal period, were employed. The inputs considered, in accordance with previous studies (Vasques and Fleming, 1998), were $\{u[k-1], u[k-2], y[k-1], y[k-2]\}$.

To determine the initial order of the splines, experiments using data belonging to the 75.4% NH were conducted with spline orders of 2, 3 and 4. It was verified that spline orders of 3 obtained the best generalisation. The final model, obtained with spline orders of 3, presented a structure of two univariate sub-model and one bivariate sub-model:

$$\hat{y}[k] = S_1\left(u[k-1]\right) + S_2\left(u[k-2]\right) + \cdots$$
$$\cdots + S_3\left(y[k-1], y[k-2]\right) \tag{6}$$

The approximation obtained, for the training data, can be seen in Fig. 5.

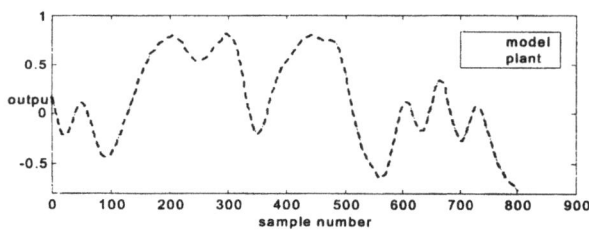

Fig. 5 – Actual and model output (training data)

The fit obtained was almost perfect, achieving a mean-square-error of 2.1×10^{-6}.

Employing this model to predict the output 1000 steps ahead, using a cascade of neural models, where each one predicts the output one step-ahead, the results obtained, for test data, are shown in Fig. 6.

It can be observed that good values of prediction were obtained until, at least, 400 steps in the future, which demonstrates the quality of the model obtained.

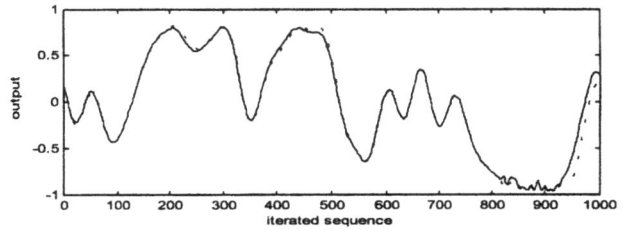

Fig. 6 – Prediction sequence (1000 steps) using test data (75.4% NH)

Afterwards, 3 other models were obtained for the other operating points (53.8%, 75.3%, 88.7%). The results obtained are summarised in Table 1. Normalised data was employed, 4000 samples for training and 1000 samples for testing. In this table, the *model* column denotes the sub-models found by the ASMOD algorithm, with {1,2,3,4} denoting the variables *{u[k-1],u[k-2],y[k-1],y[k-2]}*. The *order* column indicates the order of the splines, $\|w\|$ denotes the 2-norm of the weight vector, *x-PE* denote the error x steps-ahead (for the training set), *MSE training* and *MSE test* the Mean-Square-Error for the training and test sets, and *Comp* the complexity of the model obtained. It can be observed that the results obtained were all excellent. All the models have a similar structure, except for the 75.45% mean NH, where a bivariate sub-model is employed, therefore increasing the model complexity, and the norm of the weight vector. This increase is usually related with overfitting problems that, however, are not evident in the MSE results for the test set.

The variances of the prediction error, 20 steps ahead for the 4 models are shown in Fig. 7. Excellent values were obtained for all the operating points, slightly better results having been obtained for the 75.45% case.

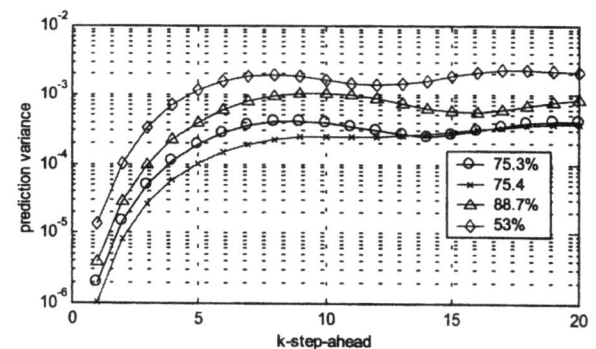

Fig. 7 – Variances of the prediction errors for 1-20 steps ahead in the future, for the 4 models, using 500 test data.

Afterwards, all the obtained models were statically validated using the following conditions, introduced in (Billings and Tao, 1991):

Table 1. Results obtained for the 4 models

Op. Point	Model	Order	‖ w ‖	MSE (training)	MSE (test)	3-PE	100-PE	Comp
53.8%	[1 2 3 4]	[2 2 2 3]	3.6	$1.1\ 10^{-5}$	$1.2\ 10^{-6}$	$8.2\ 10^{-3}$	$2.1\ 10^{-2}$	11
75.45%	[1 2 3*4]	[2 2 3 3]	$2.9\ 10^{3}$	$2.1\ 10^{-6}$	$3.2\ 10^{-7}$	$3.9\ 10^{-3}$	$1.7\ 10^{-2}$	28
75.3%	[1 2 3 4]	[2 2 2 2]	2.7	$1.7\ 10^{-6}$	$5.6\ 10^{-3}$	$1.1\ 10^{-3}$	$1.6\ 10^{-2}$	9
88.7%	[1 2 3 4]	[2 2 3 2]	5.5	$2.9\ 10^{-6}$	$5.7\ 10^{-3}$	$1.7\ 10^{-3}$	$8.7\ 10^{-3}$	10

$$\Phi_{\varepsilon\varepsilon}(\tau) = E[\varepsilon(t-\tau)\varepsilon(t)] = \delta(\tau) \qquad \forall \tau$$

$$\Phi_{u\varepsilon}(\tau) = E[u(t-\tau)\varepsilon(t)] = 0 \qquad \forall \tau$$

$$\Phi_{u'^2\varepsilon}(\tau) = E[(u^2(t-\tau) - E[u^2(t)])\varepsilon(t)] = 0 \qquad \forall \tau \quad (7)$$

$$\Phi_{u'^2\varepsilon^2}(\tau) = E[(u^2(t-\tau) - E[u^2(t)])\varepsilon^2(t)] = 0 \qquad \forall \tau$$

$$\Phi_{\varepsilon(\varepsilon u)}(\tau) = E[(\varepsilon(t)\varepsilon(t-1-\tau))u(t-1-\tau)] = 0 \quad \tau \geq 0$$

Data has been normalised to give all tests a range of ±1 and approximate 95% confidence bands at 1.96√N. All conditions were satisfied for all models except the autocorrelation test, which failed for all the models. This indicates that the network structure is right, but that coloured noise was present in the data.

Finally the results obtained with the neural networks were compared with the results obtained in (Vasques and Fleming, 1998), where a multiobjective genetic programming scheme, considering as objectives the number of polynomial terms, polynomial degree, model lag, residual variance and the k-step prediction error, was employed to determine polynomial NARX models for each operating point:

$$y(k) = \theta_0 + \sum_{i_1=1}^{n} \theta_{i_1} x_{i_1}(k) + \sum_{i_1=1}^{n}\sum_{i_2=i_1}^{n} \theta_{i_1}\theta_{i_2} x_{i_1}(k) x_{i_2}(k)$$

$$+ \ldots + \sum_{i_1=1}^{n} \ldots \sum_{i_l=i_{l-1}} \theta_{i_1} \ldots \theta_{i_l} x_{i_1}(k) \ldots x_{i_l}(k), \qquad (8)$$

where $x_i(k)$ represent lagged terms in y and u and θ_i are the model parameters.

This example was chosen as comparison as the models obtained by that approach are also nonlinear.

Table 2 illustrates the results obtained with the neural approach (first line), and the results obtained with the genetic programming technique (second line), in terms of the variance of the error (VAR) and the 3

steps-ahead prediction error (PE). The neuro approach employs 1000 test samples.

It can be observed that the neuro model achieves in all cases a much smaller error variance.

With respect to the 3-steps ahead prediction error the two schemes produce similar results, the NARX approach obtaining better results for small shaft speeds, and the neuro technique better performance for high operating points.

5. CONCLUSIONS

A preliminary study of the performance of B-splines neural networks for modelling aircraft gas-turbine dynamics was presented in this paper. Different models were trained for different operating points, presenting promising results, especially in terms of prediction, and comparing favourably with one alternative scheme.

This research will be continued, by, at a first stage, try to avoid the bias detected by the first statistical validity test. Afterwards, one single model will be identified for all the operating points, and validated accordingly. Finally, the performance of this single model will be compared with alternative approaches, referenced above.

ACKNOWLEDGMENTS

The authors wish to acknowledge Rolls-Royce plc. (specially Dr. D. Hill) and DERA (Pyestock) for their cooperation and support for the provision of the engine data.

Table 2. Comparison between neuro and genetic programming approaches

VAR 53.6%	PE 53.6%	VAR 75.45%	PE 75.45%	VAR 75.3%	PE 75.3%	VAR 88.7%	PE 88.7%
$1.6\ 10^{-5}$	$1.7\ 10^{-2}$	$2.1\ 10^{-5}$	$1.5\ 10^{-2}$	$2.6\ 10^{-5}$	$6.1\ 10^{-3}$	$2.3\ 10^{-5}$	$5.7\ 10^{-3}$
$5.9\ 10^{-3}$	$7\ 10^{-3}$	$5.7\ 10^{-3}$	$8.6\ 10^{-3}$	$3.9\ 10^{-3}$	$5.1\ 10^{-3}$	$4.7\ 10^{-3}$	$1.2\ 10^{-2}$

REFERENCES

Evans, C. (1998). Testing and Modelling Aircraft Gas Turbines, *UKACC Int. Conf. On Control 98*, IEE, UK, pp. 1361-1366

Arkov, V.Y., G. G. Kulikov, T. V. Breikin and P. J. Fleming (1998). Dynamic Model Identification of Gas Turbines, *UKACC Int. Conf. on Control 98*, IEE, UK., pp. 1367-1371

Borell, A., C. Evans, and D. Rees (1998). Identification of Aircraft Gas Turbine Dynamics using Frequency-Domain Techniques, *UKACC Int. Conf. on Control 98*, IEE, UK., pp. 1372-1378

Norton, J.P., P. T. Ladlow and D. C. Hill (1998). Identification of Large-Transient Effects in Aircraft Gas-Turbine Dynamics, *UKACC Int. Conf. on Control 98*, IEE, UK., pp. 1379-1384

Vasques, K. and P. J. Fleming (1998). Multiobjective Genetic Programming for Gas Turbine Engine Model Identification, *UKACC Int. Conf. On Control 98*, IEE, UK., pp. 1385-1390

Weyer, E. and T. Kavli (1995). The ASMOD Algorithm. Some new Theoretical and Experimental Results, *SINTEF report STF31 A95024.*, Oslo

Brown, Martin and C. Harris (1994). *Neurofuzzy Adaptive Modelling and Control*, Prentice Hall, London

Billings, S. A. and Q. M. Tao (1991). Correlation based Model Validity for Nonlinear Signal Processing Applications, *Int. J. Control,* **54**, pp. 157-194

A PARALLEL VARIABLE MEMORY BFGS TRAINING ALGORITHM

Seán Mc Loone

*Intelligent Systems and Control Research Group, School of Electrical and Electronic
Engineering, The Queen's University of Belfast, Belfast BT9 5AH.
Tel.: 01232 274535, Email: s.mcloone@ee.qub.ac.uk*

Abstract: This paper considers the parallel implementation of a novel variable memory quasi-newton neural network training algorithm recently developed by the author. Unlike existing training methods this new technique is able to optimize performance in relation to available memory. Numerically it has equivalent properties to Full Memory BFGS optimization (FM) when there are no restrictions on memory and to FM with periodic reset when memory is limited. Parallel implementations of both the Full and Variable Memory BFGS algorithms are outlined and performance results presented for a PVM target architecture. *Copyright © 2000 IFAC*

Keywords: Neural networks, parallel algorithms, training, second-order.

1 INTRODUCTION

Feedforward neural network training is usually posed as an unconstrained optimization problem where the objective is to minimise a mean-squared-error cost function

$$E(\mathbf{w}) = \frac{1}{N_v}\sum_{i=1}^{N_v}(y(\mathbf{w}, \mathbf{u}_i) - d_i)^2 \qquad (1)$$

defined on a training set

$$\tau = \{\mathbf{u}_i, d_i\}, \qquad i \in 1, N_v \qquad (2)$$

with respect to the network weights $\mathbf{w}$. Here $\mathbf{w}$ is an N_w x 1 vector, N_v is the size of the training set and $\mathbf{u}_i$, d_i and $y(\mathbf{w},\mathbf{u}_i)$ are the input vector, desired network output and actual network output for the ith training vector respectively.

Training is achieved by updating the weights using a learning rule of the form

$$\mathbf{w}_{k+1} = \mathbf{w}_k + \eta_k \mathbf{d}_k \qquad (3)$$

where $\mathbf{d}_k$ is a descent search direction and η_k is the step size or "learning rate" at the kth iteration.

The choice of descent direction, $\mathbf{d}_k$, dictates the rate of convergence achievable with a given technique and is typically a trade-off between performance and computational/memory requirements. In the Batch Back Propagation (BBP) algorithm, the original method developed for training MLPs (Rumelhart *et al.* (1986)), $\mathbf{d}_k$ is simply the negative of the cost function gradient at $\mathbf{w}_k$, that is:

$$\mathbf{d}_k = -\mathbf{g}_k \qquad (4)$$

where

$$\mathbf{g}_k = \left[\frac{\partial E(\mathbf{w})}{\partial w_1} \frac{\partial E(\mathbf{w})}{\partial w_2} \cdots \frac{\partial E(\mathbf{w})}{\partial w_{N_w}}\right]^T \qquad (5)$$

while in the Full Memory BFGS (FM) algorithm, a powerful second order technique, the search direction is given by

$$\mathbf{d}_k = -\mathbf{M}_k\mathbf{g}_k \qquad (6)$$

where $\mathbf{M}_k$ is a symmetric positive definite approximation to the inverse Hessian matrix of the cost function that is updated iteratively as follows.

$$\mathbf{s}_k = \mathbf{w}_k - \mathbf{w}_{k-1}, \qquad \mathbf{t}_k = \mathbf{g}_k - \mathbf{g}_{k-1} \qquad (7)$$

$$\mathbf{M}_k = \mathbf{M}_{k-1} + \left[1 + \frac{\mathbf{t}_k^T\mathbf{M}_{k-1}\mathbf{t}_k}{\mathbf{s}_k^T\mathbf{t}_k}\right]\frac{\mathbf{s}_k\mathbf{s}_k^T}{\mathbf{s}_k^T\mathbf{t}_k}$$

$$- \frac{\mathbf{s}_k\mathbf{t}_k^T\mathbf{M}_{k-1} + \mathbf{M}_{k-1}\mathbf{t}_k\mathbf{s}_k^T}{\mathbf{s}_k^T\mathbf{t}_k}, \quad \mathbf{M}_0 = \mathbf{I} \qquad (8)$$

The FM algorithm is typically two orders of magnitude faster than BBP but this is achieved at the expense of $O(N_w^2)$ memory requirements and computational complexity. Consequently FM cannot be used for large dimensional problems. Conjugate gradient or Memory-less BFGS algorithms, which only require $O(N_w)$ storage, can be used instead, but these are only one order of magnitude faster than BBP (McLoone and Irwin (1997). This leads to very poor utilisation of resources, especially when the available memory is only marginally less than that required for FM. Clearly much better utilisation of resources could be achieved if the performance of a training algorithm varied as a function of available memory. The variable memory (VM) BFGS algorithm recently developed by McLoone and Irwin (1999) has this property. Instead of storing and updating the inverse Hessian approximation ($\mathbf{M}_k$) at each iteration VM BFGS stores only those vectors and scalars needed to compute the product $\mathbf{M}_k\mathbf{g}_k$. This yields the following set of equations for the search direction calculation.

$$\mathbf{u}_k = \mathbf{t}_k - \sum_{j=1}^{k-1} \mathbf{f}_j(\mathbf{t}_k) \qquad (9)$$

$$n_k = \mathbf{s}_k^\mathrm{T}\mathbf{t}_k , \qquad v_k = \mathbf{t}_k^\mathrm{T}\mathbf{u}_k \qquad (10)$$

and

$$\mathbf{d}_k = -\mathbf{g}_k + \sum_{j=1}^{k} \mathbf{f}_j(\mathbf{g}_k) \qquad (11)$$

where

$$\mathbf{f}_j(\mathbf{r}) = \left[\frac{\mathbf{u}_j^\mathrm{T}\mathbf{r}}{n_j} - \left(1 + \frac{v_j}{n_j}\right)\frac{\mathbf{s}_j^\mathrm{T}\mathbf{r}}{n_j}\right]\mathbf{s}_j + \left[\frac{\mathbf{s}_j^\mathrm{T}\mathbf{r}}{n_j}\right]\mathbf{u}_j \quad (12)$$

In order to compute $\mathbf{d}_k$ at the kth iteration, the scalars v_j, n_j and vectors $\mathbf{u}_j$ and $\mathbf{s}_j$ have to be stored for all the previous iterations. Thus, the memory requirements and computational complexity of VM increase linearly with iteration number and are therefore $O(kN_w)$. Consequently when available memory is limited to $O(L_{mem}N_w)$, where $L_{mem} \ll N_w$, a situation which precludes FM, the VM alternative can be used if it is reset every time the memory limit is reached. VM is numerically equivalent to FM with reset and its performance varies between that of the Full and Memory-less BFGS algorithms as a function of the resetting frequency.

Neural networks are naturally parallel structures, therefore it is not surprising that parallel implementations of these architectures and their training algorithms have been the subject of a great deal of research over the past decade with implementations reported across the full spectrum of concurrent architectures, from Single Instruction Multiple Data (SIMD) machines such as the Connection Machine (CM) and MasPar to Multiple Instruction Multiple Data (MIMD) machines which include Transputer networks and workstation clusters. Examples include a parallel back-propagation algorithm by Besch *et al.* (1994), a Connection Machine based partial conjugate gradient algorithm by Kramer (1988), a transputer based parallel Memory-less BFGS algorithm by Lightbody and Irwin (1992) and PVM based implementation of the Full Memory BFGS algorithm by McLoone and Irwin (1997).

The intention in this paper is to extend the work of McLoone and Irwin (1997) on Full Memory BFGS to cover the new Variable Memory training algorithm. In that contribution a parallel implementation of FM incorporating matrix partitioning of the search direction calculations and training data partitioning of the cost function and gradient calculations was identified as being the most appropriate for the problem dimensions typically encountered in engineering system identification and function approximation applications. Since VM only differs from FM in the formulation of the search direction calculation this paper will only consider parallel implementation of this segment of the algorithm. Accordingly the remainder of the paper is structured as follows.

In section 2 efficient sequential and parallel implementations of the FM BFGS update equations (6 - 8) are presented and a timing analysis is used to highlight the significant contribution these equations make to the total iteration time. Section 3 compares the performance of the sequential VM and FM BFGS updates with the aid of benchmark training problems from the Proben benchmark suite (Prechelt (1994)). Two different parallel implementations of the VM algorithm are also presented in section 3. The first is based on placing the data associated with each iteration on a different processor while the second is based on partitioning individual vectors over all the processors. Finally in section 4 numerical results are presented comparing the performance of the various algorithms considered for a PVM target architecture consisting of a cluster of 4 Sun workstations.

2 FULL MEMORY BFGS

2.1 *Sequential FM BFGS update algorithm*

To obtain an efficient sequential implementation the BFGS update calculations (6 - 8) are generally expressed in the form

$$\mathbf{s}_k = \mathbf{w}_k - \mathbf{w}_{k-1} , \qquad \mathbf{t}_k = \mathbf{g}_k - \mathbf{g}_{k-1} \qquad (13)$$

$$n_k = \mathbf{s}_k^\mathrm{T}\mathbf{t}_k, \qquad \mathbf{u}_k = \mathbf{M}_{k-1}\mathbf{t}_k, \qquad (14)$$

$$v_k = \mathbf{t}_k^T \mathbf{u}_k, \qquad \phi_k = \frac{n_k^2}{n_k + v_k} \qquad (15)$$

$$\mathbf{M}_k = \mathbf{M}_{k-1} + \frac{\mathbf{s}_k \mathbf{s}_k^T}{\phi_k} - \frac{\mathbf{u}_k \mathbf{s}_k^T + \mathbf{s}_k \mathbf{u}_k^T}{n_k} \qquad (16)$$

$$\mathbf{d}_k = -\mathbf{M}_k \mathbf{g}_k \qquad (17)$$

where v_k, n_k and ϕ_k are scalars and $\mathbf{u}_k$ is a vector. This formulation allows exploitation of the symmetry in $\mathbf{M}$ so that only the lower diagonal elements of $\mathbf{M}$ need to be computed and stored, that is:

$$[m_{ij}]_k = [m_{ij}]_{k-1} + \frac{s_i s_j}{\phi_k} - \frac{u_i s_j + s_i u_j}{n_k} \qquad (18)$$

for $i=1$ to N_w, $j=1$ to i. The memory requirements and complexity if the symmetric FM BFGS update are thus

$$\mu_{SFM} \cong \frac{N_w(N_w + 7)}{2} \qquad (19)$$

$$\chi_{SFM} \cong 2N_w(4N_w + 5) \qquad (20)$$

Here μ is defined as the number of storage locations and χ is defined as the number of arithmetical operations (multiplications, divisions, additions or subtractions) required by the equations. While there are a number of $O(N_w)$ operations in the update the $O(N_w^2)$ operations dominate for large N_w.

2.2 Timing analysis

The principle components of the Full Memory BFGS training algorithm are the batch gradient calculation which has $O(N_v N_w)$ complexity, the step size calculation which has $O(N_c N_v N_w)$ complexity and the BFGS search direction calculation outlined above. (N_v and N_w are as previously defined and N_c is the average number of cost function evaluations needed by the step size estimation routine). Figure 1 shows a graph of the average percentage contribution of each of these components to the overall algorithm iteration time plotted as a function of the ratio N_w/N_v. The boundary curves have been estimated from timing results obtained for a range of MLP network and training set sizes trained using a Matlab 5® implementation of the algorithm. In all cases the remaining computations amounted to less than 0.1% of the total iteration time and have omitted. From the graph it can be seen that the line search computation is the most computationally intensive element of the training algorithm when the ratio N_w/N_v is small, but that the influence of the BFGS update increases with N_w/N_v and dominates for $N_w/N_v > 0.5$. Note that for N_w/N_v =0.2 the BFGS update and line search amount to 30% and 60% of the computation respectively but that the situation is reversed for N_w/N_v=0.35.

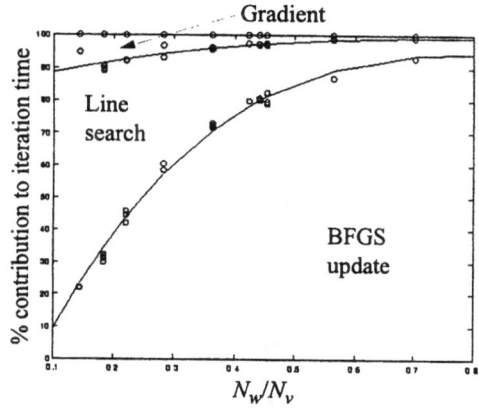

Fig. 1. Cumulative sum of the % contributions of the BFGS update, line search and gradient calculations to the total iteration time plotted as a function of the ratio of the number of weights N_w to the number of training vectors N_v.

2.3 Parallel implementation

Parallel implementation of the BFGS search direction equations can be achieved by partitioning the $\mathbf{M}$ matrix by rows so that N_w/p rows are placed on each of p slave processes. Defining the row partition of a matrix $\mathbf{A}$ and vector $\mathbf{x}$ resident on the rth processor as

$$\Delta_r \mathbf{A} = [a_{ij}], \qquad \Delta_r \mathbf{x} = [x_i]$$

$$i = \left[1 + \frac{N_w}{p}(r-1)\right] \text{ to } \left[\frac{N_w}{p}r\right] \qquad (21)$$

$$j = 1 \text{ to } N_w$$

the parallel FM BFGS update algorithm can be summarised as follows:

At the kth iteration, after the gradient has been determined, $\Delta_r \mathbf{M}_{k-1}$, $\mathbf{w}_k$, $\mathbf{w}_{k-1}$ and $\mathbf{g}_{k-1}$ are present on the rth slave processor and $\mathbf{g}_k$ is present on the master processor.

Step 1: The slave processors receive $\mathbf{g}_k$ from the master processor and compute $\mathbf{s}_k$ and $\mathbf{t}_k$.

Step 2: Equations

$$\Delta_r \mathbf{u}_k = \Delta_r \mathbf{M}_k \mathbf{t}_k,$$
$$\Delta_r v = \Delta_r \mathbf{t}_k^T \Delta_r \mathbf{u}_k \qquad (22)$$

are evaluated in parallel and the results transmitted to the master processor where they are used to obtain v_k and $\mathbf{u}_k$, that is:

$$\mathbf{u}_k = \left[\Delta_1 \mathbf{u}_k^T \,.. \,\Delta_p \mathbf{u}_k^T\right]^T, \quad v_k = \sum_{r=1}^{p} \Delta_r v_k \quad (23)$$

Step 3: The slave processors receive v_k and $\mathbf{u}_k$ from the master processor and compute the scalars n_k and ϕ_k.

$$n_k = \mathbf{s}_k^T \mathbf{t}_k, \qquad \phi_k = \frac{n_k^2}{n_k + v_k} \qquad (24)$$

Step 4: Each slave process then updates its portion of the $\mathbf{M}$ matrix ($\Delta_r \mathbf{M}_k$) using equation (18).

Step 5: Finally the slave processes compute $\Delta_r \mathbf{d}_k$, $r=1$ to p in parallel and transmit the results to the master processor where they are combined to give $\mathbf{d}_k$. that is:

$$\Delta_r \mathbf{d}_k = \Delta_r \mathbf{M}_k \mathbf{g}_k, \qquad (25)$$

$$\mathbf{d}_k = \left[\Delta_1 \mathbf{d}_k^T \quad .. \quad \Delta_p \mathbf{d}_k^T \right]^T \qquad (26)$$

The memory and computational requirements of this implementation of the FM BFGS update are given by

$$\mu_{\text{PFM}} \cong N_w \left(\frac{N_w}{p} + 3 \right) \qquad (27)$$

$$\chi_{\text{PFM}} \cong \frac{2N_w}{p}(6N_w + 2p + 1) \qquad (28)$$

but this is at the expense of communication overheads (τ) which are approximately given by

$$\tau_{\text{PFM}} \cong 2[N_w(p+1) + p]\tau_c + 4p\tau_o \qquad (29)$$

where τ_c and τ_o are the mean transfer time per floating point number and mean communication initiation time respectively.

The symmetry in $\mathbf{M}$ cannot be exploited when employing standard row partitioning as given in equation (21). However if the partitioning of $\mathbf{M}$ is defined as

$$\Delta_r \mathbf{M}^* = [m_{ij}] \qquad (30)$$

$$\left\{ i = \left[1 + \frac{N_w}{p}(r-1) \right] \text{ to } \left[\frac{N_w}{p} r \right], j = 1 \text{ to } i \right\}$$

$$\cup \qquad (31)$$

$$\left\{ i = \left[\frac{N_w}{p} r + 1 \right] \text{ to } N_w, j = \left[1 + \frac{N_w}{p}(r-1) \right] \text{ to } \left[\frac{N_w}{p} r \right] \right\}$$

symmetry can be exploited leading to reduced memory requirement and complexity. The difference between the partitioning strategies is illustrated graphically in Figure 2. Partitions $\Delta_r \mathbf{M}^*$ have fewer elements than $\Delta_r \mathbf{M}$, but contain the same information because of symmetry. The reduced memory usage and complexity arising from this modification are

$$\mu_{\text{PSFM}} \cong \mu_{\text{PFM}} - \frac{N_w}{2p}\left(\frac{N_w}{p} - 1 \right) \qquad (32)$$

$$\chi_{\text{PSFM}} \cong \chi_{\text{PFM}} - \frac{4N_w}{p}\left(\frac{N_w}{p} - 1 \right) \qquad (33)$$

Note that the reductions are only significant when the partitions N_w/p are large.

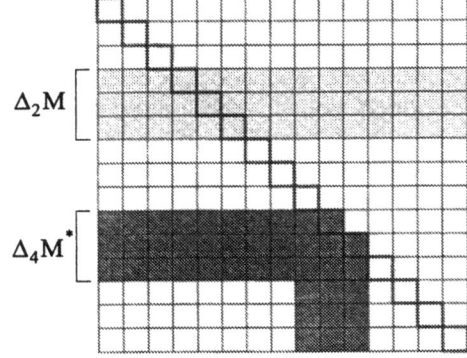

Fig. 2. Graphical representation of matrix partitioning strategies $\Delta_r \mathbf{M}$ and $\Delta_r \mathbf{M}^*$ applied to a 15 x 15 matrix divided over 5 processors.

3 VARIABLE MEMORY BFGS UPDATE

3.1 Performance of the sequential VM algorithm

The memory requirements and complexity of the VM BFGS update equations (9) to (12) at the kth iteration are given by

$$\mu_{\text{VM}} \cong 2k(N_w + 1) + N_w \qquad (34)$$

$$\chi_{\text{VM}} \cong 2k(7N_w + 6) + N_w \qquad (35)$$

Comparing these expressions with the corresponding expressions for the Full Memory BFGS update (equations (19) and (20)) it can be shown that:

$$\mu_{\text{VM}} < \mu_{\text{SFM}} \text{ if } k < \frac{N_w}{4} \qquad (36)$$

$$\chi_{\text{VM}} < \chi_{\text{SFM}} \text{ if } k < \frac{4N_w}{7} \qquad (37)$$

$$\sum_{k=1}^{N_k} \chi_{\text{VM}} < \sum_{k=1}^{N_k} \chi_{\text{SFM}} \text{ if } N_k < \frac{8N_w}{7} \qquad (38)$$

Equations (36) and (37) define the conditions under which VM uses less memory and has shorter iteration times than FM, while equation (38) indicates the number of iterations N_k that can be computed before the total training time for VM exceeds that of FM. In practice when using FM BFGS neural network training algorithms termination occurs in significantly less than N_w iterations. Therefore by equation (38) VM BFGS will always be computationally superior to FM

BFGS. If the VM algorithm is reset every L iterations where $L << N_w$ mean iteration times will be much quicker than those of FM, but convergence will generally require more iterations. In many cases the convergence versus iteration time trade off is such that VM with reset will outperform VM without reset. To illustrate this point a 10 neuron MLP was trained for, two test problems from the Proben benchmark suite (Prechelt (1994)) (*build1a* and *flare1a*) using the FM algorithm and various flavours of VM. Training was halted in each case when an error goal of 0.2% was achieved or the number of iterations exceeded 1000. The number of iterations and total training time for each algorithm are recorded in table 1 below. As expected VM is superior to FM in both problems. This is most evident in *flare1a* where the BFGS update represents a significant portion of the computation (i.e. $N_w/N_v > 0.5$). Note also that VM with a resetting factor of 70 outperforms VM in problem *build1a*.

Table 1: No. of iterations and training times for FM, VM and VM with periodic reset

Alg.	build1a N_w=183, N_v=2100		flare1a N_w=283, N_v=530	
	No. of iterations	Time (secs)	No. of iterations	Time (secs)
FM	148	144.7	146	148.8
VM	148	122.9	146	48.0
VM5	1000+	628.1	1000+	227.6
VM10	1000+	662.9	1000+	238.6
VM20	479	350.2	297	77.1
VM30	237	181.2	206	55.0
VM40	197	155.2	187	51.4
VM50	168	134.2	182	52.0
VM60	153	121.2	178	51.2
VM70	149	118.8	172	50.3

3.2 Parallel VM BFGS update implementations

Since the computational complexity and memory requirements of the Variable Memory BFGS update are $O(kN_w)$, parallel implementation is possible either by partitioning the data in terms of k or in terms of N_w.

Partitioning in terms of k. Here, the data associated with each iteration is stored on a different processor so that at the kth iteration $\mathbf{u}_j$, $\mathbf{s}_j$, n_j and v_j, j=1 to k-1, are distributed over the p parallel processors with the jth components resident on the φ_j^{th} processor where φ_j is defined as

$$\varphi_j = (j\text{-}1) \bmod p + 1 \qquad (39)$$

Using this approach the $\mathbf{f}_j(.)$ expressions in equations

(9) and (11) can be determined in parallel. The algorithm proceeds as follows.

After completing the gradient calculation at the kth iteration $\mathbf{w}_k$, $\mathbf{w}_{k-1}$ and $\mathbf{g}_{k-1}$ are present on the slave processors and $\mathbf{g}_k$ is present on the master processor.

Step 1: The slave processors receive $\mathbf{g}_k$ from the master processor and compute $\mathbf{t}_k$.

Step 2: Each slave process evaluates the partial sum $\sum \mathbf{f}_j(\mathbf{t}_k)$ for the data ($\mathbf{u}_j$, $\mathbf{s}_j$, n_j and v_j) currently stored on it and transmits the result to the φ_k^{th} processor.

Step 3: The φ_k^{th} processor collects the partial sums from the parallel processes and uses them to compute $\mathbf{u}_k$. Vector $\mathbf{s}_k$ and scalars n_k and v_k are also computed and stored along with the $\mathbf{u}_k$ for future use.

Step 4: Partial sums $\sum \mathbf{f}_j(\mathbf{g}_k)$ are determined in parallel and transmitted to the φ_k^{th} processor where the search direction $\mathbf{d}_k$ is computed using eqt. (11).

The memory requirements, computational complexity and inter processor communication overheads are approximately given by

$$\mu_{\text{PVM}k} \cong \frac{2k}{p}(N_w + 1) + N_w \qquad (40)$$

$$\chi_{\text{PVM}k} \cong \frac{N_w}{p}(14k + 6p + 2p^2) + \frac{12k}{p} \qquad (41)$$

$$\tau_{\text{PFM}k} \cong 3pN_w\tau_c + 3p\tau_o \qquad (42)$$

This partitioning strategy results in a factor p increase in storage capacity, hence if k_s full BFGS updates can be accommodated when using the sequential algorithm, the partitioned algorithm can compute pk_s updates. However, computationally the strategy is very poor since parallelisation is incomplete and processing power is under utilised.

Partitioning in terms of N_w: In this approach individual vectors are partitioned over all the processors leading to the following parallel algorithm:

At the kth iteration, $\Delta_r\mathbf{u}_j$, $\Delta_r\mathbf{s}_j$, n_j and v_j, j=1 to k-1 are distributed over the p parallel processors. The gradient $\mathbf{g}_k$ is present on the master processor and $\mathbf{w}_k$, $\mathbf{w}_{k-1}$ and $\mathbf{g}_{k-1}$ are available on each slave process.

Step 1: Each slave processor receives its segment of $\mathbf{g}_k$ ($\Delta_r\mathbf{g}_k \rightarrow$ processor r) and computes $\Delta_r\mathbf{t}_k$ and $\Delta_r\mathbf{s}_k$.

Step 2: Partial sums $\Delta_r\mathbf{u}_j^T\Delta_r\mathbf{t}_k$ and $\Delta_r\mathbf{s}_j^T\Delta_r\mathbf{t}_k$, j=1 to k-1 are computed by the slave processes and transmitted to the master processor where they are summed to give $\mathbf{u}_j^T\mathbf{t}_k$ and $\mathbf{s}_j^T\mathbf{t}_k$, j=1 to k-1.

Step 3: The slave processes receive $\mathbf{u}_j^T\mathbf{t}_k$ and $\mathbf{s}_j^T\mathbf{t}_k$, j=1 to k-1 from the master processor and hence compute $\Delta_r\mathbf{u}_k$. $\Delta_r\mathbf{u}_k$ and $\Delta_r\mathbf{s}_k$ are stored for future use.

Step 4: The slave processes compute the partial sums $\Delta_r n_k$ and $\Delta_r v_k$ and transmit them to the master processor where they are combined to give n_k and v_k. These quantities are then returned to the slave processes where they are stored for future use.

Step 5: Following the procedure outlined in step 2 the slave processes determine $\mathbf{u}_j^T\mathbf{g}_k$ and $\mathbf{s}_j^T\mathbf{g}_k$, j=1 to k and hence compute $\Delta_r\mathbf{d}_k$. Finally the $\Delta_r\mathbf{d}_k$ partitions are transmitted to the master process where they are combined to give $\mathbf{d}_k$.

The memory, complexity and communication costs associated with this implementation are

$$\mu_{PVMN_w} \cong 2k\left(\frac{N_w}{p}+2\right)+N_w \qquad (43)$$

$$\chi_{PVMN_w} \cong \frac{N_w}{p}(14k+1)+4k(p+3) \qquad (44)$$

$$\tau_{PFMN_w} \cong (2N_w+8kp)\tau_c + 8p\tau_o \qquad (45)$$

Comparing the expressions for both implementations it can be seen that provided the ratio N_w/p is large the second algorithm is superior. The trade-off is a slight reduction in memory capacity because the scalars n_k and v_k are needed on all processors.

4 SIMULATION RESULTS

To assess the performance of the parallel FM and VM BFGS updates discussed in the paper, speed-ups relative to the sequential FM BFGS algorithm have been determined for PVM implementations on a cluster of 4 Sun workstations. The results, which are based on the benchmark problems and training conditions described in section 3, are given in table 2.

Table 2: Parallel FM and VM BFGS speeds-ups relative to the sequential FM BFGS update.

Training Algorithm	Speed Up Relative to FM BFGS			
	build1a		*flare1a*	
	p=2	p=4	p=2	p=4
PFM	1.14	1.61	1.24	2.04
PSFM	1.33	1.70	1.46	2.18
VM	*1.41*	*1.41*	*3.20*	*3.20*
PVMk	1.68	1.56	4.42	4.68
PVMN_w	2.15	2.55	5.17	6.61

The speed-ups observed are consistent with predictions. The parallel FM BFGS algorithm which exploits symmetry (PSFM) out performing the version that doesn't (PFM) while the vector based partitioning strategy (PVMN_w) gives the best results for the VM algorithm. While the speeds-ups obtained are significant they fall well short of the theoretical maximum. This is mainly due to the influence of the communication overheads which are amplified by the relatively slow data transmission rates of the PVM architecture employed.

5 CONCLUSIONS

This paper has proposed parallel implementation strategies for the search direction calculation in the FM and VM BFGS training algorithms. In addition to facilitating larger problems sizes the proposed algorithms provide significant speed-ups relative to their sequential counterparts.

6 REFERENCES

Besch M. & Pohl H.W. (1995), "Flexible data parallel training of neural networks using MIMD-computers", In: *Third Workshop on Parallel and Distributed Processing*, Sanremo, Italy.

Kramer A.H. & Sangiovanni-Vincentelli (1988), "Efficient parallel learning algorithms for neural networks", *Advances in Neural Information Processing Systems*, Vol.1, Pages 75-89.

Lightbody G. & Irwin G.W. (1992). A parallel Algorithm for Training Neural Network Based Nonlinear Models. In: *Proc. 2nd IFAC Workshop on Algorithms and Architectures for Real-time Control*, pp. 99-104.

McLoone S.F. & Irwin G.W. (1999), 'A Variable memory Quasi-Newton Training Algorithm', *Neural Processing Letters*, Vol.9(1), pp. 77-89.

McLoone S.F. & Irwin G.W. (1997), "Fast parallel off-line training of multilayer perceptrons", *IEEE Trans. Neural Networks*, Vol.8, No.3, pp. 646-653.

Prechelt L. (1994), "Proben 1 - A set of neural network benchmark problems and benchmarking rules", Technical Report 21/94, Fakultät für Informatik, Universität Karlsruhe, 76128 Karlsruhe, Germany, Sept 1994. Available by anonymous FTP at ftp.ira.uka.de in /pub/neuron/proben1.tar.gz.

Rumelhart D.E, G. Hinton & R. Williams (1986). "Learning internal representations by error propagation". In: *Parallel Distributed Processing* (D.E Rumelhart, J.L. McClelland, (Ed.)), Vol.1, pp. 318-364. MIT Press.

HIGH PERFORMANCE ALGORITHM FOR TRACKING TRAJECTORIES OF ROBOT MANIPULATORS.

Juan C. Fernández * **Enrique Arias** **
Vicente Hernández *** **Lourdes Peñalver** ****

* *Departamento de Informática*
Universidad Jaume I
12.071-Castellón (Spain)
Phone: +34-964-345680; Fax: +34-964-345848
e-mail: jfernand@inf.uji.es
** *Departamento de Informática*
Universidad Castilla- La Mancha
02071- Albacete (Spain)
Phone: +34-967-599200; Ex: 2460
e-mail: earias@info-ab.uclm.es
*** *Dept. de Sistemas Informáticos y Computación*
Universidad Politécnica de Valencia
46.071-Valencia (Spain)
Phone: +34-96-3877356; Fax: +34-96-3877359
e-mail: vhernand@dsic.upv.es
**** *Dept. de Ingeniería de Sistemas y Computadores*
Universidad Politécnica de Valencia
46.071-Valencia (Spain)
Phone: +34-96-3877000 Ext: 5725; Fax: +34-96-3877579
e-mail: lourdes@disca.upv.es

Abstract: The paper is concerned with the application of quadratic optimization for motion control to the feedback control of robotic manipulators using a high performance computing approach. The implementation of this optimal control requires, at each sample instant, the computation of all the terms of the dynamic equation. For solving this problem different approaches have been used to reduce the computational cost of the Lagrange-Euler formulation. This formulation permits us to establish matrix structures with important properties (symmetry, repeated rows, etc.) that can be adequately exploited. Simulation results on a six-links manipulator are presented. *Copyright ©2000 IFAC*

Keywords: Robotics, dynamic models; optimal control; inverse dynamics control; real-time.

1. INTRODUCTION

Using knowledge of the dynamic systems, a mathematical model is developed in the linear optimal control design. For optimal control design an accurate linear system (Kirk, 1970; Lewis and Syrmos, 1995) or a "feedback linearization" of nonlinear systems (Slotine and Li, 1984; Lewis *et al.*, 1993) can be used.

Several authors have addressed the application of optimal control techniques to the nonlinear robotic manipulator. These approaches often combine feedback linearization and optimal control techniques (Johansson, 1990). An optimal design method has been recently proposed by (Kim *et al.*, 1999) where a neural network implementation is also included.

To apply these optimal applications it is necessary to know the robot dynamics. Many advanced manipulator control schemes are based on inverse dynamics calculation (Good *et al.*, 1985; Tarn *et al.*, 1984; Verdier *et al.*, 1989). There are also different approaches based on adaptive inverse dynamics (Balestrino and Zinober, 1984; Berghuis *et al.*, 1993; Craig, 1986; Slotine and Li, 1987).

In this paper, we describe the implementation of an optimal design, using the method described in (Kim *et al.*, 1999), that integrates optimal control techniques and high performance computing techniques. The main problem here is the need to implement algorithms to obtain the dynamic equation in real-time. Various robot formulations have been proposed. The Lagrange-Euler formulation has a well structured and systematic representation that accomodates different control applications. The recursive Lagrangian formulation has a better computation time but destroys the structure of the equation. The Newton-Euler formulation is based on a set of highly recursive equations which makes it very difficult to use in control applications. In this paper the Lagrange-Euler formulation is employed. This formulation provides explicit state equations for the robot dynamics that can be used to analyze and design advanced control schemes (e.g. adaptive control). Although this formulation incurs in a heavy computational cost for the forward and inverse dynamics problems it is possible to reduce this computational cost. An implementation avoiding arithmetic operations involving null elements (matrix products and additions) has been developed, obtaining in this way similar performance than using the Newton-Euler formulation (Fernández, 1999).

The paper is organized as follows. Section 2 describes the dynamic equation of robot manipulators and introduces different data structures. The optimal control is presented in section 3. In section 4 a high performance sequential algorithm is introduced. The experimental results are presented in section 5. Finally, conclusions and remarks are given in section 6.

2. DYNAMIC EQUATIONS OF RIGID MANIPULATORS

The dynamic equation of rigid manipulators with n arms in matrix form is

$$\tau = D(q)\ddot{q} + h(q, \dot{q}) + c(q), \qquad (1)$$

where τ is the $n \times 1$ vector of nominal driving torques, q is the $n \times 1$ vector of nominal generalized coordinates, $\dot{q}$ and $\ddot{q}$ are the $n \times 1$ vectors of the first and second derivatives of the vector q, respectively, D is the inertia matrix, $h(q, \dot{q})$ is the vector of centrifugal and Coriolis forces and $c(q)$ is the vector of gravitational forces.

Given a desired manipulator trajectory $q_d(t) \in R^n$ the tracking error is defined by

$$e(t) = q_d(t) - q(t) \qquad (2)$$

and the instantaneous performance measure is defined as

$$r(t) = \dot{e}(t) + \Gamma e(t) \qquad (3)$$

where Γ is the constant gain matrix which will be determined later from the user-specified performance index.

The robot dynamics (1) may be written in terms of the instantaneous performance measure as

$$D(q)\dot{r}(t) = -h(q, \dot{q})r(t) - \tau(t) + h(x) \qquad (4)$$

where the robot nonlinear function is

$$h(x) = D(q)(\ddot{q}_d + \Gamma\dot{e}) + \\ + h(q, \dot{q})(\dot{q}_d + \Gamma e) + c(q). \qquad (5)$$

This function $h(x)$ captures all the unknown dynamics of the robot manipulator. Let the control input torque be defined as

$$\tau(t) = h(x) - u(t) \qquad (6)$$

with $u(t)$ an auxiliary control input to be optimized later. The closed-loop system becomes

$$D(q)\dot{r}(t) = -h(q, \dot{q})r(t) + u(t). \qquad (7)$$

This is the error system wherein the instantaneous performance measure dynamics is driven by an auxiliary control input.

From the dynamic equation (1) the elements of the $n \times n$ symmetric positive definite inertia matrix $D(q)$ are given by

$$d_{ij} = \sum_{k=max\{i,j\}}^{n} tr(U_{jk} J_k U_{ik}^T), \qquad (8)$$

for $i, j = 1 : n$. In this expression $tr(\cdot)$ represents the trace of a matrix and J_k, $k = 1 : n$, is the inertia tensor related to the link k. To obtain D it is necessary to compute U_{ij}, $i, j = 1, \ldots, n$, given by

$$U_{ij} = \frac{\partial {}^0 A_j}{\partial q_i} = \begin{cases} {}^0 A_{i-1} Q_i {}^{i-1} A_j, & i \le j \\ 0, & i > j \end{cases} \qquad (9)$$

where Q_i is a constant matrix that allows us to calculate the partial derivative related to the link i. A block upper triangular matrix structure can be defined, where each block is one of the matrices U_{ij}

$$U = \begin{bmatrix} U_{11} & U_{12} & \cdots & U_{1n} \\ & U_{22} & \cdots & U_{2n} \\ & & \ddots & \vdots \\ & & & U_{nn} \end{bmatrix}. \qquad (10)$$

To obtain the matrix U it is necessary to compute matrices $^jA_i \in R^{4\times 4}$. From the Denavit-Hartenberg convention (Fu et al., 1988),

$$^jA_i = {}^jA_{i-1} \, {}^{i-1}A_j, \qquad (11)$$

where

$$^{i-1}A_i = \begin{bmatrix} cos\theta_i & -cos\alpha_i sin\theta_i & sin\alpha_i sin\theta_i & a_i cos\theta_i \\ sin\theta_i & cos\alpha_i cos\theta_i & -sin\alpha_i cos\theta_i & a_i sin\theta_i \\ 0 & sin\alpha_i & cos\alpha_i & d_i \\ 0 & 0 & 0 & 1 \end{bmatrix} \quad (12)$$

in the case of a revolution joint, or

$$^{i-1}A_i = \begin{bmatrix} cos\theta_i & -cos\alpha_i sin\theta_i & sin\alpha_i sin\theta_i & 0 \\ sin\theta_i & cos\alpha_i cos\theta_i & -sin\alpha_i cos\theta_i & 0 \\ 0 & sin\alpha_i & cos\alpha_i & d_i \\ 0 & 0 & 0 & 1 \end{bmatrix} \quad (13)$$

for the case of a prismatic joint. Then the following block upper triangular matrix can be defined

$$A = \begin{bmatrix} ^0A_1 & ^0A_2 & \cdots & ^0A_n \\ & ^1A_2 & \cdots & ^1A_n \\ & & \ddots & \vdots \\ & & & ^{n-1}A_n \end{bmatrix}. \qquad (14)$$

The components of the $n \times 1$ vector of centrifugal and Coriolis forces, $h(q, \dot{q})$, could be expressed as $h_i(q, \dot{q}) = \dot{q}^T H_i \dot{q}$, $i = 1 : n$, where $H_i = [h_{ijk}]$ is an $n \times n$ matrix given by

$$h_{ijk} = \sum_{l=max\{i,j,k\}}^{n} tr(U_{kjl} J_l U_{ii}^T) \qquad (15)$$

for $j, k = 1 : n$. These matrices define the matrix structure

$$H = \begin{bmatrix} H_1 & H_2 & \cdots & H_n \end{bmatrix}^T. \qquad (16)$$

To obtain H, it is necessary to compute matrices $U_{kjl} \in R^{4\times 4}$, which represent the second derivative terms. U_{kjl} is given by

$$U_{kjl} \equiv \frac{\partial U_{jl}}{\partial q_k} =$$

$$= \begin{cases} ^0A_{j-1}Q_j \, ^{j-1}A_{k-1}Q_k \, ^{k-1}A_l, & j \le k \le l \\ ^0A_{k-1}Q_k \, ^{k-1}A_{j-1}Q_j \, ^{j-1}A_l, & k \le j \le l \\ 0, & j > l \text{ or } k > l. \end{cases} \quad (17)$$

Note that $U_{kjl} = U_{jkl}$. As a consequence matrices H_i are symmetric. From matrices U_{kjl}, the following matrix structure can be defined

$$DU = \begin{bmatrix} DU_1 & DU_2 & \cdots & DU_n \end{bmatrix}^T, \qquad (18)$$

where

$$DU_i = \begin{bmatrix} 0 & \cdots & 0 & U_{i1i} & U_{i1i+1} & \cdots & U_{i1n} \\ 0 & \cdots & 0 & U_{i2i} & U_{i2i+1} & \cdots & U_{i2n} \\ \vdots & \vdots & \vdots & \vdots & \vdots & \vdots & \vdots \\ 0 & \cdots & 0 & U_{iii} & U_{iii+1} & \cdots & U_{iin} \\ 0 & \cdots & 0 & 0 & U_{ii+1i+1} & \cdots & U_{ii+1n} \\ \vdots & \vdots & \vdots & \vdots & 0 & \ddots & \vdots \\ 0 & \cdots & \cdots & \cdots & \cdots & \cdots & U_{inn} \end{bmatrix}. (19)$$

Finally, the components of the $n \times 1$ vector of gravitational forces are given by

$$c_i = \sum_{j=i}^{n} (-m_j g^T U_{ij} \, {}^jr_j) \qquad (20)$$

for $i = 1 : n$. In this expression ^jr_j is the position of the center of mass of link j with respect to the origin of the coordinates of the link j.

3. OPTIMAL CONTROL OF ROBOT MANIPULATORS

From the robot dynamic equation described in section 2, the control aim is to keep the dynamic response of the robot according to some established operation criteria. The control problem can be considered as a tracking problem. A feedback control is presented in (Kim et al., 1999) to make the robot track a reference trajectory. If some constraints are added as a functional to be optimized, then this is an optimization problem. Thus, the motion control objective is to follow a desired trajectory $q_d(t)$ and $\dot{q}_d(t)$ with suitable small position errors $e(t)$ and velocity errors $\dot{e}(t)$. The position error $e(t)$ is expressed in terms of the instantaneous performance index (3) as

$$\dot{e}(t) = -\Gamma e(t) + r(t). \qquad (21)$$

Combining (21) and (4), the following system is obtained

$$\dot{z} = \begin{bmatrix} \dot{e} \\ \dot{r} \end{bmatrix} = \begin{bmatrix} -\Gamma & I \\ 0 & -D^{-1}(q)H(q,\dot{q}) \end{bmatrix} \begin{bmatrix} e \\ r \end{bmatrix} +$$

$$+ \begin{bmatrix} 0 \\ D^{-1}(q) \end{bmatrix} u(t) \qquad (22)$$

or using an equivalent shorter notation

$$\dot{z} = A(q,\dot{q})\tilde{z}(t) + B(q)u(t) \qquad (23)$$

with $A(q,\dot{q}) \in R^{2n\times 2n}$, $B(q) \in R^{2n\times n}$ and $\tilde{z}(t) \in R^{2n\times 1}$.

If the robot dynamic is known exactly, then a quadratic performance index $J(u)$ can be formulated as follows

$$J(u) = \int_{t_0}^{\infty} L(\tilde{z}, u) dt \qquad (24)$$

with the Lagrangian

$$L(\tilde{z}, u) = \frac{1}{2} \tilde{z}^T(t) Q \tilde{z}(t) + \frac{1}{2} u^T(t) R u(t) =$$

$$= \frac{1}{2} \begin{bmatrix} e^T & r^T \end{bmatrix} \begin{bmatrix} Q_{11} & Q_{12} \\ Q_{12} & Q_{22} \end{bmatrix} \begin{bmatrix} e \\ r \end{bmatrix} + \frac{1}{2} u^T R u. \quad (25)$$

where $R = R^T > 0 \in R^{n \times n}$ is the control weight matrix and $Q = Q^T > 0 \in R^{2n \times 2n}$ is the state weight matrix.

An optimal control input $u^*(t)$ minimizes the quadratic performance index (24) if there exists a function $V = V(\tilde{z}, t)$ satisfying the Hamilton-Jacobi-Belman (H-J-B) equation

$$\frac{\partial V(\tilde{z}, t)}{\partial t} + \min_{u} H(\tilde{z}, u, \frac{\partial V(\tilde{z}, t)}{\partial \tilde{z}}, t) = 0 \ , (26)$$

where the Hamiltonian of optimization is

$$H(\tilde{z}, u, \frac{\partial V(\tilde{z}, t)}{\partial z}, t) = L(\tilde{z}, u) + \frac{\partial V(\tilde{z}, t)}{\partial \tilde{z}} \dot{\tilde{z}}. (27)$$

In this case $V = \frac{1}{2} \tilde{z}^T P(q) \tilde{z}$, where

$$P(q) = \begin{bmatrix} K & 0 \\ 0 & D(q) \end{bmatrix}, \ K = K^T \in R^{n \times n} \quad (28)$$

The optimal control input $u^*(t)$ that minimizes (24) under the constraints in (22) is given by

$$u^*(t) = -R^{-1} B^T P(q) \tilde{z} =$$

$$= -R^{-1} r(t) = -R^{-1} \{ \dot{e}(t) - \Gamma e(t) \}. \quad (29)$$

However, a suitable choice of the weigth matrices allows to obtain the matrices K and Γ (Kim *et al.*, 1999) without solving any Riccati differential equation. If weight matrices Q and R are chosen such that

$$Q = \begin{bmatrix} Q_{11} & Q_{12} \\ Q_{12}^T & Q_{22} \end{bmatrix} > 0, \ R^{-1} = Q_{22}, \quad (30)$$

with $Q_{12} + Q_{12}^T < 0$, then K and Γ can be determined by

$$K = K^T = -\frac{1}{2}(Q_{12} + Q_{12}^T) > 0 \quad (31)$$

$$\Gamma^T K + K \Gamma = Q_{11}. \quad (32)$$

Note that (32) represents a non-symmetric Lyapunov equation.

4. HIGH PERFORMANCE SEQUENTIAL ALGORITHM

In order to obtain the terms of the dynamic equation with a low computational cost, an approach is presented based on the following considerations:

- It is possible to optimize the matrix products involved, because the matrices (12), (13) and Q_j have a considerable number of null elements.
- It is possible to use different properties of the matrices, such as symmetry, repeated rows, etc., in order to reduce the computational cost.
- It is possible to reduce the cost of the trace operation.

The calculation of the different terms of the dynamic equation is presented below for the case of a robot with revolution joints.

The diagonal blocks of matrix A are given by (12) and (13). The rest of blocks of A are obtained from (11). Taking these expressions into account, matrices $^j A_i$, for $j = 0 : n - 2$ and $i = j + 2 : n$, have the following structure

$$^j A_i = \begin{bmatrix} a_{11} & a_{12} & a_{13} & a_{14} \\ a_{21} & a_{22} & a_{23} & a_{24} \\ a_{31} & a_{32} & a_{33} & a_{34} \\ 0 & 0 & 0 & 1 \end{bmatrix}. \quad (33)$$

From (9), the expression of matrix U is the following:

- First row of matrix U, U_{1j}, for $j = 1 : n$

$$U_{1j} = \begin{bmatrix} u_{11} & u_{12} & u_{13} & u_{14} \\ u_{21} & u_{22} & u_{23} & u_{24} \\ 0 & 0 & 0 & 0 \\ 0 & 0 & 0 & 0 \end{bmatrix}. \quad (34)$$

- Rest of the rows of matrix U, U_{ij}, for $i = 2 : n$, $j = i : n$

$$U_{ij} = \begin{bmatrix} u_{11} & u_{12} & u_{13} & u_{14} \\ u_{21} & u_{22} & u_{23} & u_{24} \\ u_{31} & u_{32} & u_{33} & u_{34} \\ 0 & 0 & 0 & 0 \end{bmatrix}. \quad (35)$$

Using expression (17), matrix DU has the following structure:

- Matrix DU_1, U_{1ij}, for $i = 1 : n$, $j = i : n$

$$U_{1ij} = \begin{bmatrix} du_{11} & du_{12} & du_{13} & du_{14} \\ du_{21} & du_{22} & du_{23} & du_{24} \\ 0 & 0 & 0 & 0 \\ 0 & 0 & 0 & 0 \end{bmatrix}. \quad (36)$$

- Matrix DU_i, U_{ikj}, for $i = 2 : n$, $k = i : n$, $j = k : n$

$$U_{ikj} = \begin{bmatrix} du_{11} & du_{12} & du_{13} & du_{14} \\ du_{21} & du_{22} & du_{23} & du_{24} \\ du_{31} & du_{32} & du_{33} & du_{34} \\ 0 & 0 & 0 & 0 \end{bmatrix}. \quad (37)$$

Expressions (8) and (15) are necessary to obtain matrices D and H respectively. In these expressions matrices $Y_{ij} = J_j U_{ij}^T$ need to be computed. The expression of matrix Y is the following:

- First row of matrix Y, Y_{1j}, for $j = 1 : n$

$$Y_{1j} = \begin{bmatrix} y_{11} & y_{12} & 0 & 0 \\ y_{21} & y_{22} & 0 & 0 \\ y_{31} & y_{32} & 0 & 0 \\ y_{41} & y_{42} & 0 & 0 \end{bmatrix}. \quad (38)$$

- Rest of the rows of matrix Y, Y_{ij}, for $i = 2 : n$, $j = i : n$

$$Y_{ij} = \begin{bmatrix} y_{11} & y_{12} & y_{13} & 0 \\ y_{21} & y_{22} & y_{23} & 0 \\ y_{31} & y_{32} & y_{33} & 0 \\ y_{41} & y_{42} & y_{43} & 0 \end{bmatrix}. \quad (39)$$

To calculate the trace of a matrix product it is only necessary to obtain the diagonal elements of that product.

5. EXPERIMENTAL RESULTS

The experiments have been carried out on two different platforms:

- MIPS R12000-300MHz processor.
- Intel 440BX-200MHz processor.

Two different implentations have been developed:

- A nonoptimized implementation. In this case, BLAS and LAPACK standard linear algebra routines have been used for computations.
- An optimized implementation. In this case, the operations have been done taking into account the null elements to minimize the number of needed operations. A detailed analysis of the operations has been developed.

Using previous studies, the algorithms were executed for a PUMA with six links. The experimental results are summarized in the following tables. In table (1) the execution time (in milliseconds) of the sequential optimized and nonoptimized algorithms are shown. The execution time in the nonoptimized case is three times greater than the optimized version.

Table (2) shows that the results obtained by the optimized version on the R12000 are quite similar to those obtained on the Intel 440BX.

Table (3) shows the results obtained executing one iteration of the algorithm to compute the optimal control for traking trajectories. Table (3)

Table 1. Execution time (milliseconds) for the optimized and nonoptimized implementations of the algorithm to compute the dynamic equation on the Intel 440BX.

Matrix	Time optimized	Time nonoptimized
J	0.010	0.015
A	0.063	0.089
U	0.018	0.096
D	0.070	0.204
DU	0.156	0.253
H	0.088	0.657
c	0.014	0.016
Total	0.419	1.330

Table 2. Execution time (milliseconds) for the optimized implementation of the algorithm to compute the dynamic equation on the Intel 440BX and the R12000 processors.

Matrix	Intel 440BX	R12000
J	0.010	0.009
A	0.063	0.041
U	0.018	0.024
D	0.070	0.085
DU	0.156	0.069
H	0.088	0.128
c	0.014	0.002
Total	0.419	0.358

summarizes the results of the sequential optimized version on the Intel 440BX and on the R12000.

Table 3. Execution time (milliseconds) of the sequential optimized implementation to compute one iteration of the control algorithm on the Intel 440BX and the R12000 processors.

Platform	Execution Time
Intel 440BX	0.531
R12000	0.423

The following input data have been considered for the experimental results in the tracking problem:

- Denavit-Hartenberg parameters.

ρ	a	α	d
0.0	0.0	0.0	0.0
0.0	0.0	$\pi/2$	0.0
0.0	0.71	0.0	0.0
0.0	0.85	$\pi/2$	0.125
0.0	0.0	$\pi/2$	0.0
0.0	0.0	$\pi/2$	0.0

- Kind of links. Array of dimension equal to the number of links, with values 0 if rotation link and 1 if prismatic link.
- Initial values of positions, velocities and accelerations.

$$q = \begin{bmatrix} 0.25 \\ 0.5 \\ 0.0 \\ 0.75 \\ 0.0 \\ 0.75 \end{bmatrix}, \quad \dot{q} = \begin{bmatrix} 0 \\ 0 \\ 0 \\ 0 \\ 0 \\ 0 \end{bmatrix}, \quad \ddot{q} = \begin{bmatrix} 0 \\ 0 \\ 0 \\ 0 \\ 0 \\ 0 \end{bmatrix}.$$

- Desired positions, velocities and accelerations.

$$q_d = \begin{bmatrix} 1.5 \\ 1.0 \\ 1.0 \\ 2.0 \\ 1.0 \\ 2.0 \end{bmatrix}, \quad \dot{q}_d = \begin{bmatrix} 0 \\ 0 \\ 0 \\ 0 \\ 0 \\ 0 \end{bmatrix}, \quad \ddot{q}_d = \begin{bmatrix} 0 \\ 0 \\ 0 \\ 0 \\ 0 \\ 0 \end{bmatrix}.$$

- Rotational angles

$$Kr_i = \begin{bmatrix} 0 & 0 & 0 & 0 \\ 0 & 1.1547 & 0 & 0 \\ 0 & 0 & 1.1547 & 0 \\ 0 & 0 & 0 & 0 \end{bmatrix}, \quad i = 1:6$$

- Gravity centers

$${}^i r_i = \begin{bmatrix} 1 & 0 & 0 & 1 \end{bmatrix}^T, \quad i = 1:6$$

- State weight matrix Q

$$Q = \begin{bmatrix} Q_{11} & Q_{12} \\ Q_{12}^T & Q_{22} \end{bmatrix}],$$

$$Q_{11} = \begin{bmatrix} 10 & 2 & 2 & 2 & 2 & 2 \\ 2 & 10 & 2 & 2 & 2 & 2 \\ 2 & 2 & 10 & 2 & 2 & 2 \\ 2 & 2 & 2 & 10 & 2 & 2 \\ 2 & 2 & 2 & 2 & 10 & 2 \\ 2 & 2 & 2 & 2 & 2 & 10 \end{bmatrix}],$$
$$Q_{12} = -I_6, \quad Q_{22} = 30I_6.$$

From the initial and desired values of positions, velocities and accelerations (q, q_d, $\dot{q}$, $\dot{q}_d$, $\ddot{q}$ and $\ddot{q}_d$), and the final time of simulation, the trajectory is defined by the following polynomial

$$qt_{(i)}(t) = a_{0(i)} + a_{1(i)}t + a_{2(i)}t^2 + a_{3(i)}t^3 + a_{4(i)}t^4 + a_{5(i)}t^5,$$

where coefficients can be obtained by solving the following system of equation

$$\begin{bmatrix} q_{(i)}^T \\ \dot{q}_{(i)}^T \\ \ddot{q}_{(i)}^T \\ q_{d(i)}^T \\ \dot{q}_{d(i)}^T \\ \ddot{q}_{d(i)}^T \end{bmatrix} = \begin{bmatrix} 1 & 0 & 0 & 0 & 0 & 0 \\ 0 & 1 & 0 & 0 & 0 & 0 \\ 0 & 0 & 2 & 0 & 0 & 0 \\ 0 & t_{f(i)} & t_{f(i)}^2 & t_{f(i)}^3 & t_{f(i)}^4 & t_{f(i)}^5 \\ 0 & 1 & 2t_{f(i)} & 3t_{f(i)}^2 & 4t_{f(i)}^3 & 5t_{f(i)}^4 \\ 0 & 0 & 2 & 6t_{f(i)} & 12t_{f(i)}^2 & 20t_{f(i)}^3 \end{bmatrix} \begin{bmatrix} a_{0(i)}^T \\ a_{1(i)}^T \\ a_{2(i)}^T \\ a_{3(i)}^T \\ a_{4(i)}^T \\ a_{5(i)}^T \end{bmatrix}$$

In this system, $a_{j(i)}$, $j = 0, \ldots, 5$ is an array of dimension $1 \times n$ whose components are the coefficients corresponding to each joint. The references and positions of the manipulator variables are showed in figures (1) and (2).

6. CONCLUSIONS

One of the most important problems to achive real-time control of a robot manipulator is to compute the dynamic equation. In this paper a high performance sequential implementation to compute the dynamic equation based on the Lagrange-Euler formulation is presented. Also, an algorithm for tracking trajectory using a Hamilton-Jacobi-Bellman optimization has been implemented. The results show that the most important computational cost is due to the dynamic equation.

The results obtained show the good performance of the optimized version (three times faster than the nonoptimized version) using an operation-level optimization (1).

Table (2) and (3) show that similar results are obtained using a low cost platform (Intel 440BX) or a MIPS R12000 processor.

Figures (1) and (2) show how the links track the established reference.

Our future work will include the analysis of the performance of the optimized sequential algorithm on parallel distributed and shared architectures.

7. REFERENCES

Balestrino, A. D. and A. S. L. Zinober (1984). Nonlinear adaptative model following control. *Automatica, 20*,(5), 559–568.

Berghuis, H., R. Ortega and H. Nijmeijer (1993). A robust adaptive controrller for robot manipulators. *IEEE Trans. Robotics and Automation.* **10**,(6), 825–830.

Craig, J. (1986). Adaptative control of mechanical manipulators. In: *Int. Conf. on Robotics and Automation (San Francisco).*

Fernández, Juan Carlos (1999). Simulación Dinámica y Control de Robots Industriales Utilizando Computación Paralela. PhD thesis. Univ. Politécnica de Valencia. Dept. Sistemas Informáticos y Computación, Valencia (Spain).

Fu, K. S., R.C. González and C. S. G. Lee (1988). *Robótica: Control, detección, visión e inteligencia artificial.* McGraw-Hill.

Good, M. C., L.M. Sweet and K. L. Strobel. (1985). Dynamic models for control system design of integrated robot and drive systems. *ASME Journal of Dynamics Systems, Measurement and Control* **107**,, 53–59.

Johansson, R. (1990). Quadratic optimization of motion coordination and control. *IEEE Trans. Automat. Control* **35**,(11), 1197–1208.

Kim, Young H., Frank L. Lewis and Darren M. Dawson (1999). Intelligent optimal control of robotic manipulators using neural networks. *Automatica*.

Kirk, D. (1970). *Optimal Control Theory: An Introduction*. Prentice-Hall. Englewood Cliffs, NJ.

Lewis, F. L. and V.L. Syrmos (1995). *Optimal Control*. Addison-Wesley. New York.

Lewis, F. L., C.T. Abdallah and D.M. Dawson (1993). *Control of Robot Manipulators*. Macmillan. New York.

Slotine, J. J. and W. Li (1984). *Applied Nonlinear Control*. Prentice-Hall.

Slotine, J. J. and W. Li (1987). On adaptative control of robot manipulators. *Internat. J. Robotics Research*. **6**,(3), 49–59.

Tarn, T., A.K. Bejczy, A. Isidori and Y.L. Chen (1984). Nonlinear feedback in robot arm control. *Proc. of the 23rd IEEE Conf. on Decision and Control*.

Verdier, M., M. Rouff and J.G. Fontaine. (1989). Nonlinear control robot: A phenomenological approach to linearization by static feedback. *Robotica* **7**, 315–321.

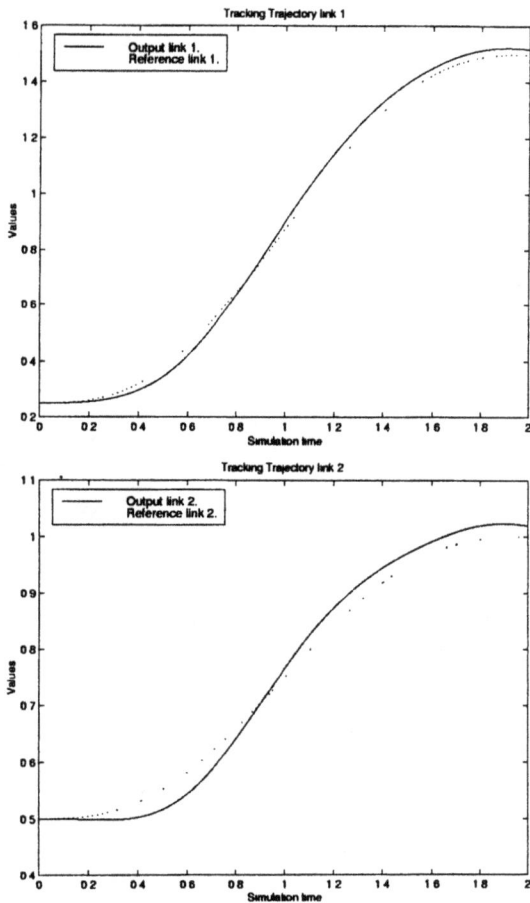

Fig. 1. Output and reference for joints 1 to 2

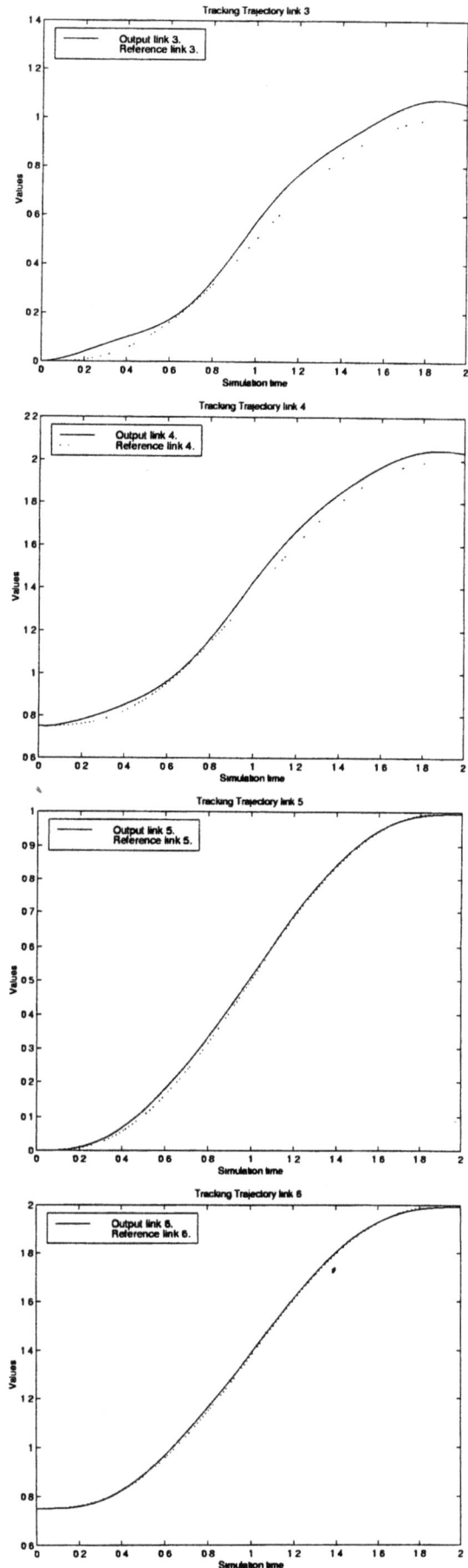

Fig. 2. Output and reference for joints 3 to 6

141

ADAPTIVE QoS MANAGEMENT SYSTEM FOR AUTONOMOUS VEHICLES [1]

Houcine Hassan, Alfons Crespo, José Simó

*Dept. Computer Engineering, Universidad Politécnica de Valencia, Camino de Vera
s/n - 46071 Valencia (SPAIN). e-mail:{husein, acrespo, jsimo}@disca.upv.es*

Abstract: Many researchers in mobile robotic systems have stated that the robotic architecture
must explicitly recognise multiple performance criteria and need to use various empirical
metrics to evaluate the utility of the architecture for the tasks accomplished by the vehicle.
The most recent hybrid architectures proposed in the field of complex mobile robotic
architectures have shown the necessity of internal monitoring systems that dynamically
measure vehicle quality of service (QoS) and adjusting it for the best vehicle fit. The paper
presents an adaptive system for managing the vehicle QoS. The proposed system allows
guaranteeing the requirements of the vehicle performance parameters and improving them for
a better vehicle QoS results. *Copyright© 2000 IFAC*

Keywords: mobile robots, robot navigation, real-time systems, adaptive systems, adaptive
algorithms

1 INTRODUCTION

Many researchers in mobile robotic systems (Arkin 98, Michaud 99, Hexmoor 99, Kortenkamp 98, Gat 98) have stated that the robotic architecture must explicitly recognise multiple performance criteria and need to use various empirical metrics to evaluate the utility of the architecture for the tasks accomplished by the vehicle. Researchers in the field of mobile robotic architecture are looking for internal monitoring systems that dynamically measure vehicle quality of service (QoS) and adjusting it for the best vehicle fit

Besides guaranteeing the functional and temporal requirements of the vehicle, the real-time robotic architecture developed (Hassan 1999) introduces enough flexibility and adaptation in the mobile system to guarantee and improve the quality of service. The proposed architecture allows adapting the vehicle system to the environmental conditions. This focuses on the ability of the system to select the appropriate activity to be executed (depending on the available time) and to change its behaviour (depending on the environment information).

In the real-time community it could be found a variety of QoS architectures, most of them are specifically for multimedia and telecommunications applications (Liu 1997, Shankar 1999) that are not adequate for mobile robotic systems. Indeed, in robotic systems the premises

are quite different, as it is not enough with determining the QoS of the individual components of the architecture. Vehicle systems are always submitted to unpredictable and dynamic environments, which influence the system load. Sometimes there is a need of processing high amount of information due to the environment (i.e. high quantity of obstacles). In such conditions it is necessary to change the vehicle dynamics for accomplishing the QoS requirements. Furthermore, the QoS architectures studied focus the adaptation only on the levels of quality of the flexible tasks (imprecise computation). However in robotics systems any task, either reactive (critical) and deliberative (imprecise) tasks could have influence on the global QoS of the vehicle.

In this paper the adaptive QoS management system (AQoSS) integrated on a flexible planner of a real-time hybrid architecture is presented. The real-time architecture assures the functional and temporal requirements of the robot by means of a flexible real-time system. The AQoSS guarantee the vehicle performance constraints and adapts the system to improve the global vehicle quality of service.

The paper is organised as follows, after the introduction to the main aspects that motivated the research, section 2 describes the real-time task model and the system scheduling supporting the execution of the vehicle plans. One of the main components of the flexible planner, the adaptive QoS management system, is detailed in section

---

[1] This work has been partially funded by the Spanish Government grant
CICYT TAP99-1226-C02

3. An application example illustrating the effect of the AQoSS on improving the quality performance is presented in section 4 The QoS feasibility analysis and simulations are shown in section 5. Section 6 presents the evaluation of the proposed AQoSS architecture. Finally, the conclusions and the future work are outlined in section 7.

2 REAL-TIME HYBRID ARCHITECTURE

The hybrid mobile robotic architecture proposed consists of a set of distributed behaviour entities communicating with a centralised arbiter. The behaviour entities define the task-specific knowledge for the domain. Each behaviour entity runs completely independently and asynchronously, using sensor data and processing algorithms. They provide commands to the arbiter, each at their own rate and according to their own time constraints. The kind of information represented by the commands relates to speed, trajectory, turn, etc. Each behaviour also has a motivation parameter reflecting its importance in performing the control action. Currently, a flexible planner module is responsible for configuring the motivations, depending on the vehicle state and the environment. The arbiter is then combines the behaviour commands, taking into account their motivation weight, in order to generate the actions that cope with the vehicle guidance. The different components of the architecture are: *Behaviours, Fusion algorithms, Arbiters, Sensors and the Intelligent planner.* For further detail of the behaviour architecture report to (Buendia 1999, Hassan 1999).

The proposed real-time architecture has to cope with the guarantee of the functional and temporal requirements of the hybrid mobile robotic model. But it also introduces a high degree of flexibility and adaptation to the mobile system by integrating in the flexible planner a QoS management system for assuring and enhancing the quality performances required by the autonomous vehicle. The task model is organised in order to cope with the execution of the reactive and deliberative components of the hybrid architecture.

2.1 Task model

The architecture is organised in two planning levels. The critical level gives support to the reactive behaviours, and the optional level supports the deliberative behaviours. The tasks of different natures that coexist in the architecture are:

- Fixed periodic tasks: represent the reactive behaviours of the model.

- Variable periodic tasks: implement reactive behaviours that depend on some application parameter. Such as, the *obstacle_avoidance* behaviour which in turn depends on the vehicle speed.

- Optional tasks: support the execution of deliberative

tasks. These are soft aperiodic tasks.

The fixed periodic task set models the periodic activities of the system. These tasks are contain the following parameters:

$$T_i^f = (C_i, T_i, P_i, \phi_i)$$

where Ci is the worst case execution time, T_i is the period, P_i is the priority, and ϕ_i is the task offset.

The variable periodic task subset is characterised by:

$$T_i^v = (C_i(k), T_i(k), P_i, \phi_i)$$

In this case, the parameters *Ci* and *Ti* depend on the parameter *k*. And the priority is, consequently, reconfigured dynamically.

The variable parameter *k* can, in a robotic system, be the robot speed. If the robot speed is slow, the period of these tasks can be longer, and so reducing the *CPU* load and allowing more optional computation. If the robot speed increases, the period should be reduced to guarantee the response time. These tasks have some specific features such as the variable computation time. If the number of objects in the environment is low, the computation time is shorter than when there are more objects to be recognised. Both aspects are complementary. When the number of objects detected in the environment is large, the computational requirements increase and, consequently, the robot speed must be reduced and the periods have to be increased. On the other hand, if the number of objects decreases, then the computational requirements decrease, so the robot can increase speed and reduce the task periods. The goal of the system is to adjust the system load to achieve the maximum robot speed and process the maximum optional load.

Moreover, the tasks are split in two parts: initial and final. The initial part computes the answer with sufficient quality. The final part sends the final answer to the actuator. However, the answer can be improved during the optional load – thus obtaining a higher quality system response. To allow this, all periodic tasks can be structured into two tasks with the following features: the period and priority are the same for both parts; the computation time corresponds to the computation of each part; and the offset of the final part is delayed to allow for refinement of the answer.

The optional tasks considered in the system are of multiple versions outlined in the imprecise computation model (Liu 91, Audsley 91). For the multiple version task set, each task To_i is composed of different deepening levels. The first level generates a minimum quality response of the task. Increasing the level, the quality of the response is also increased. This task set is characterised by:

$$T_i^o = (LL_{i,j}, \mathrm{Im}_i).$$

Where $L_{i,j}$ is the estimated computation time of the jth level of the ith task and Im_i is the importance of the task.

The group of tasks and activities of the system are basically handled by two types of schedulers that are structured in two levels. The first level guarantees the execution of the reactive tasks while the second level assures the planning of the operations regarding the predetermined quality response. The scheduled tasks of the critical level are periodic tasks. The priority assignation is based on deadline monotonic scheduling, and the schedulability analysis applied are those outlined for fixed priority pre-emptive systems (Lehoczky, 90). In the second level, the optional tasks are executed in slack time by the slack server (Lehoczky 92, Davis 93).

3 ADAPTIVE QOS MANAGEMENT SYSTEM

The adaptive QoS administration system for autonomous vehicles is designed to provide a platform that allows to analyse, to guarantee and to improve the global vehicle QoS. The AQoSS has to assure the accomplishment of the performances required by the vehicle mission planner. The other AQoSS main objective is to enhance the parameter performances in order to increase the vehicle QoS.

The proposed AQoSS is composed by the following modules:
- Performance supervision system (PSS)
- Execution monitoring system (EMS)
- Decision making system (DMS)

The Figure 1 shows the architecture of the adaptive QoS management system. The PSS look after the execution of the performances required in the current mission by sending orders, that must be taken by the DMS to satisfy it (i.e. increase map level). On the other hand the EMS guarantees the execution of high level planning tasks required by the mission planner, by emitting orders to the DMS.

The DMS analyzes the impact of the orders received in the execution of the mission, based on the motivations and on the heuristic strategies and then taking the opportune decisions that better results provide to the mission.

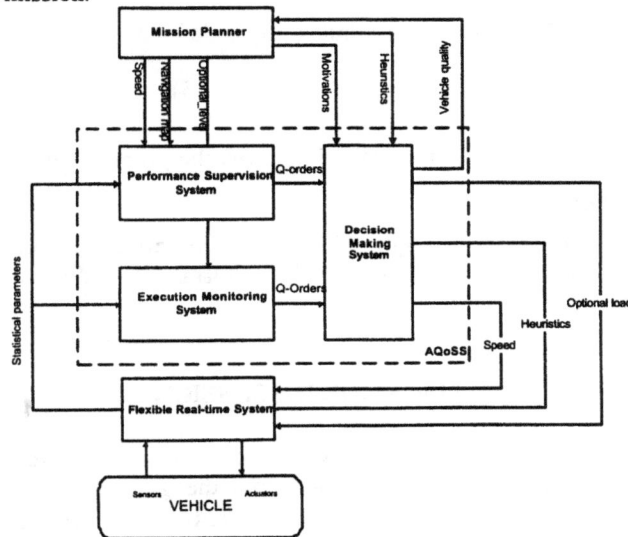

Fig. 1. Architecture of the vehicle adaptive QoS management system

3.1 QoS parametrisation

The AQoSS uses two indexes. One is the guarantee index $\delta_S^C(P)$ for assuring the parameter performance requirements. The second is the quality index for enhancing the parameter performance qualities.
The guarantee index is defined in the equation (1)

$$\delta_S^C(P) = \left(S, P, v_p^\delta, m_c\right) \qquad (1)$$

- P Performance parameter.
- S is the system that regulates the guarantee of the parameter P.
- v_p^δ it is the function value of the parameter P. It is defined as the ratio between the executed value of the parameter and the required value of this parameter
- m_c is the motivation of the behavior c that controls the parameter P. It determines the degree of importance of this parameter in the current vehicle plan.

The AQoSS achieves the performance requirements of a certain parameter P whenever it is accomplished that $\delta_S^C(P) = 1$. In any other case, the system will unfulfill the requirements of the mission planner.
The quality index parameter is characterised in the equation (2).

$$q_S^C(P) = \left(S, P, v_p^q, m_c\right) \qquad (2)$$

All the parameters are the same as in the equation (1) except the function value v_p^q. v_p^q is the ratio between the executed value of the parameter P and its maximum value.
The AQoSS provides higher quality results of a certain parameter P, as near, as $q_S^C(P)$ is to 1 and vice-versa.

3.2 QoS management subsystems

Before detailing the AQoS system it is necessary to characterize the parameter requirements of the mission planner (MP). The MP analyses the environment and the objectives that have to fulfill. Then, it decides which are the parameter values that have to be guaranteed. Thereafter, the mission with the required parameters is sent to the AQoSS. The mission is defined as:

$$Mission_k = \left(L_r, L_{d1}, v, m_b, H_j\right) \qquad (3)$$

L_r is the reactive load, L_d is the deliberative load, v is the speed, m_b is the behaviour motivation, H_j is the heuristic strategy.

The behaviour motivation is assigned to each behaviour by the MP system. The motivation value is the relevance that a given behaviour has in the mission. The MP assigns the motivation value based on the environment and on its objectives. (i.e. when the environment is overloaded, the obstacle avoidance behaviour will have the maximum motivation value).

On the other hand, the flexible real-time system has a second scheduling level where the deliberative load is executed based on some heuristic criteria. (i.e. When the vehicle is on a free obstacle area, a balance heuristic is useful. This heuristic consists on balancing the system resources between the local maps and the global maps.)

Performance supervision system (PSS): This system is responsible of the guarantee and of the improvement of the speed references, indicated by the mission planner. As well as it executes the levels of the sonar maps required for the navigation. As shown in the Figure 1, the PSS receives as input the speed reference and the minimum level of the navigation maps.

On the other hand, it knows if it has enough CPU time for the execution of the map levels since the real time kernel informs it. The PSS receives speed petitions V_{req}, and it is aware of the executed speed V_{exec} and of the maximum speed of the vehicle V_{max}. Consequently it knows if the speed performance requirements are fulfilled: $\delta_{PSS}^{MG}(V) = \frac{V_{exec}}{V_{req}}$, as well as if the speed quality could be improved $q_{PSS}^{MG}(V) = \frac{V_{exec}}{V_{max}}$. The speed commands that generates this module to the DMS module are composed by the $\delta_{PSS}^{MG}(V)$, $q_{PSS}^{MG}(V)$, and the request of increase, decrease, or maintain the parameter value.

The procedure followed by the PSS for accomplishing the performance requirements of the navigation maps is similar to the speed. The map commands are composed by $\delta_{PSS}^{OA}(M)$, $q_{PSS}^{OA}(M)$ and the map level requested.

Execution monitoring system (EMS): The mission planner indicates which are its forecasts of time that will dedicate to planning tasks (i.e path planning). This information is transmitted to the monitoring system, by means of the *optional_level* parameter, for its guarantee and possibly improvement. The EMS knows how long it can be dedicated to this type of tasks, thanks to the information of the statistical parameters provided by the real time kernel. The EMS is able to know which is the degree of fulfillment of the time of planning that is requested by the mission planner by means of the formula $\delta_{EMS}^{PP}(NO) = \frac{NO_{exec}}{NO_{req}}$.

On the other hand, the quality that measures the planning level is defined as the ratio among the optional time executed and the available time in the system $q_{EMS}^{PP}(NO) = \frac{NO_{exec}}{NO_{available}}$.

While the planning performance requirements are not guaranteed the performance quality is not analyzed. In order to insure the planning guarantee index the EMS requests orders to the DMS. Once insured the guarantee index $\delta_{SME}^{PC}(NO)$ it could be improved the quality index

$q_{SME}^{PC}(NO)$. Again the commands sent to the DMS includes guarantee index, quality index, and the selected action (increase, decrease or maintain the planning level).

Decision Making system (DMS): The DMS decides what parameters of the system it is necessary to modify for the fulfillment of the requirements of the mission planner. It receives from the two previous systems, the PSS and the EMS the commands that each one proposes for the guarantee and the improvement of the performance parameters that they supervise.

The DMS analyzes the orders and taking the decisions in consequence (i.e. to increase speed, to reduce planning level). The analysis phase of the orders is carried out based on the behavior motivations involved in the plan. Taking into account the behavior motivations associated to the parameter, the DMS decides what parameters to modify. The behaviors with more repercussion in the plan will have more motivation value and it will have the biggest influence in the adopted decision. When the performance parameter of the more important behaviour is satisfied, the DMS selects the next more important parameter's behaviour.

On the other hand, the DMS has some heuristic strategies that are based on the motivations that distribute differently the load (i.e. it balanced load, to prioritize the most important process).

The parameter that informs about the degree of performance requirement fulfillment is the global guarantee index parameter and is defined as

$$\Omega_v = \sum_{k=1}^{k=n} m_k \, \delta_S^K(P) \qquad (4)$$

The degradation parameter shows the degree of non-fulfillment of the requirements requested by the mission planner and is $D = 1 - \Omega_v$. It should be necessary to guarantee the global index, Ω_v before improving the quality. The global quality index Q_v is defined as:

$$Q_v = \sum_{k=1}^{k=n} m_k \, q_S^K(P) \qquad (5)$$

Once insured Ω_v, the DMS improves the global quality of the vehicle Q_v.

An application example illustrating the effect of the AQoSS on enhancing the vehicle performance quality is presented in section 4.

4 APPLICATION DISCUSSION

The application will show with more clarity how the AQoS management system performs the guarantee and improvement of the parameter qualities. A typical scenario where the vehicle has to move is shown in the

Fig. **2**. The squares represents the obstacles and the thick line is the trajectory that the vehicle describes from the initial position I to the final goal O. This trajectory is generated in real-time by the co-operation of the *move_to_goal* and the *obstacle_avoidance* behaviours.

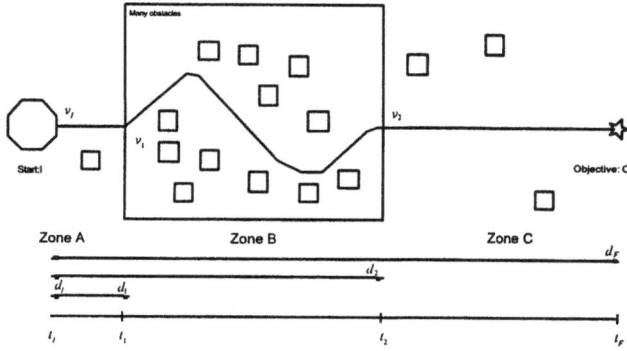

Fig. 2. Application example

Two are the performance parameter requirements of the vehicle that the QoS management system has to fulfil for the optimal quality vehicle guidance. First, the vehicle has to reach its objective at time t_F and not later. Second, the level of proximity to the obstacles has to be respected. The latter requirement is dependent on the map quality parameters (Object recognition rate and Degree quality of the fusion algorithms). The former objective will depend on the vehicle speed.

The vehicle starts the execution of the path at instant t_I. It could be coming from completing another objective. As it doesn't detect in its surroundings any obstacle and for attaining the position of the objective the mission planner decides that the speed v will be $v_I = \dfrac{d_F - d_I}{t_F - t_I}$.

Arriving to t_1 the *obstacle_avoidance* behaviour detects the presence of many obstacles. For processing the environment correctly the system has not enough time. This means that the mission planner detects that at the current speed the requirements of object recognition will fail. To prevent this situation, the mission planner will find the correct speed allowing adequate object recognition. Let α be the increment factor time due to object processing time needs. Hence, the task periods would decrease in order to allow the execution of the new maps. The new task periods becomes $T_{i,1} = \alpha \cdot T_{i,0}$, what implies that the vehicle speed decreases and will be $v_1 = \dfrac{1}{\alpha} \cdot v$. The AQoSS has to guarantee this speed and try to improve it, while guaranteeing the new map level needs.

Exiting the central full object zone, the environment is less loaded. It could maintain the current speed, if the vehicle had not any arriving time constraint t_F. Once more, the mission planner has to check if at the current speed the response time required t_F is fulfilled. Consequently, the minimum required speed is calculated as: $v_2 = \dfrac{d_F - d_2}{t_F - t_2}$. While t_2 is obtained as: $t_2 = t_1 + \dfrac{\sum\limits_{i}^{n} s_i}{v_1}$,

being s_i the number of segments in the central region. If $v_1 < v_2$ then the mission planner asks the AQoSS to reconfigure the real-time system to the speed v_2. Otherwise the v_1 speed is maintained. In both cases the AQoSS will try to improve the speed parameter.

For the correct vehicle guidance, the next processes are needed:

- *Obstacle_avoidance* behaviour: based on the sonar maps, decides the control actions (turns) for guiding the vehicle by avoiding the obstacles in its path.
- *Move_to_goal* behaviour: Move towards the direction of the current objective by generating turns control action.
- *Odometer* process: gets the estimated position of the vehicle based on the encoder.
- *Local_min_map* process (ultrasounds): builds a minimum environment map that allows the object detection.
- *Turn_arbiter*: controls the steering wheels of the vehicle. When more than one task generates a turn command, the action with greatest motivation is applied. In the example, the *Obstacle_avoidance* and the *Move_to_goal* behaviour compete for the steering wheels control.
- *Speed_arbiter*: controls the vehicle speed. The action of the task with greatest motivation is applied to the controller.

The reactive processes of the application are shown in Table 1. These processes are dependent on the vehicle speed. Different speeds have been considered: $[0, 4m/s]$.

For all these speeds the reactive load is schedulable. Hence, the planner could switch, in run-time, from one speed to another without sacrificing the schedulability. The relationship between speed and periods is: $T = k \cdot \dfrac{1}{v}$ where k is a constant.

The attributes appearing in Table 1 correspond to the speed $v_0 = 2m/s$. It is simple to calculate the attributes for the other speeds, taking into account the last equation.

Table 1. Reactive attributes of the system.

Reactive Task	$Cm_{i,0}$	$Cf_{i,0}$	$D_{i,0}$	$T_{i,0}$
Behaviours				
Obstacle_avoid	2	1	200	200
Move_to_goal	2	1	225	225
Fusion				
Odomtric_process	1	1	50	50
Local_min_map	3	1	100	100
Arbiters				
Turn_arbiter	1	1	250	250
speed_arbiter	1	1	275	275

The variables of a reactive task i for the initial vehicle speed v_0 are: $Cm_{i,0}$ is the mandatory computation time, $Cf_{i,0}$ is the action computation time, $D_{i,0}$ is the deadline and $T_{i,0}$ is the period of task.

The Table 2 represents the attributes of the flexible tasks: *Local_Map* and *Global_Map*.

The *Local_Map* task depends on the number of samples per revolution performed by the sonar system. Different quality maps have been obtained depending on these samples (Benet 1998). The computational cost of the map is proportional to the number of samples. For the *Global_Map* process, different quality levels have been considered depending on the fusion technique used (cummulative value, dead-reckonning, etc.) (Duckett, 1997). Each level identifies the degree of confidence obtained from fusing the odometer and the infrared data with the sonar map.

Table 2. Optional attributes of the unbounded tasks

Deliberative Task	Quality versions	$L_{i,j}$	$Im_{i,0}$
Local_Map	LM₁ (10 samples)	30	max
	LM₂ (20 samples)	60	
	LM₃ (80 samples)	105	
	LM₄ (120 samples)	133	
	LM₅ (180 samples)	342	
Global_Map	GM₁	32	min
	GM₂	58	
	GM₃	90	
	GM₄	265	
	GM₅	375	

The detailed analysis performed by the QoS management system is shown in the section 5.

5 QOS FEASIBILITY ANALYSIS AND SIMULATIONS

The interest is focused on the analysis of the adaptation capacity of the AQoSS behind the performance requirements of the mission planner (MP). Furthermore, it is examined the AQoSS spread of improving the quality response of the performance parameters required by the MP. For the example of the Figure 2, since the vehicle has to arrive in time to the objective and has to avoid the obstacles in its trajectory, the performance metrics that the vehicle requires to be fulfilled are the speed and the navigation maps. The analysis is focused on these two parameters.

The Figure 3 show a simulation of how is adapted the system to the requirements of the mission planner. It can be seen that initially, the vehicle speed is 2m/s. In 100 ms it is changed to a 4m/s by the AQoSS. And at t=1250ms the speed changes once again to 1m/s.

The levels of the maps that could be executed are dependent on the vehicle speed. The best maps are needed after the instant 1250 ms, when the vehicle is entering the most obstacle-loaded zone. Indeed, it can be seen in the simulation, that the best maps are executed after t=1250ms.

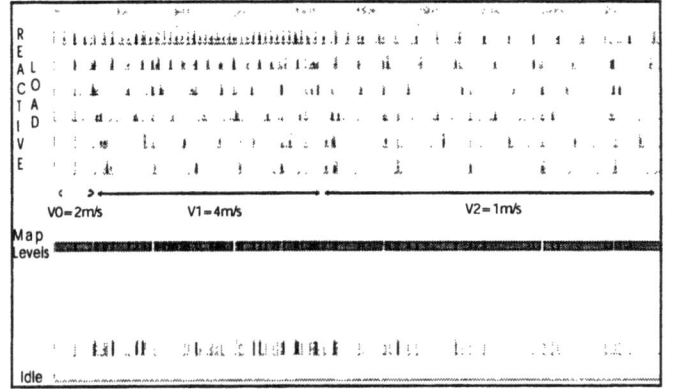

Fig. 3. Simulation of the application

The planner analyses the environment ant its objectives and then it transmits the required parameter values to be guaranteed to the AQoSS. The AQoSS analyses the system, assures the parameter performances required and try to improve them, as the system resources become available.

At t=0:

Mission Planner

Environment analysis few obstacles →Map level 1
Objective analysis reach objective at t=15s→v=2m/s

$$Mission_0 = (L_r, M = M_1, v = 2m/s, m_{MTG} = 0.7, m_{OA} = 0.3)$$

AQoSS

PSS
Guarantee

$$\delta_{SSP}^{OA}(M) = \frac{62}{62} = 1 \wedge \delta_{SSP}^{MTG}(V) = \frac{2}{2} = 1$$

→ Parameters guaranteed

PSS
Improvement

$$q_{SSP}^{OA}(M) = \frac{62}{398} = 0.16 \wedge q_{SSP}^{MTG}(V) = \frac{2}{4} = 0.5$$

→ Quality commands to DMS
(V): increase the speed.
(M) increase the map level

DMS
Analysis

The speed (0.7) is more motivated than the map (0.3)

Decision

increase the speed $q_{SSP}^{MTG}(V) = \frac{4}{4} = 1$

Increase the map $q_{SSP}^{OA}(M) = \frac{118}{398} = 0.3$

At t=1250 ms

Mission Planner

Environment analysis many obstacles →Map level 3
Objective analysis cross with maximum security→v=1m/s

$$Mission_1 = (L_r, M = M_3, v = 1m/s, m_{MTG} = 0.2, m_{OA} = 0.8)$$

AQoSS

PSS
Guarantee

$$\delta_{SSP}^{MTG}(V)=\frac{4}{1}\neq 1;\ \delta_{SSP}^{MTG}(V)=\frac{3}{1}\neq 1;$$

$$\delta_{SSP}^{MTG}(V)=\frac{2}{1}\neq 1;\ \delta_{SSP}^{MTG}(V)=\frac{1}{1}=1$$

$$\delta_{SSP}^{OA}(M)=\frac{195}{195}=1\rightarrow$$

Parameters guaranteed in 100ms

PSS
Improvement

$$q_{SSP}^{OA}(M)=\frac{195}{398}=0.5\wedge q_{SSP}^{MTG}(V)=\frac{1}{4}=0.25$$

$\rightarrow$ Quality commands to DMS:
(*M*) increase the map level &
slow speed.
(*V*) Increase the speed

DMS
Analysis

The map (0.8) is more motivated
than the speed (0.2)

Decision

Increase the map level
$$q_{SSP}^{OA}(M)=\frac{398}{398}=1$$
increase the speed to 1.2m/s
$$q_{SSP}^{MTG}(V)=\frac{1.2}{4}=0.3$$

At t=12000 ms:

Mission Planner

Environment analysis   no obstacles $\rightarrow$Map level 1
Objective analysis   reach objective soon$\rightarrow$v=3m/s

$$Mission_2 =(L_r, M=M_1, v=3m/s, m_{MTG}=0.8, m_{OA}=0.2)$$

AQoSS

PSS
Guarantee

$$\delta_{SSP}^{MTG}(V)=\frac{1.2}{3}\neq 1;\ \delta_{SSP}^{MTG}(V)=\frac{2}{3}\neq 1;$$

$$\delta_{SSP}^{MTG}(V)=\frac{3}{3}=1$$

$$\delta_{SSP}^{OA}(M)=\frac{62}{62}=1\rightarrow$$

Parameters guaranteed in 50ms

PSS
Improvement

$$q_{SSP}^{OA}(M)=\frac{62}{398}=0.16\wedge q_{SSP}^{MTG}(V)=\frac{3}{4}=0.75$$

$\rightarrow$Quality commands to DMS:
(V) Increase the speed
(M) increase the map level

DMS
Analysis

The speed (0.8) is more
motivated than the map (0.2)

Decision

increase the speed $q_{SSP}^{MTG}(V)=\frac{4}{4}=1$

increase the map level

$$q^{OA}{}_{SSP}(M)=\frac{118}{...}=0.3$$

The two objectives of the vehicle have been
accomplished correctly and their quality results have been
improved as possible without the intervention of the
mission planner.

6 EVALUATION RESULTS

After analyzing the protocol allowing the guarantee and
the improvement of the performances, it will be evaluated
the influence of the adaptive QoS system on the
enhancement degree reached by the two concerned
parameters: the speed and navigation map.

The Figure 4 shows the results of the AQoSS effect on
the guarantee and on the improvement of the speed
performance metric.

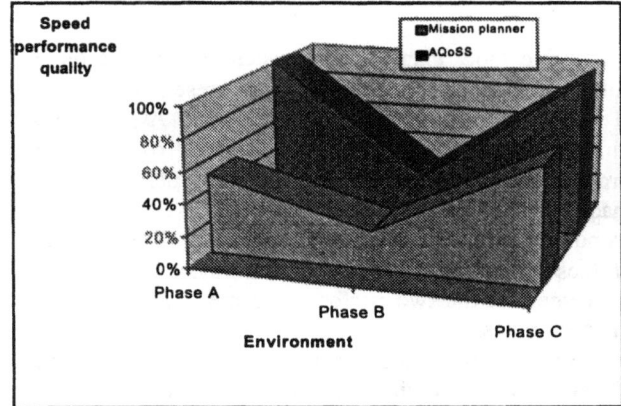

Fig. 4. AQoSS effect on improving the speed quality

In the first action of the AQoSS the speed has been
improved substantially from 50% to its maximum
capacity because the environment is few loaded. When,
in the central region, more obstacles appear, slower is the
speed. However the AQoSS increase the quality of the
MP speed constraint: 25% to 30%. At the last phase C,
there are less obstacles, hence the MP speed is improved
by the AQoSS from 75% to the maximum speed.

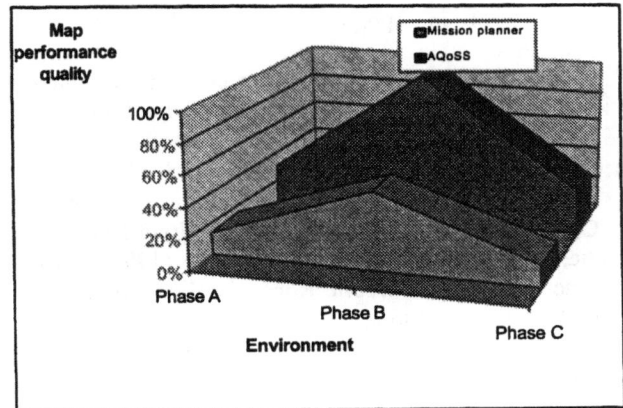

Fig. 5. AQoSS effect on improving the Map quality

In the Figure 5 are compared the guaranteed MP performance parameter requirements of the navigation maps and how are they improved by the AQoSS. With few obstacles on the environment, the MP map requirements are poorer. Once guaranteed the AQoSS decide to ameliorate the quality map in the double:15% to 30%. Recall that the vehicle is navigating at its maximum speed. When the map requirement becomes more important, on the phase B, the AQoSS increase the quality of the map parameter significantly from 50% to 100%. At the end of the travel the MP has a little need of map levels, so as it requires a minimum quality: 16%. However the AQoSS as it has enough resources and no other priorities it invest them on the map quality refinement:30%.

7 CONCLUSIONS

An adaptive QoS management system integrated on a flexible real-time hybrid architecture is presented.
The different subsystems composing the AQoSS are detailed. A special emphasis on the study of the QoS guarantee and enhancement of the vehicle performances is made.
A robotic application example served to illustrate the usefulness of the adaptive QoS system on the enhancement of the two performance metrics: speed and navigation maps.
Future work will pursue the following objectives: The analysis of the heuristic strategy impact on the quality performances. To elaborate more sophisticated rules on the decision making system based on the motivations.
Finally, from the implementation point of view, the system must be embedded under a real-time operating system (RT-Linux).

REFERENCES

Arkin, R. C. (1998). "Behaviour based Robotics" Cambridge MIT Press 1998.

Audsley, N.C., Burns, A., Richardson M.F. and Wellings, A.J. (1991). Incorporating Unbounded Algorithms into Predictable Real-Time Systems. Real-Time Systems Research Group. Department of Computer Science. University of York, UK. Report number RTRG/91/102.

Benet, G. Blanes, F. Martínez, M. Simó, J. (1998). A Multisensor Robot Distributed Architecture. IFAC Conference INCOM'98. Metz-Nancy. June 1998.

Buendia, F. Hassan, H. Simó, J. Crespo, A. (1999) Real-time Systems for Mobile Robot Applications based on Behavioural Model. 23 IFAC/IFIP Workshop in Real-time Programming. Saarland Germany. May 1999.

Davis, R.I., Tindell, K.W., and Burns, A. (1993). Scheduling Slack Time in Fixed Priority Pre-emptive Systems. Proc. Real-Time Systems Symposium, Raleigh-Durham, North Carolina, December 1-3, 222-231

Duckett T. Nehmzow U. (1997). Quantitative analysis of mobile robot localisation systems. British conference on autonomous mobile robotic systems. Manchester 1997.

Gat E. et al.(1998) "Three Layer Architectures" Artificial Intelligence and Mobile Robots. The MIT Press. 1998

Hassan, H. Buendia, F. Simó, J. Crespo, A, Benet, G.(1999) "A Real-time Model of Flexible Tasks for Mobile Robotic Applications based on a Behavioural Architecture." 7th. IEEE International Conference on Emerging Technology in Factory Automation. Barcelona'99.

H. Hexmoor, M. Lafary, M. Trosen, 1999. "Towards Empirical Evaluation of Agent Architecture Qualities", In Agent Theoresi, Architectures, and Languages (ATAL-99), Orlando, Florida.

Kortenkamp D. Et al.(1998) "Three NASA Application Domains for Integrated Planning, Scheduling and Execution.". IEEE Symposium on Intelligence in Automation and Robotics. 1998

Lehoczky J. (1990). Fixed Priority Scheduling for Periodic Tasks Sets with Arbitrary Deadlines. Proc. Of the 11th Real-Time Systems Symposium. December, 201 -- 209, IEEE Computer Society Press.

Lehoczky J., and Ramos-Thuel S. (1992). An Optimal Algorithm for Scheduling Soft-Aperiodic Tasks in Fixed-Priority Preemptive Systems. Proc. Real-Time Systems Symposium, Phoenix, Arizona, December 2-4, 110 -- 123, IEEE Computer Society Press.

Liu, J.W.S., Lin, K.J.L, Shih, W.K., Yu, A.C., Chung, J.Y., and Zhao, W. (1991) Algorithms for Scheduling Imprecise Computations. Computer IEEE May 1991.

Liu, J.W.S., Nahrstedt, K. Hull, D., Chen, S., Baochum, Li..(1997) EPIQ QoS Characterization, Internal Report, Dept. Computer Science, Univ. of Illinois, Urbana-Champaign.

Michaud F. (1999) "Managing Robot Autonomy and Interactivity Using Motives and Visual Communication" International Conference on Autonomous Agents. Seattle 1999.

Shankar M, De miguel M, Liu J. (1999) An End-to-End QoS Management Architecture, RTAS'99.

COMMUNICATIONS STRUCTURE FOR SENSOR FUSION IN DISTRIBUTED REAL TIME SYSTEMS

J.L.Posadas, P.Pérez, J.E.Simó, G.Benet, F.Blanes

Departamento de Informática de Sistemas y Computadores (D.I.S.C.A.)
Universidad Politécnica de Valencia, Spain
{ jposadas,pperez,jsimo,gbenet,pblanes}@disca.upv.es

Abstract: This paper describes a communication system called SC (*Sistema de Comunicaciones*) that is suitable for real-time systems with distributed sensory architecture. This system has been implemented in the YAIR robot, an autonomous robot with intelligent sensors that produces different measurements about the environment and its position within it. To guarantee good response times, the sensory modules are connected using the CAN bus. The robot's main controller executes a control algorithm that can be decomposed into smaller parts using a communications server. Thus, the execution can be shared between different processors connected through an Ethernet network. *Copyright © 2000 IFAC.*

Keywords: Mobile robots, Distributed computer control systems, Intelligent instrumentation, Sensor fusion, Fieldbus.

1. INTRODUCTION

YAIR[1] (Blanes, et al., 1998; Gil, 1997; Simó, et al., 1997) is an autonomous robot that was built for the experimental study of reactive systems, sensor fusion and distributed computing. This paper describes a communication system suitable for hybrid systems. These systems have a reactive level based on computing and communication under real-time constraints. A deliberative level without real-time constraints also exists, but a good mean response time must still be guaranteed.

The communications system presented includes two communication models: one model is vertical and based on the CAN bus – a fieldbus that enables real-time features; the second model is hybrid-horizontal and supported by a distributed blackboard system (SC) (Posadas, et al., 1997). The SC software enables the main robot controller (Windows NT based) to communicate transparently through different channels: CAN, ethernet, DDE, RS232, and so on. The coupling between these two models is possible

using an application interface. The SC behaviour has been verified executing an application (a probabilistic data fusion algorithm) that uses the space/time tagged sensory information broadcast from different modules to decide the optimal trajectory of the robot, and avoid obstacles found during its walk.

The goals of the YAIR development are the following (fig. 1):

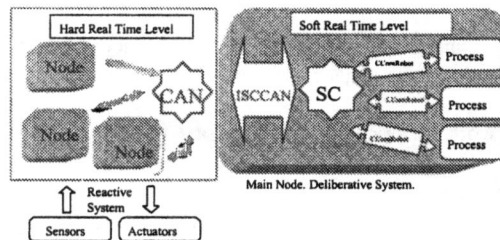

Fig. 1: Communication System Structure

❑ Development of intelligent, microprocessor-based, sensor and actuator modules. The sensors developed must supply the robot with diverse information about the robot's environment.
❑ Integration of the whole robot system using CAN as the main communication bus.

---

[1] YAIR stands for Yet Another Intelligent Robot, and is currently being developed under CICYT grant TAP98-0333-C03-02 from the Spanish Government

- Development of a distributed objects structure over the bus CAN for sensory data access.
- Construction of a reactive system over the previous distributed objects architecture. This reactive system will allow the robot to avoid obstacles without high-level planning or local environment maps.
- Addition to the system of a deliberative level and sensor fusion level, by adding a Pentium processor that can access the distributed objects (sensor readings) by accessing the CAN bus.
- Provision of the necessary software support so that any Windows application (local or remote) can access the communication system with all the necessary communications hidden (including CAN).
- Development of an application that uses the SC to accomplish several functions: sensor reading requests, sensor fusion and local map representation, objective specification (co-ordinates to be reached by the robot), high-level task planning, and so on.

In the following sections, the developed work and the results obtained together with the response times are described. Firstly, the global architecture of YAIR and its main sensors is described. Secondly, the CAN level of distributed objects (sensor readings) is presented. Thirdly, the implemented SC and its real-time features are also described. Finally, the results of the tests carried out are presented, together with the conclusions.

2. YAIR ROBOT ARCHITECTURE

The backbone of YAIR is the CAN bus (Bosch, 1991), a fieldbus initially developed for the automotive industry that is actually being used in numerous technological areas, specially in mobile robotics, due mainly to its reliability and versatility. Its medium access mechanism, its multimaster capability, and the ability to detect transmission errors make it suitable for distributed real-time systems. Sensor modules and computing nodes use the bus to share the sensory information.

In this work, YAIR has five sensory nodes, and a control node that manages the external communications and executes the main control application program (Fig. 2). Sensory nodes have reckoning capability, meaning that they pre-process the information and reduce the main processor load. This pre-processed data (sensations) is put on the bus, making it accessible to the remaining nodes. (i.e.: the speed vector computed by the motion controller can be used by the ultrasonic sonar module to point the sonar head towards the displacement direction).

Figure 2 shows the global architecture of YAIR. The main nodes can be concisely described as follows:

- Motion controller and odometry reckoning node. This module consists of three processors: a supervisor controller that handles the communications tasks as well as managing the other two processors in the node; and two HCTL1000 motion controllers, one for each motor in the robot. These two motion controllers manage synchronously the motion of each robot wheel and compute the respective odometry. Finally, this node also includes a bumper ring to avoid direct collisions and generate emergency messages.
- Infrared node. Consists of a single processor that supervises the 16 IR detectors, detects objects and estimates, in a range from 10cm to 1m, the distance from them and the robot. It can be operated by polling, or in autonomous scan mode. Also, it can generate emergency messages and distance messages.

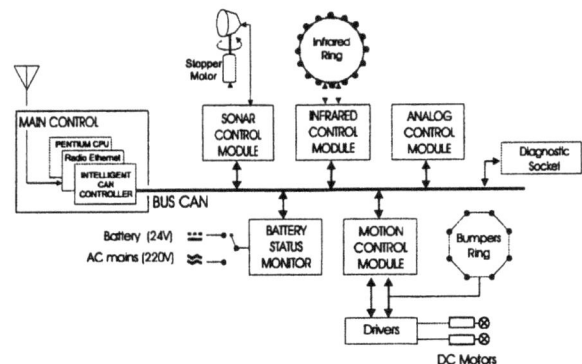

Fig. 2: YAIR architecture

- Ultrasonic sonar node. This module also provides range information about environment objects, but it can reach up to 4 metres in distance. The sonar module can generate maps in two different ways: simple time-of-flight measurements, and deconvolution algorithms. This former approach requires co-operation with a DSP processor.
- Main control node. This is an embedded PC under Windows-NT. It has a digital colour camera, a floating point DSP board, a full CAN interface, and radio ethernet link that provides external IP communication. Currently, this module runs a control application whose main purpose is fusing sensor data to obtain an escape vector for collision avoidance (Braintenberg, 1984)
- Auxiliary nodes. Some examples of this type of module include: battery status monitoring, stereo audio microphones for sound tracking, temperature measurements, etc.

3. DISTRIBUTED CAN OBJECTS SYSTEM

Communications between processes running in CAN nodes use shared memory. So, when a process needs to know data from a sensor, it only needs have to obtain the memory pointer where the information is. Using this address, the process will gain access to data calling CAN interface functions.

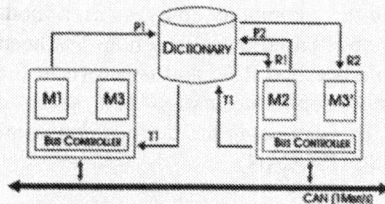

Fig. 3: Distributed CAN object system

The communication CAN system is structured using the following abstractions:

- ❑ Node: a node is a computer (physical entity) which can do two possible tasks. It can make numerical operations or control transducers (sensors) i.e.; central computer or motion controller. The system can have until 32 nodes.
- ❑ Handlers: logical entities that divide a node in different software units. Each handler executes a specific task that the node needs. For example, tasks could be controlling the infrared or communications. The system can have until 256 handlers (NM).
- ❑ Communication objects (COB): communication objects are the shared variables. A shared variable is used in the system to obtain data from a sensor, or to set points in an actuator. For example, each one of the 200 radial measures that the ultrasonic sensor obtains can be a communication object. When a process needs to know data from a sensor it needs only to gain access to the shared variable or COB. Each handler can control up to 256 variables (NV). So, the system can have a maximum of NM*NV=2^{16} shared variables.
- ❑ Pointers to the variables (*COB_ID´s*): the pointers are the memory addresses of the shared variables or communication objects. A pointer has two different parts:
  - • Handler: it is the process that shares a memory block.
  - • Offset: it is the index inside the memory block.
- ❑ COB_ID dictionary (fig. 3): a database that establishes a mapping between the COB_ID´s and the CAN identifiers. Using the COB_ID (address of the variable) a process looks in the dictionary and obtains the associated CAN identifier. Using this identifier the process will send a CAN message to read or write the

variable. Each node in the system holds its own local copy of the dictionary.

The access to shared variables through handlers is implemented by a uniform interface. This interface implements three functions:

- ❑ XXX_Execute, to read/write a variable.
- ❑ XXX_Read, to read a variable.
- ❑ XXX_Request, to write a variable.

Where XXX is the reference of the handler. These functions obtain, or send, the communication objects that are transmitted through the CAN bus.

4. DISTRIBUTED BLACKBOARD SYSTEM (SC)

High-level access to distributed data in WinNT processes has been provided by the development of a system (SC) that hides communication details behind a uniform bind-notification interface (Fig. 4).

The SC makes an internal representation of the data objects using a distributed blackboard architecture (Penny, 1989). The data structure that forms the blackboard is continually updated with the changing values of the objects through the SC established channels.

The SC system requires a program instance to be executed in each computer belonging to the configuration. The programs instance will communicate with each other to control and update the distributed data. As a result, each computer has a partial copy of the blackboard. It is important to emphasise that the processes, independently of their location, have only to execute local accesses to the corresponding SC program instance in order to contact all the system.

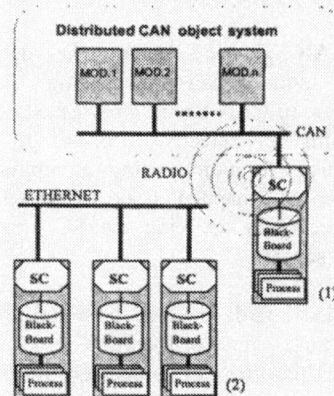

Fig. 4: Distributed blackboard structure.

The SC system establishes the necessary communications to transmit the information between the computers and to ensure that the copies of the distributed blackboard are consistent.

The system is event driven. In this way, it is possible to associate the execution of code with specific events (for instance an event could be a change on the blackboard). Thus, the processes, which are bound with SC objects, automatically receive the values they need from blackboard objects.

The YAIR robot is equipped with a Pentium processor which has a physical connection with the CAN bus (Figure 4, node 1). Processes that provide the robot a deliberative behaviour and a sensorial fusion level can be executed in every node of the IP network (Fig. 4).

5. SC-CAN GATEWAY

The distributed blackboard generated by the SC software is extensive to the data in the CAN network. Each computer node in the CAN network serves data to its running processes through the homogeneous SC software interface. The gateway software ISCCAN performs specific translations between CAN protocol and SC data.

The ISCCAN gateway supports communication of the CAN raw data, as well as the mapped mode that consists of a bi-directional mirroring of CAN identifiers and objects in the distributed blackboard. The mapped mode allows processes running in every node in the IP network access to the CAN information through the SC software and the defined notification scheme.

Two kind of problems arise in implementing the data chain linking low-level processes running in CAN nodes and high-level processes running in computer nodes:

❑ Data format conversions and serialisation coherence.
❑ Semantic guided data filtering.

The ISCCAN gateway solves the data format conversion and serialisation using ASCII-Hex representation of CAN binary streams. SC distributes these streams for selective processing. Processes translate this information using a supplied object toolbox. The SC mapped mode allows the use of defined filtering by applying the SC general bind-notification scheme.

The SC blackboard level is intended to provide communications for deliberative soft real-time processes. This kind of processes must manage communication overloads of 20ms introduced by the SC+ISCCAN system.

Typically, deliberative processes are related to sensory integration, data fusion and map building. In this case, when temporal and spatial sensory fusion is essential, time properties must be attached to sensor data and control actions. The time property attached to each SC blackboard object is in the form of a time firewall – a register that accumulates all communication overloads. From this information, each process using data computes the t_{use}-$t_{observation}$ difference and uses it in data integration tasks.

6. TESTING PROTOTYPE

To validate the communication system, a module for reactive control based on a local map has been built. This module is called REC and is shown in Figure 5. It obtains sensory information and sends control actions to the robot using the defined communication structure (SC+ISCCAN).

The REC test bed has been designed to stress the communication system in order to evaluate its performance. Reactive control actions for avoiding obstacles are computed from local map data that consist of a bundle of vectors. Each vector offers information regarding obstacle's proximity, as well as time and probability properties used in data fusion. Periodically, the local map is transformed by translations and rotations based on dead-reckoning data. Local map information is also used to direct the ultrasonic sensor head by focussing data sampling on useful information for the model. Figure 5 shows the graphical representation of the information in the model. On the left of the screen are some fusion indexes (i.e. probability used in the bayesian algorithm) and the central part shows the current local map.

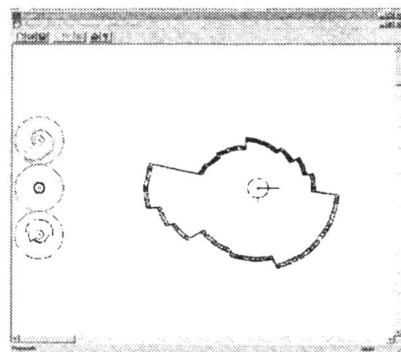

Figure 5. Reactive control based on local map.

During the test, the REC process was running in a node outside the CAN network and communicated through a wireless IP network and the described SC+ISCCAN facilities. We obtained good communications performance running REC tasks with the following periodicity.

❑ Obtain odometric information: 100ms
❑ Send control action: 100ms
❑ Obtain ultrasonic information: 300ms
❑ Obtain infrared information: 300ms

These control periods are slow and force the robot to move slowly during the test.

7. REAL TIME RESTRICTIONS.

The YAIR architecture combines two domains differing in real-time restrictions. Soft real-time restricted processes, as before described, can run in every node connected to the IP network. Reasoning in time is allowed by the attachment of information about the communication overload.

Hard real-time restrictions, defined by the real world interaction, are managed by a CAN network of local processors. Each processor runs specific algorithms using CAN distributed data. The use of shared resource, the CAN bus, is critical in the schedulability of the set of tasks. Worst-case scenarios are analysed in the following section.

8. RESPONSE TIME ANALYSIS.

This section describes the analysis of read/write temporary costs on sensorial distributed variables. This analysis enables us to study the system schedulability.

Distributed variable access time is split into the following factors:

$$T_{acc} = T_{comp} + R_{pet} + T_{proc} + R_{resp} \qquad (1)$$

- ❑ T_{comp}: Computation time, time from task activation until CAN controller queues messages. This depends on operating system and communication layers.
- ❑ R_{pet} y R_{resp}: CAN latencies for request and response. These latencies depend mainly on the workload over each message identifier, and that means in CAN, over each priority.
- ❑ T_{proc}: Processing time or local access time. This depends on the distributed variable type. Internal variables are traditional distributed variables at robot nodes, for example, motion control type (PID control position or P control velocity), motivation coefficients in robot behaviours. (tracking or avoiding obstacles). On the other hand, external or sensor variables involve sensing, that is to say, digital signal processing. Reading variables means acquiring and processing signals; and writing variables means setting point actuators. Access time analysis of these variables depends mainly on the mapped sensor. Transducer nodes are one-task systems and their analysis consists in calculating processing time and workload peaks during the process.

The first factor T_{SC} is the longest time between task activation and the time the message can take part in arbitration. Since T_{SC} is included into CAN latency, it is not studied here.

The analysis of the second factor is comparable to the schedulability test by fixed priorities. (Tindell, et al., 1994). Timing analysis makes the following simple assumptions:

- ❑ Transmission periods T_m are known and constant.
- ❑ Maximum jitters J_m are also known.
- ❑ Message m has a bounded size (s_m bytes).
- ❑ CAN controllers have prioritised communication queues.
- ❑ The length and identifiers of all messages are known.

The longest time from task activation until message m gains control of the bus R_m can be calculated using the following extended schedulability test equation, assuming that environment map variables (infrared or ultrasonic) are segmented in up to 37 CAN frames. In this case, the problem is equal to the worst-case response time of the last transmitted frame of the message. The last transmitted frame of the message m is denoted by $\hat{f}_m$. Thus, we have the burst response time as:

$$R_m = J_m + w_{m\hat{f}} + C_{m\hat{f}} < T_m \qquad (2)$$

where,

$$w_{m\hat{f}} = wh_{m\hat{f}} + ws_{m\hat{f}} + wl_{m\hat{f}} \qquad (3)$$

where, $wh_{m\hat{f}}$, $ws_{m\hat{f}}$, $wl_{m\hat{f}}$ are latency increments of higher, the same, and lower priority frames than the last frame of message m. These values can be calculated using the recursive method described in (Gil and Pont, 1996). Thus the following equation 4 can be used.

$$w_{m\hat{f}}^{n+1} = \sum_{\forall j \in HP(m)} \sum_{\forall g \in F(j)} \left(\left\lceil \frac{w_{m\hat{f}}^n + J_j + \tau_{bit}}{T_j} \right\rceil C_{jg} \right) + \\ \sum_{\forall h \in F(m); h \neq \hat{f}} C_{mh} + \sum_{i=1}^{min(p,n_m)} b_i \qquad (4)$$

,where:

n_m = number of frames of message m (depends on protocol).

C_{mf} = physical transmission time of frame f.

F(m), set of all frames coming from the message m.

$$B = (b_1, ..., b_p)_{b_i = S_{jg}; j \in lp(m); g \in F(j); b_i \geq b_{i+1}} \qquad (5)$$

Table 1 Response time analysis (time units in ms).
[1] COB_ID is the physical identifier.
[2] C_{id} and B_{id} the maximum length and transmission times of these messages.

Message	COB_ID[1]	C_{id}(length)[2]	T_{id}	J_{id}	B_{id}[2]	W_{id}	R_{id}
IR-Dist(WR)	0x023	135(4)	10,0	0,0	0,548	0.959	1,096
Odom(RD)	0x024	135(2)	8,0	0,0	0,274	0.959	1,096
US-Dist(RD)	0x044	135(1)	50,0	6,0	0,135	0.959	7,096
IR_Dist(RD)	0x061	135(1)	10,0	0,0	0,135	1,096	1,233
IR_Dist(RD)	0x064	135(1)	10,0	2,0	0,135	1,233	3,370
US-Dist(WR)	0x081	135(1)	50,0	0,0	0,135	1,370	1,507
Odom(WR)	0x082	135(2)	8,0	0,0	0,274	1,781	1,918
IR-Dist(WR)	0x084	135(4)	10,0	0,0	0,548	2,603	2,740

$$p = \sum_{\forall j \in lp(m)} \sum_{\forall g \in F(j)} \left\lceil \frac{w_{m\hat{j}}^{n} + J_j + \tau_{bit}}{T_j} \right\rceil \quad (6)$$

The third factor T_{proc} or access time depends on variable type: internal or external. Since the former are mapping sensors and actuators, a specialised analysis about each external variable must be carried out. So, it is quicker to construct a local map using infrared sensor data than construct it from ultrasonic data.

9. ANALYSIS RESULTS.

Two processes are executed to validate the low-level communication system. A local version of the reactive control application described above is running in the main processor. The motion control module runs another obstacle avoiding algorithm.
CAN analysis latencies require fixed transmission period times and efficient CPU scheduling to guarantee this supposition. Consequently, in the following analysis, the deadline processes are supposed to be guaranteed. The definition of these analysed applications is:

❑ Local reactive application, threads:
  • Obtain odometric information: 8 ms
  • Obtain ultrasonic information: 50 ms
  • Obtain infrared information: 10 ms
The jitter of these threads, listed in table 1, depends exclusively on scheduling in WinNT.
❑ Obstacle avoiding algorithm:
  • Obtain infrared information: 10 ms

10. CONCLUSIONS

A communication system suitable for real-time systems with distributed sensory architecture is described. This system has been implemented in the YAIR robot, an autonomous robot with intelligent sensors that produces different measurements about the environment and its position within it. Low-level

communication based on the CAN bus has been presented. Also, the SC+ISCCAN combination solves the high-level data diffusion in the distributed blackboard system. Finally, timing analysis has been conducted showing real-time capabilities of both, low and high-level domains.

11. REFERENCES

Blanes, F., G. Benet, M. Martinez, J.Simó. (1998). Grid Map Building from Reduced Sonar Data. IFAC International Symposium on Intelligent Autonomous Vehicles. IAV'98. Madrid.

Gil, J.A (1997). A CAN Architecture for an Intelligent Mobile Robot, 65-70. SICICA-97.

Simó, J., A. Crespo, J.F. Blanes (1997). Behaviour Selection in the YAIR Architecture. Proceedings of IFAC Conference on Algorithms and Architectures for Real Time Control. Vilamoura, Portugal. AARTC'97.

Posadas, J.L., J.Simó, F.Blanes (1997). Un modelo para el desarrollo de aplicaciones distribuidas. El Servidor de Comunicaciones. Jornadas españolas de Automática (JA'97). Gerona 1997.

Bosch (1991) "CAN Specification 2.0A". © Robert Bosch GmbH.

Braitenberg, V. (1984). "Vehicles: Experiments in Synthetic Psychology". Cambridge. MIT Press.

Penny, H. (1989) "Blackboard Architectures and Applications". Edited by V. Jagannathan, Rajendra Dodhiawala, Lawrence S. Baum.

Tindell, K.W., H. Hansson, A.J. Wellings (1994) "Analysing Real-Time Communication: Controller Area Network (CAN)". In F. Jahanian and K. Ramaritham, editors, Proc. 15 th RealTime Systems Symposium, pages 259--263. IEEE Computer Society Press.

Gil, J.A., A.Pont. (1996) "Guaranteeing Message Block Transfer Latencies in CAN". Internal Report DISCA. UPV. Valencia (Spain).

ADAPTIVE GENERALIZED PREDICTIVE CONTROL ALGORITHM AUTOMATIC PARALELIZATION USING MAPS ENVIRONMENT

H. A. Daniel*[1] and A. E. Ruano**

*Unidade de Ciências Exactas e Humanas, Universidade do Algarve
and Institute of Systems & Robotics*
* Email: hdaniel@ualg.pt   Tel: +351 289 800950   Fax: +351 289 818560
** Email: aruano@ualg.pt   Tel: +351 289 800912

Abstract: Parallelization of real time control algorithms is a problem that the control engineer must consider to meet tighter specifications in terms of plant sampling time. However, using conventional tools, the development time of an efficient parallel algorithm is much higher than its sequential equivalent. If the algorithm can be represented in a matrix form, the Matrix Automatic Parallelization System – MAPS - puts in the hands of the control engineer the power of parallel processing at the cost of the sequential programming model. To illustrate this environment, the automatic parallelization of an Adaptive Generalized Predictive Control algorithm is employed. *Copyright © 2000 IFAC*

Keywords: Parallel Algorithms, Multiprocessing Systems, Digital Signal Processors, Adaptive Control, Predictive Control.

1. BRIEF DESCRIPTION OF THE MAPS ENVIRONMENT

The concept of parallelizing matricial algorithms by allocating sets of rows of the matrix operands in different processors, which execute the same code, allows the parallel algorithm to achieve high levels of efficiency (Daniel and Ruano, 1997; Piedra, 1991). However, when using a message passing parallel model, the programmer must develop routing processes, so that data may be broadcasted through the parallel network (Daniel and Ruano, 1997). Since the efficiency of the parallel algorithm, which can be seen as the ratio between the time in which the processors are computing the algorithm and the time required to broadcast data, strongly depends on the routing strategy, a considerable part of the development time is spent in optimizing this strategy. Furthermore, different network topologies must be investigated, so that the most appropriate one, in

terms of execution time and efficiency, may be found. This way, the development time of efficient parallel applications, with the conventional tools available, is many times superior to the development time of a sequential application. Another problem with the common parallel programming model is that the parallelization of these algorithms is strongly dependent on the hardware used, so portability and upgradability are also time consuming steps in the development process.

The MAPS environment was first introduced in (Daniel and Ruano, 1999a). For a full description of the toolbox, please refer to that paper. Under this framework, developing a parallel algorithm requires only a sequential description of this algorithm and the desired network topology, using appropriate tools from this toolbox, thus speeding up the development time of parallel applications towards levels near the sequential development time.

---

[1] Helder Daniel acknowledges the financial support of "Programa Praxis XXI" for this work

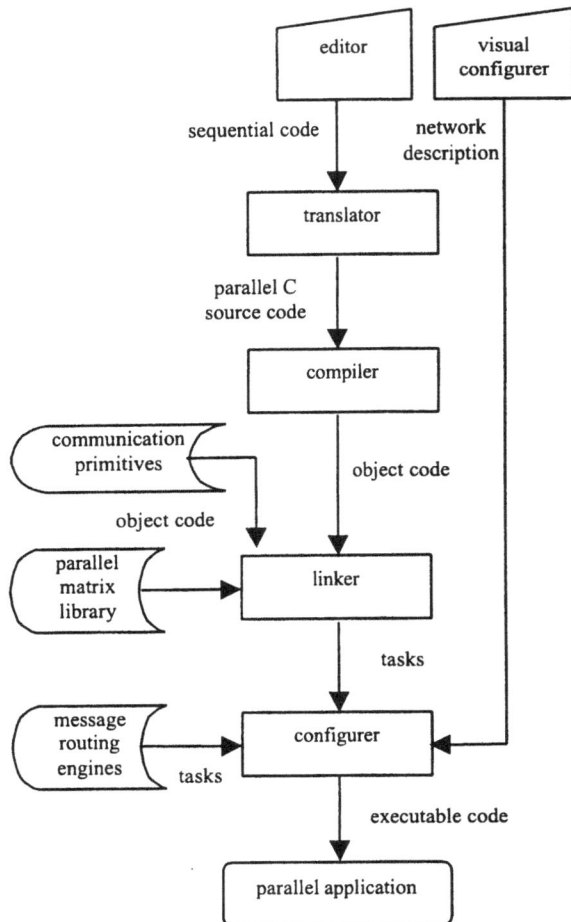

Fig.1. Automatic generation of parallel executable code

The fluxogram in Fig. 1 shows the process of automatic generation of parallel executable code. This process begins in the translator, where the sequential code is converted into ANSI C source code with extensions for parallel processing. This code is identical for every processor in the network. However, two parameters, specified in the visual configurer, are different for every processor. These parameters are the processor number, which is used to identify it to the message routing engines, and the number of rows of the operand matrices allocated in the processor local memory, which allows the amount of data to process to be balanced for different types of processors. The parallel C source code is mainly formed of calls to a library of parallel matrix functions, which will be linked in a following step, and two communicating primitives to send and receive data from any other processor in the network. Code to measure the performance of the algorithm is also included. In the next step the parallel C source code is sent to a compiler suitable for the hardware used. This compiler, the linker and the configurer are tools available on the market. The compiler generates the object code, which is then linked with the parallel matrix library and an additional library that includes the communication primitives. These primitives allow

that the application can access an inferior level of abstraction. This level interfaces the application with the hardware in such a way that, at the application level, the hardware is transparent. So, whatever the hardware used in the network, at the application level the communication primitives are all that is needed to broadcast messages across the network. Finally the configurer reads the network description and selects the message routing engine appropriated for the target network. Currently a *Simulink* block diagram is used to design the network.

An example of a heterogeneous network description, using *Simulink*, is presented in Fig. 2. This model also allows that future changes in network topology, as long as the hardware is supported, require only the programmer to redraw the network description in the visual configurer. To support this hardware independence, the application is structured in hierarchic abstraction levels as shown in Fig 3. The bottom level, or the hardware level, consists in the network of processing elements - PEs - which can be built of any kind of processors supported. Next there is a level that interfaces the generated parallel C code, at the programming level, with the hardware, called kernel. Currently this level of abstraction supports networks built of Digital Signal Processors ADSP2106x and *TMS320C40*, and the *T80x Transputer*.

The host interface, as well as the calculation processes, automatically generated by the translator tool, forms the programming level. The host interface process can be seen as an extension of the kernel, allocated in the root processor, so that the host resources are available to the network. Finally the host level, which is allocated in the host computer, handles the interface on the host side.

2. MATHEMATICAL FORMULATION OF AGPC

The GPC algorithm was introduced in the past decade (Clarke, et al., 1987a, 1987b) and has subsequently

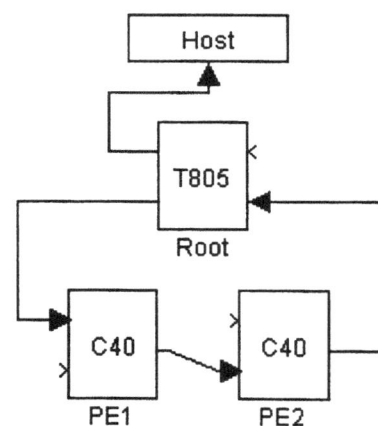

Fig. 2. Network design in Simulink

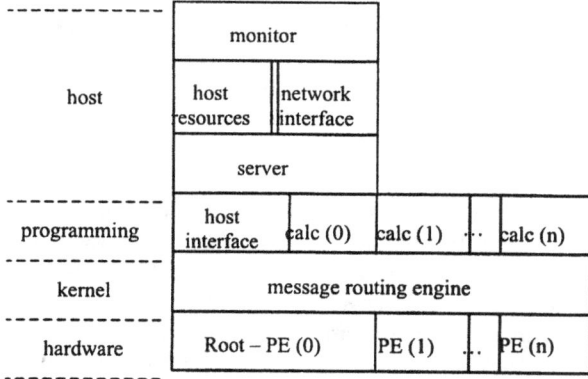

Fig. 3. Implementation model

been proved to be superior to other self-tuning algorithms such as the ones based on pole-placement and generalized minimum variance. In fact the GPC algorithm conserves its robustness even when the plant dead-time and model order are unknown (Aström and Wittenmark, 1989). However the cost of achieving such robustness is a substantial cost in computational overhead, suggesting the use of parallel processing to speed it up, therefore making it a valuable proposition for the control of plants with tighter specifications in terms of sampling time.

As the adaptive GPC algorithm needs to know the plant transfer function, a first stage in the algorithm must be an RLS estimator. The estimated parameters are then injected in the predictor who computes the predicted plant output and the control signal with a subsequent minimisation of a suitable loss function. The block diagram of this algorithm is presented in Fig. 4.

2.1 Recursive least squares algorithm

Let the plant model be a CARIMA representation:

$$A(q^{-1})y(k) = q^{-d}B(q^{-1})u(k-1) + \frac{C(q^{-1})}{\Delta}e(k), \quad (1)$$

where $y(k)$ is the plant output at time step k, $u(k-1)$ is the plant input at time step $k-1$, $e(k)$ is an uncorrelated random sequence, Δ is the difference operator $1 - q^{-1}$, d is the time delay and $A(q^{-1})$, $B(q^{-1})$, $C(q^{-1})$ are polynomials in the backward shift operator q^{-1}:

$$A(q^{-1}) = 1 + a_1 q^{-1} + \ldots + a_{na} q^{-na}$$
$$B(q^{-1}) = b_0 + b_1 q^{-1} + \ldots + b_{nb} q^{-nb}$$
$$C(q^{-1}) = 1 + c_1 q^{-1} + \ldots + c_{nc} q^{-nc}$$

The unknown parameters vector and the regression or data vector are represented by:

$$\theta^T = \left[-a_1,\ldots,-a_{n_a}, b_0,\ldots,b_{n_b}, c_1,\ldots,c_{n_c}\right]$$
$$\mathbf{x}^T(k) = \left[y(k-1),\ldots,y(k-n_a),u(k-1),\ldots\right.$$
$$\left.\ldots,u(k-n_b-1),e(k-1),\ldots,e(k-n_c)\right]$$

$\mathbf{P}(k)$ is the covariance matrix, initialised as the Identity matrix. The following set of equations defines a standard RLS estimator, optimized in terms of computational operations:

$$\varepsilon(k) = y(k) - \mathbf{x}^T(k)\theta(k-1) \quad (2)$$
$$\mathbf{K}_b(k) = \mathbf{P}(k-1)\mathbf{x}(k) \quad (3)$$
$$\mathbf{K}(k) = \frac{\mathbf{K}_b(k)}{\lambda + \mathbf{x}^T(k)\mathbf{K}_b(k)} \quad (4)$$
$$\mathbf{P}(k) = \frac{\mathbf{P}(k-1) - \mathbf{K}(k)\mathbf{K}_b(k)^T}{\lambda} \quad (5)$$
$$\theta(k) = \theta(k-1) + \mathbf{K}(k)\varepsilon(k) \quad (6)$$

2.2 GPC Predictor

The GPC predictor minimisation criteria is given by:

$$J(N_1, N_2, N_u) = E\left[\sum_{j=N_1}^{N_2}\delta(j)[\hat{y}(k+j\,|\,k) - w(k+j)]^2 \right.$$
$$\left. + \sum_{j=1}^{N_u}\lambda(j)[\Delta u(k+j-1)]^2\right], \quad (7)$$

where $E[\ldots]$ is the expectation operator, $\hat{y}(k+j\,|\,k)$ is an optimal j steps ahead predictor using data available until time step k, N_1 is the minimum cost horizon, N_2 is the maximum cost horizon, N_u is the control horizon, $\lambda(j)$ and $\delta(j)$ are weighting sequences (($\lambda(j)$ is considered to be constant and $\delta(j) = 1$), and $w(k+j)$ is the future reference trajectory. To obtain the output j steps ahead, the Diophantine equation (8) needs to be solved:

$$C(q^{-1}) = E_j(q^{-1})\tilde{A}(q^{-1}) + q^{-j}F_j(q^{-1}), \quad (8)$$

where $\tilde{A}(q^{-1}) = \Delta A(q^{-1})$, deg $(E)_j = j - 1$, deg $(F)_j = n_a$.

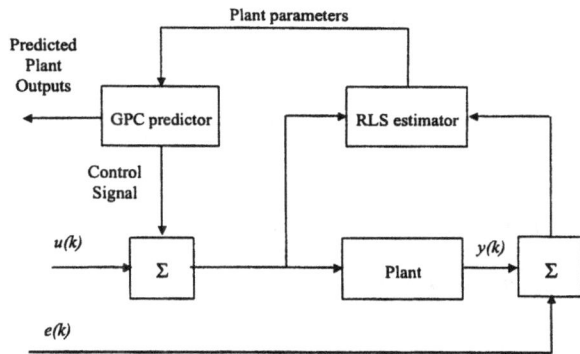

Fig. 4. Block diagram of a plant controlled by a GPC predictor

E_j (q^{-1}) and F_j (q^{-1}) can be obtained dividing C (q^{-1}) by $\tilde{A}$ (q^{-1}) until the remainder can be factored as q^{-j} F_j (q^{-1}), where E_j (q^{-1}) is the quotient or recursively (Clarke, et al., 1987a, 1987b; Camacho and Bordons, 1994). Considering now the following sequence of j steps ahead predictions:

$$\hat{y}(k+d+1|k)=G_{d+1}(q^{-1})\Delta u(k)+F_{d+1}(q^{-1})y(k)$$
$$\hat{y}(k+d+2|k)=G_{d+2}(q^{-1})\Delta u(k+1)+F_{d+2}(q^{-1})y(k)$$
$$\vdots \qquad\qquad (9)$$
$$\hat{y}(k+d+N|k)=G_{d+N}(q^{-1})\Delta u(k+N-1)+F_{d+N}(q^{-1})y(k)$$

where G is obtained with:

$$G_j = E_j . B, \qquad j = 1,...,N. \qquad (10)$$

According to (Aström and Wittenmark, 1989) equations (9) can be written as:

$$\mathbf{y} = \mathbf{G}\Delta\mathbf{u} + \mathbf{f}, \qquad (11)$$

where:

$$\Delta\mathbf{u}^T = \left[\Delta u(k),\Delta u(k+1),\cdots,\Delta u(k+N-1)\right],$$

$$\mathbf{y}^T = \left[\hat{y}(k+d+1|k),\hat{y}(k+d+2|k),\cdots,\hat{y}(k+d+N|k)\right]$$

$$\mathbf{G}=\begin{bmatrix} g_0 & 0 & \cdots & 0 \\ g_1 & g_0 & \cdots & 0 \\ \vdots & \vdots & \ddots & \vdots \\ g_{N-1} & g_{N-2} & \cdots & g_0 \end{bmatrix}, \begin{matrix} g_n = n^{th}\text{coefficient} \\ \text{of } G_{d+N} \\ n = 0,...,N-1 \end{matrix}$$

$$\mathbf{f}=\begin{bmatrix} (G_{d+1}(q^{-1})-g_0)\Delta u(k)+F_{d+1}(q^{-1})y(k) \\ (G_{d+2}(q^{-1})-g_0-g_1q^{-1})\Delta u(k)+F_{d+2}(q^{-1})y(k) \\ \vdots \\ (G_{d+N}(q^{-1})-g_0-g_1q^{-1}-\cdots-g_{N-1}q^{-(N-1)})\Delta u(k)+F_{d+N}(q^{-1})y(k) \end{bmatrix},$$

and $N=N_2-N_1+1= N_u$.

$\Delta\mathbf{u}$, the minimum solution of (7), can be computed as:

$$\Delta\mathbf{u} = (\mathbf{G}^T \mathbf{G} + \lambda \mathbf{I})^{-1} \mathbf{G}^T (\mathbf{w} - \mathbf{f}) \qquad (12)$$

where:

$$\mathbf{w} = \left[w(k+d+1),w(k+d+2),\cdots,w(k+d+N)\right]$$

and the control signal is given by:

$$u(k) = u(k-1) + \Delta\mathbf{u}(1) \qquad (13)$$

3. IMPLEMENTATION USING THE TOOLBOX

The parallel implementation of the AGPC algorithm was mapped over several architectures, with the same topologies, since point to point communication between processors was a goal. Figure 2 illustrates this topology. This figure represents the only heterogeneous architecture tested. From this point on it will be referenced as Het (p), where p represents the number of processors in the network. So, Het (1) represents a network with only one T805, while Het (2) represents a network with both one T805 and one C40. Of course, Het (3) stands for the network in the figure. Also three homogeneous architectures where used, referenced as T8 (p) if built with T805s at 25 MHz, C4 (p) if built with C40s at 50 MHz and finally Sharc (p) if built with ADSP21060 Sharcs running at 40 MHz. All the architectures were built with up to three processors, except for this last one, which was built with up to two processors only. It is also important to notice that architectures Het and T8, with only one processor, degenerate in the same architecture.

As mentioned above, the sequential code that the application programmer uses to express the algorithm, is architecture independent. This code should be familiar to everyone used to express math calculation in a sequential language. For instance, the code that expresses the predictor core, which requires the solution of matrix equations 11 to 13, might look like these:

```
y = G * u   + f;
u = inv (G' * G + 1 * I) * G' * (w - f);
u1 = u2 + u (1);
u2 = u1;
```

Due to problems of space, it is impossible to show the complete development process of this algorithm using the MAPS environment. This step, however, is as simple as it was shown.

4. PERFORMANCE OF THE PARALLEL AGPC ALGORITHM

The performance of the automatic generated parallel

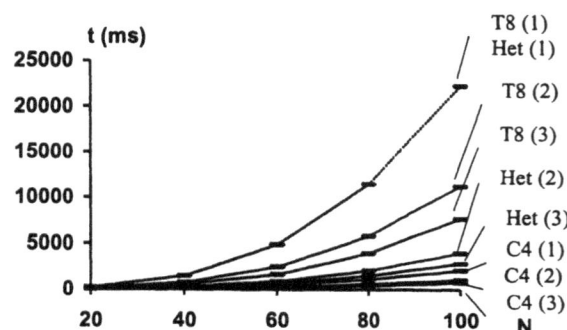

Fig. 5. Execution times of the AGPC algorithm (I)

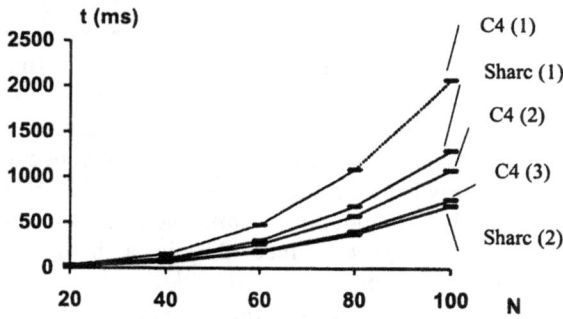

Fig. 6. Execution times of the AGPC algorithm (II)

AGPC algorithm was measured assuming that the plant is a 2nd order CARIMA model. A predictive horizon, N, from 20 to 100 steps ahead was used. Fig. 5 represents the AGPC execution time for networks T8, C4 and Het with one, two, and three processors. As this scale is not appropriate to compare architectures C4 and Sharc, its performance is shown in figure 6. It is obvious that the T8 architecture is always the slowest, whatever the number of processors employed, while the Sharc is the fastest. This was expected since the ADSP21060 is much faster than the T805 and faster than the C40. Further more a network with only two ADSP21060 is faster than a network with three C40s. Since the Heterogeneous network Het (1) is the same as T8 (1) both plots are superimposed. By adding one C40 processor to this architecture, Het (2) is built. As in this case the computational load is distributed between a T805 and a C40 processor it was expected that this architecture would have better performance than C4 (1). However, the communication costs between the C40 and the T805, with floating point format conversion, are approximately 2.5 times slower than between two T805s and 21.3 times slower than between two C40s (Daniel and Ruano, 1999b), so a bottleneck occurs reducing the performance. In an independent investigation (Tokhi et al., 1995), slightly better communication times are presented, however this analysis remains valid. It was also expected that Het (3) would be much faster than Het (2), and at least faster than C4 (1), since 1 more C40 is used to compute the algorithm, however this is not observed. This can also be explained by the

bottleneck introduced by the C40 – T8 link, and also, as the message routing engines were designed to be as much ANSI C compatible as possible, they are not taking advantage of the capabilities of the T805. This way the T805, rather than communicating with both C40 concurrently, is serving each one at a time. Thus, as the computational load in each processor decreases, when using Het (3) rather than Het (2), the amount of time spent in communications increases. A common form of evaluation of a parallel algorithm is efficiency. This measures the percentage of time in which p processors are actually computing an algorithm rather than communicating data between them. It can be obtained by the following equation:

$$Eff\,(p) = \frac{Exec.\,time\,for\,1\,processor}{(Exec.\,time\,for\,p\,processors)\,.\,p}100\% \quad (14)$$

When using heterogeneous architectures, however, as different types of processors have different performances, equation (14) is not valid. Lets consider a network built of p processors and that function $T(K(x), R(x))$ gives the execution time required for processor type $K(x)$ to solve the algorithm over $R(x)$ rows. Considering an optimum load balancing and that communication time is null, the optimum parallel computational time is given by:

$$T_o = max[T(K(1), R_o(1)), \cdots, T(K(p), R_o(p))] \quad (15)$$

Then the efficiency may be obtained by the ratio between the optimal time and the actual measured time, T_m:

$$Eff(p) = \frac{T_o}{T_m}100\% \quad (16)$$

These measures are presented in figures 7 and 8. The most efficient architecture is the T8. Architecture C4 also has a good efficiency when computing large data sets, followed by architecture Sharc. The heterogeneous architecture obtains the poorest efficiency. Note that, for the same architecture, efficiency decreases as the number of processors in the network increases. On the other hand, although it

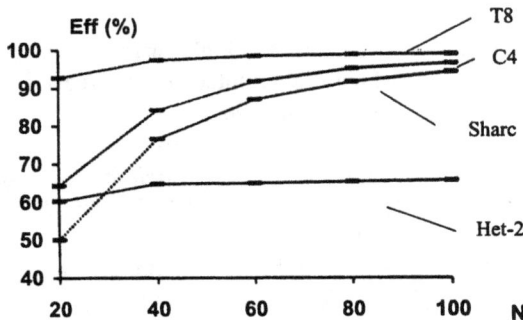

Fig. 7. Efficiency of the AGPC parallel algorithm with 2 processors

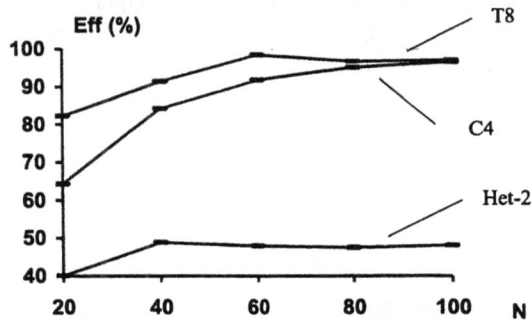

Fig. 8. Efficiency of the AGPC parallel algorithm with 3 processors

161

may seem strange that the faster processor is the most inefficient one, that can be explained with the help of the ratio between the capabilities of computation and communication:

$$\frac{\text{Run - time quantum}}{\text{Communication quantum}} \qquad (17)$$

where run-time quantum is the time spent in computing one floating point operation and communications quantum the time spent in communicating one float data type (4 byte word). This way, the higher this ratio is, the more efficient the architecture is. So, from the 3 types of processors used, the T805 processor is the one with the highest ratio, followed by C40 and finally the ADSP21060. This means that, in spite of the lower run-time quantum of the ADSP21060, the communications quantum is not small enough to maintain a ratio equivalent to the T805 ratio, leading to a loss in efficiency when compared to this processor. This can also explain why the heterogeneous architecture has such a poor efficiency, since the run-time quantum can not be smaller than the C40's, and the communication quantum is larger than the T8's, as already mentioned.

As a final remark, it should be noted that the performance of the parallel AGPC algorithm presented here is better than the one presented in (Daniel and Ruano, 1999a). This does not mean that the version of the MAPS environment presented in that paper is different than the one presented here, only that the parallel matrix library has been updated. In fact, only one function of this library was optimized, the parallel matrix inversion. However, matrix inversion is a time consuming stage in the AGPC algorithm, so optimizing this stage is widely reflected in the overall performance.

The algorithm used to perform a matrix inversion is based on a parallel version of the Gauss-Jordan method for solving linear systems of equations (Daniel and Baltazar, 1994) adapted to invert matrix operands (Daniel and Ruano, 1996). However, the earlier version used memory addressing in a less efficient way. Also the memory space required was redundant and implied a strategy of memory manipulation that reduced the performance. This optimizations leads to an increase in the overall performance of the AGPC algorithm in a range from 25% to 45% depending on the architecture.

5. CONCLUSIONS

The toolset described in this paper adds the easy of use of sequential programming environments to the power of parallel processing. As already discussed, an optimization of the matrix algebra sub-programs,

can drastically improve the overall performance of the generated parallel algorithms. So, to achieve peak performance, a low level implementation of this library, targeted for every supported processor is required. Also, additional work may be done at kernel level in order to optimize it for heterogeneous architectures where the interfacing between different processors has an important cost, such as the Het architecture. However, since this is a hardware question, one can not expect miracles. Finally, the toolset can be improved adding a simulator tool, which allows the programmer to investigate the performance prior to the implementation in the actual hardware.

REFERENCES

Aström, K. J., and B. Wittenmark (1989). *Adaptive Control*. Addison - Wesley.

Camacho, E. F. and J. C. Bordons (1994). *Model Predictive Control in the Process Industry*. Springer Verlag

Clarke, D. W., C. Mohtadi and P. S. Tuffs (1987a). Generalized Predictive Control - Part I. The Basic Algorithm. In: *Automatica, 23*, 137 - 148

Clarke, D. W., C. Mohtadi and P. S. Tuffs (1987b). Generalized Predictive Control - Part II. Extensions and Interpretations. *Automatica,* **23**, 149 - 160

Daniel, H. and S. Baltazar, S. (1994). *Paralelização do método de Gauss-Jordan de resolução de sistemas de equações lineares* (in Portuguese). UAL Internal Report.

Daniel, H. and A. E. B. Ruano (1996). Parallel Implementation of an Adaptive Generalized Predictive Control Algorithm. *Proc 2nd International Meeting on Vector and Parallel Processing (VECPAR 96)*

Daniel, H. and A. E. B. Ruano (1997). Adaptive Generalized Predictive Control Algorithm Implemented over an Heterogeneous Parallel Architecture. In: *Proceedings of the IFAC International Workshop on Distributed Computer Control System 97*, pp. 185 – 190

Daniel, H. e A. E. B. Ruano (1999a). Automatic parallelization of matricial algorithms. *Proc 14th IFAC World Congress (IFAC 99)*. Vol. Q, 453-458

Daniel, H. e A. E. B. Ruano (1999b). Performance comparison of parallel architectures for real-time control. *Microprocessors and Microsystems*, 23 (1999), 325-336

Piedra, R.M. (1991). *A parallel Approach for Solving Matrix Multiplication on the TMS320C4x DSP*. Texas Instruments, Inc.

Tokhi, M. O., M. A. Hossain, M. J. Baxter e P. J. Fleming (1995). Heterogeneous and homogeneous parallel architectures for real-time active vibration control. *IEEE Proc Control Theory Appl.*, Vol. 142, No. 6, 625-632

MEMORY MANAGEMENT AND COMMUNICATION ON HOMOGENEOUS AND HETEROGENEOUS PARALLEL SYSTEMS

Ventura P. * Ruano M. Graça *
Cardoso e Cunha J. **

** GPSBio/ADEEC, Unidade de Ciências Exactas e
Humanas, Universidade do Algarve, Campus de Gambelas,
8000 Faro, Portugal
** Departamento de Informática, FCT, Universidade Nova
de Lisboa, Quinta daTorre, 2825-114 Caparica, Portugal*

Abstract: This paper reports a study held on the inter-processor communication
performance of homogeneous and heterogeneous parallel systems when different
data structure allocation schemes are implemented, regarding internal and/or external processor memory. To evaluate the performance parameters some case-study
algorithms were implemented on homogeneous and heterogeneous architectures.
Due to algorithm-machine dependency, hardware and software features have to be
considered. Where the access to external memory is not efficient, some internal
memory buffering methods are analysed. A comparison of the results obtained is
presented, enabling establishment of memory management references, regarding
the parallel architectures employed. *Copyright© 2000 IFAC*

Keywords: architectures, digital signal processors, parallel processing, memory
banks, communication channels, performance analysis

1. INTRODUCTION

Recent technological evolution allows computational resolution of increasingly complex problems, involving larger volumes of data. The viability of real-time implementation of this type of
problems frequently requires the use of parallel
and/or distributed systems. One of the factors
affecting the performance of the global system is
the hardware architecture selected and its ability
to perform efficiently the problem. On a parallel
processing environment, distribution of the computational load across the processors is a task involving features such as, the network topology, the
relative computation and communication capabilities of the processors, as well as algorithm task

partitioning and consequent scheduling of these
tasks to processors (Tokhi and Hossain, 1995b). It
is important to consider all these features on the
implementation of an algorithm on either a homogeneous or a heterogeneous parallel platform.

Previous works on this field considered performance evaluation on distinct parallel architectures of complex and highly demanding signal-
processing and control algorithms (Tokhi and
Hossain, 1995b; Tokhi et al., 1995; Tokhi and
Hossain, 1995a). The present study evaluates the
influence of different types of data structure allocation schemes on internal and/or external memory. The memory of a C program is divided into
four sections, code, static storage, stack and heap.

Mapping of these sections to physical memory (internal or external) can have a significant impact on the parallel system performance. Knowledge about the characteristics of a parallel systems influences algorithm design. So, the major goal of the study hereby reported is concerned with the evaluation of the tradeoffs between specific parallel systems and certain types of algorithms, when memory management and communications are considered.

2. HARDWARE

The parallel architectures in study, concerned homogeneous and heterogeneous architectures. Three different type of processors were employed: T805 Transputers (T8), Texas Instruments TMS320C40 (C40) and ADSP-21060 Sharc (Sharc). In general, the behaviour of a parallel system reflects the nature of its processors, being essentially a description of the processors relevant features and of the parallel architecture itself.

2.1 *Processing elements*

The T8 (INMOS Limited, 1990) is a 32-bit processor with 25 MHz clock speed, 4 Kbytes on-chip RAM and capable of 4.3 million floating-point operations per second (MFLOPS). The T8 includes a 64-bit floating-point unit and four serial communication links. The links operate at speeds of 20 Mbits/sec, achieving data rates of up to 1.7 Mbytes/sec unidirectionally or 2.3 Mbytes/sec bi-directionally.

The C40 (Texas Instruments, 1995) is a 32-bit DSP processor with 50 MHz clock speed, 2 blocks of on-chip RAM of 4 Kbytes each and is capable of 50 MFLOPS. For inter-processor communication includes six parallel communication links with 28 Mbytes/sec asynchronous transfer rate. It has separate internal program, data and DMA coprocessor buses for support of massive concurrent I/O of program and data throughput.

The Sharc (Analog Devices, 1995) is a 32-bit DSP processor with 40 MHz clock speed, a central processing unit (CPU) peak performance of 120 MFLOPS and includes 2 blocks on-chip SRAM of 256 Kbytes each. This processor possesses six links for point-to-point communication with others processors and operating concurrently and bi-directionally. The links has a maximum communication rate of 40 Mbytes/sec. This processor has three internal buses connected to its dual-ported memory, the program memory (PM) bus, data memory (DM) bus and I/O bus. The I/O bus allows concurrent data transfers between both

memory block and the links ports. With this dual-ported structure, accesses to internal memory by the processor and the I/O processor are independent and transparent to one another.

2.2 *Homogeneous and heterogeneous architectures*

Three homogeneous parallel architectures were considered, employing T8´s, C40´s and Sharc´s. The configuration of the homogeneous systems consisted of three processors, the root processor acting as a router of input/output operations, with the other two processors being actually responsible for the algorithm calculus.

Each processor has local memory and the links are used for communication with the other processors in the network. The network topology was chosen on the basis of the algorithm structure.

The heterogeneous system included T8's and C40, where a T8 was employed as the root processor, and the two other processing nodes were T8 and C40 processors (Figure 1). Again, the network topology was chosen considering the algorithm structure.

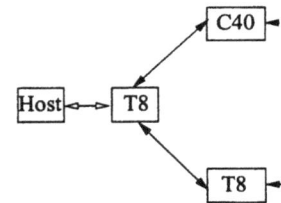

Fig. 1. Block diagram of the heterogeneous parallel system

3. SOFTWARE

Software support is essential for the development of efficient programs. Characteristics such as disk and memory accesses, input and output activities, compilation time, CPU time, and operating system overhead must be considered (Kumar, 1994). A good memory management, an efficient partitioning and mapping of tasks and the selection of a suitable compiler can affect the performance of a system. The selection of a compiler is a critical factor, since the compiler is responsible for the transformation of a high level language into a hardware dependent implementation, which differs from compiler to compiler, and presents different performance on different processors. Because of that, the considered algorithms were implemented in the same compiler family the 3L Parallel C.

The ideal performance of a parallel architecture demands a perfect match between the capability of the architecture and the program behaviour. The hardware features can be optimised with new technologies and efficient resource management. However, program behaviour is difficult to predict, since it suffers the influence of factors such as data structures and communication between processors.

Communication among processors is one of the main features influencing the performance of a parallel system. The amount of data, the frequency at which they are transmitted, the speed of their transmission and the route they take, are relevant factors on the communication among processors of a parallel system (Tokhi and Hossain, 1995b). The last two factors depend on the hardware, being the transmission speed an important discussion point since the memory zone where the data structure is mapped can influence the communication.

4. ALGORITHMS

The algorithms considered on this study were the cross-correlation (CC) algorithm (Proakis and Manolakis, 1988), the least mean square (LMS) adaptive filter (Haykin, 1998) and matrix back-propagation (MBP) (Anguita $et\ al.$, 1994), algorithms typically used in signal processing and control applications.

4.1 Cross-correlation Algorithm

The cross-correlation measures the similarity between two waveforms. Consider two signals with finite energy, $x(n)$ and $y(n)$. The cross-correlation of the two signals is defined as:

$$r_{xy}(l) = \sum_{n=-\infty}^{\infty} x(n)y(n-l); l = 0, \pm1, ... \quad (1)$$

or, equivalently, as

$$r_{xy}(l) = \sum_{n=-\infty}^{\infty} x(n+l)y(n); l = 0, \pm1, ... \quad (2)$$

The index l is the time shift, and the subscripts xy indicate the sequences being correlated.

4.2 LMS Algorithm

The LMS is one of the most successful adaptive algorithms. The LMS may be formulated as follows:

$$W_{k+1} = W_k - 2e_k\mu X_k \quad (3)$$

and,

$$e_k = y_k - W_k^T X_k \quad (4)$$

where W_k and X_k are the weight and the input signal vectors at time k respectively, e_x is the error signal, y_k is the current contaminated signal sample and μ is the constant controlling the stability and rate of convergence.

4.3 MBP Algorithm

MBP is a variation of the Back-Propagation. The algorithm can be defined in three phases as follows:
Feed-Forward

$$S(l) = b(l).1^T \quad (5)$$

$$S(l) = b(l-1)W(l) + S(l) \quad (6)$$

$$S(l) = f(S(l)) \quad (7)$$

Error Back-Propagation

$$\Delta(l) = \frac{2}{PN_L}(T - S(l)) \quad (8)$$

$$\Delta(l) = \Delta(l)g[S(l)] \quad (9)$$

$$\Delta(l) = \Delta(l+1)W^T(l+1) \quad (10)$$

$$\Delta(l) = \Delta(l)g[S(l)] \quad (11)$$

Weight updating

$$W(l) = \eta(S^T(l-1)\Delta(l) + W(l) \quad (12)$$

$$b(l) = 1^T\Delta(l) + \Delta b(l) \quad (13)$$

$S(l)$ is the neuron output matrix, T represents the network output matrix, $W(l)$ corresponds to the neuron weights matrix, $\Delta(l)$ is the propagated errors matrix and $b(l)$ is the neuron bias matrix. Besides these matrixes, there are references to functions f, the neuron activation function, and to g the first-order derivative of f. Wherever 1^T appears is a $mx1^T$ or 1^Txm operation.

5. IMPLEMENTATION

Considering that parallel systems are built upon processors with on-chip memory, message passing and data parallelism were the programming models used. The parallelisation of each algorithm was done according to Foster methodology. The algorithm parallelization considered three phases:

1. Identification of parallelism in the algorithm, choosing domain or functional partitioning,

2. Partition of the algorithm into tasks,

3.Allocation of tasks to processors.

To investigate the algorithms real-time implementation performance, the referred aspects have been efficiently explored in order to minimize inter-processor communications and to enable equally distributed computation load (Foster, 1995). Experiments were carried out varying the following parameters:

1. mapping stack, heap and static storage to internal and external memory,

2. communication of data structures in internal and external memory,

3. buffering techniques for data structures in external memory and its influence in the communication between processors.

The initial algorithm study was done with the default compiler memory allocation. Previously, for those compilers which allow optimisations in the memory management, other options of assigning memory to code, static storage, stack and heap were considered.

6. RESULTS

To evaluate the impact upon inter-processor data transmission of internal or external data location, some experiences were performed.

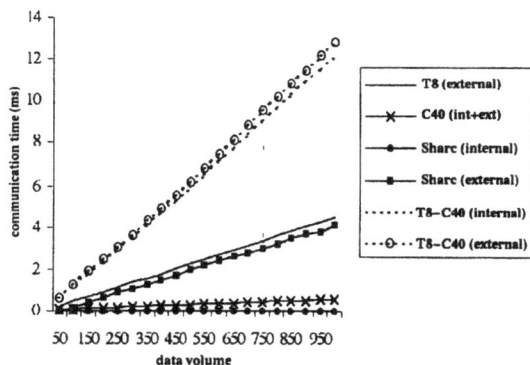

Fig. 2. Data communication times (ms) on homogeneous and heterogeneous systems

Figure 2 shows the communication times of 1000 data float elements on homogeneous and heterogeneous architectures with two processors. In the case of homogeneous architecture with two Sharc's, it is evident that the internal memory data transmission is 10 times faster than external transmission. For this amount of data, the homogeneous architecture with T8's didn't have sufficient on-chip memory to allocate data, so in the figure only appears the external memory data transmission time. In the platform with two C40

processors, it is not visible any significant difference on the transmission time of data in internal and external memory.

On the heterogeneous architecture composed by one T8 and a C40 (T8-C40), the inter-processor transmission time is the worst because of the hardware links among different type of processors and the software conversion from IEEE to ITI format (and vice-versa). The C40 processor has enough on-chip memory for this data set, but considering data mapped to external memory some differences are observed because of the slow transmission time among different types of processors.

Table 1 shows a comparison of the behaviour of the three homogeneous architectures in the implementation of cross-correlation algorithm with data structures in internal and external memory. The implementation of cross-correlation used two waveforms, of 2500 samples each. It is obvious that neither in the homogeneous architectures of T8's nor the C40's ones have enough internal memory to hold all elements.

Table 1. Cross-correlation algorithm execution times (ms) onto homogeneous architectures

Parallel Arch.	Physical Memory	Static Storage	Stack Storage	Heap Storage
2 x T8	external	12504,2	12504,2	12504,2
2 x C40	local	883	884	883
	global	2260	2244	2245
2 x Sharc	int-b0	707	707	706
	int-b1	394	394	394
	external	416	416	416

On the other hand, there is the Sharc processor with 2 blocks of 256 Kbytes each. In this condition, all data could be contained in internal memory. However, to show how the way of mapping data to memory blocks could influence the algorithm performance, other configurations will be considered.

The T8's homogeneous architecture presents the same execution time independently of the area where the data structure is placed. None of the stack, heap or static storage sections feet in internal memory because of their large dimensions.

On the C40´s architecture, the execution time for data structures in external memory depends on data being in local or global area. As can be seen in Table 1, the implementation of cross-correlation in local memory is about 150% more efficient than in global memory. When data is declared in stack or heap space the execution times are smaller than considering static storage. This is due to the extra cycle needed to set the data pointer (DP) to the correct page before accessing the global variable.

Cross-correlation had been implemented in the Sharc's homogeneous architecture, always considering code segment in the first block of internal memory (int-B0). In cases where this block includes others segments, the algorithm have the worst performance. Maximum efficiency is achieved when one block in internal memory contains instructions, while the other (int-B1) contains only data. The independent PM and DM buses allow simultaneous access to instructions and data from both memory blocks. In this situation the optimising was almost 80%. Another way to obtain good results is to store data in external memory and the code in internal memory. In this implementation the computation time is the same that was observed with the previous best solutions. The extra time was caused by the communication of data in external memory.

To investigate the behaviour of the architectures on the implementation of LMS algorithm, a finite impulse response (FIR) filter structure of 200 weights was used, the parameter μ was fixed to 0.04 and the execution of the algorithm occured over 1000 iterations. Under these condition of data volume, it was possible to study the internal memory access for T8's and C40's homogeneous architectures.

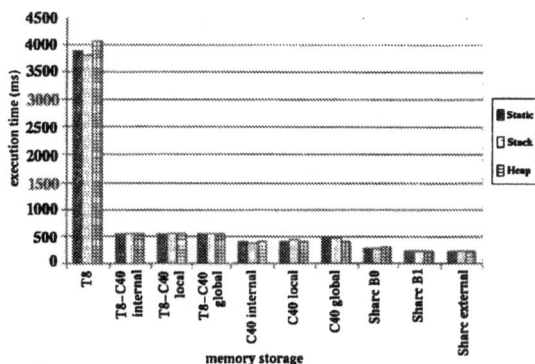

Fig. 3. Execution times (ms) obtained when implementing the LMS on the architectures

In comparasion with the implementation of cross-correlation algorithm, the T8's homogeneous architecture shows some different executions times for data declared in the heap, stack and static storage (figure 3). The most efficient option is the implementation with data mapped in the stack, benefiting from the stack being allocated in internal storage. The T8 processor has only 4 Kbytes of internal memory; this space is too small to include all the sections.

Considering the C40's architecture, the criteria observed with the cross-correlation are also applied to the LMS. In this case, it is possible to analyse the accessing time to internal memory. Here, the program and data memories lie in two separate spaces in internal memory. The best execution time in internal memory was obtained with data in stack storage. On the other hand, with external memory the best choice is to allocate data in the heap.

In the performance analysis of LMS in Sharc's homogeneous architecture, figure 3 shows that internal and external memory accessing times are similar, except when static and heap segments exist in the same block together with the code. Here, no difference was observed between execution time for data in second block or external memory. This is due to the small data volume involved in the experience.

The heterogeneous systems presented equal execution times independently of the memory allocation. The C40 processor is faster than the T8 one, computation load distribution to T8 processor had to be less than to C40, in order to reduce the execution time. All the system performance depends on the slower T8 processor.

The study held on systems with inefficient interprocessor data transmission of external data location was done to evaluate the buffering techniques improvements in the algorithm performance. In cases where on-chip data structures have to be accessed by processes on others processors, the best option was to allocate elements in internal memory. Sharc's homogeneous architecture was in those conditions, as observed in figure 2. To analyse the impact of such technique, the MPB algorithm performed the calcule of 16 bits XOR considering internal, external and hybrid storage.

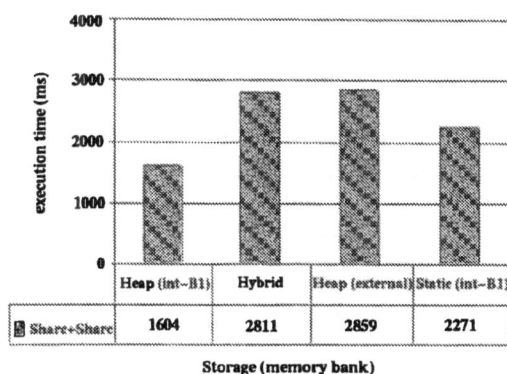

Storage (memory bank)	Heap (int-B1)	Hybrid	Heap (external)	Static (int-B1)
Sharc+Sharc	1604	2811	2859	2271

Fig. 4. Execution times (ms) obtained on the Sharc's architecture when implementing the MBP algorithm

Figure 4 compares execution times of the MBP algorithm implemented in Sharc's homogeneous architecture, considering that the main structures were on the heap. When data was allocated to external memory, the algorithm execution time was almost the double comparing with data in second block of internal memory (int-B1). The implementation of data as global variables in internal memory was analysed to evaluate the success of hybrid storage for those structures accessed by other processes.

The graphic in figure 4 show that no significant improvements were obtained. The execution times of data allocated to static storage allows to justifie the bad results obtained with the hybrid storage. Considering in internal memory only the structures that were accessed by other processors, this hybrid solution showed no impact in the algorithm performance.

7. CONCLUSION

This paper has explored the improvement in performance of parallel architectures in implementing cross-correlation, LMS and MBP algorithms. The inter-processor communication performance and computational performance have been investigated considering diferents memory storages. It has been proved that using processors such as T8 and C40, where the internal memory capacity is low (less than 10 Kbytes (INMOS Limited, 1990; Texas Instruments, 1995)) the transmission of structures in external memory is optimised. However, using the ADSP-21060 Sharc, presenting an internal memory capacity of 512 Kbytes (Analog Devices, 1995), the transmission from external memory is not efficient. When data structures mapped to external memory are used, the communication among processors presents a performance reduction of about 10 times regarding the case of communication of data structures in internal memory.

For processors with internal memory raised capacity, the assigment of data structure to external memory represents a loss of efficiency in inter-processors communication, but no significant reduction is obseved in the computation. In the Sharc's homogeneous architecture, the experiments performed demonstrate that code and data must be always in separated internal memory blocks.

For C40's homogeneous architectures, the best performance was observed with data mapped in the stack or heap. The access time to internal and local off-chip memory was similar, but the access to global off-chip area presented a great difference.

In the case of T8's homogeneous architectures, the processor had low internal memory capacity, allowing allocation to internal memory only to problems with small dimension. In this case, data structures stored in the stack were recommended.

The heterogeneous architecture had show a dependency from the slowest processor, the T8. If the algorithm shows some inter-processor communication, a tendency to maintain the execution independent from the data structures storage will be observed.

This investigation had revealed that hybrid storage implementation based on data transmission show no greatimprovements in performance as was expected, since the computation performance in external memory was similar to the computation in internal memory.

8. REFERENCES

Analog Devices (1995). *ADSP-2106x SHARC User's Manual.* Analog Devices.

Anguita, Davide, A. Da Canal, W. Da Canal, A. Falcone and A.M. Scapolla (1994). The distributed implementation of the back-propagation.

Foster, Ian (1995). *Designing and Building Parallel Programs - Concepts and Tools for Parallel Software Engineering.* Addison-Wesley Publishing Company.

Haykin, Simon (1998). *Neural Networks - a comprehensive foundation.* Prentice Hall.

INMOS Limited (1990). *IMS B004 user guide and reference manual.* INMOS Limited.

Kumar, Vipin (1994). *Introduction to Parallel Computing - Design and Analysis of Algorithms.* The Benjamin/Cummings Publishing Company, Inc.. USA.

Proakis, J. G. and D. G. Manolakis (1988). *Introduction to Digital Signal Processing.* Macmillan Publishing Company. USA.

Texas Instruments (1995). *TMS320 Floating-Point DSP Optimising C Compiler - User's Guide.* Texas Instruments. USA.

Tokhi, M.O. and M.A. Hossain (1995*a*). Cisc, risc and dps processors in real-time signal processing and control. *Microprocessors and Microsystems.*

Tokhi, M.O. and M.A. Hossain (1995*b*). Homogeneous and heterogeneous parallel architectures in real time signal processing and control. *Control Eng. Practice* **3**(12), 1675–1686.

Tokhi, M.O., M.A. Hossain, M.J. Baxter and P.J. Fleming (1995). Homogeneous and heterogeneous parallel architectures for real-time active vibration control. *IEEE Proc. Control Theory Appl.*

TOWARDS THE SIMPLIFIED COMPUTATION OF TIME-FREQUENCY DISTRIBUTIONS FOR SIGNAL ANALYSIS

F. García Nocetti, J. Solano González, E. Rubio Acosta and E. Moreno Hernández

Departamento de Ingeniería de Sistemas Computacionales y Automatización, Instituto de Investigaciones en Matemáticas Aplicadas y en Sistemas, Universidad Nacional Autónoma de México, P.O. Box 20-726, Del. A. Obregón, 01000 México D.F., México

Abstract: Conventional methods for signal analysis utilise the Fourier Transform to estimate the spectral response of a signal. However this current practice suffers from poor frequency resolution when estimating non-stationary signals. This paper presents some alternative methods based on time-frequency distributions such as the Wigner Ville, the Choi Williams, the Bessel and the Born Jordan from a Cohen's class point of view. For each case, a continuous and discrete distribution is formulated, a criterion for determining the interaction between the spectral components of the signal is given and the simplified discretised expression for the calculation of the distribution is proposed that can produce a reduction of at least half of the computations realised when using the original time-frequency distribution definition. A general parallel architecture for the parallel computation of the distribution is also proposed. *Copyright © 2000 IFAC*

Keywords: Signal Analysis, Spectral Estimation, Time frequency distribution, Computational Methods, Parallel Computation.

1. INTRODUCTION

A classic method for spectral estimation is the so-called Fourier Transform. However, its use is limited to stationary signals giving as a result poor frequency resolution when estimating non-stationary ones.

There are other spectral estimation methods (Kay, 1988) such as the Periodogram, the autoregressive method, the mobile average and the minimum variance spectral estimation. The performances of these methods also depend on the use of stationary signals and some of them (such as the autoregressive method) utilise short time-segments in order to consider the signal under study as stationary.

Other types of spectral estimators, called time-frequency distributions, have been developed. Unlike conventional methods these distributions are not limited to the use of stationary signals (Cohen, 1989). Despite of this important advantage, the number of calculations involved in obtaining the spectral estimation increases substantially compared to the traditional methods. Therefore, it is desirable to simplify the formulation of the distributions in such a way that the computations involved can be reduced without any loss in the spectral resolution.

On the other hand, there are a great variety of time-frequency distributions. It would be very useful to develop an analysis criterion such that can provide a tool for selecting the optimum time-frequency

distribution according to the features of the signal under consideration.

This paper deals with these issues. Section 2 gives the time-frequency definitions. Section 3 defines a comparison criterion based on the crossing terms weighting factors. Section 4 presents a reduction in the computational complexity of the time-frequency distributions and section 5 proposes a general parallel processing scheme to further reduce the time used in evaluating the distributions.

2. TIME-FREQUENCY DISTRIBUTIONS

This section formulates the so-called Cohen's class for the time-frequency distributions. It defines some concepts such as auto-term, crossing term and crossing term weighting factor in a distribution. It also develops the time-frequency distributions that will be analysed in this paper, that is, the Wigner Ville, the Choi Williams, the Bessel and the Born Jordan distributions.

2.1. The Cohen's Class

The Cohen's class in terms of time frequency distributions (Cohen, 1989), can be formulated as follows.

Let the time-frequency distribution kernel be defined as $\phi(\theta,\tau)$. This kernel will define the particular characteristics of each time-frequency distribution. Let the auto-correlation domain kernel $\psi(t,\tau)$ be defined as the Fourier transform of $\phi(\theta,\tau)$ (from θ to t, considering τ as constant). Let the generalised time-indexed auto-correlation function be defined as

$$R_x'(t,\tau) = \frac{1}{2\pi} \int_{-\infty}^{\infty} \psi(t-\mu,\tau) x\left(\mu+\frac{\tau}{2}\right) x^*\left(\mu-\frac{\tau}{2}\right) d\mu \quad (1)$$

Then, the Cohen class for the time-frequency distributions with kernel $\phi(\theta,\tau)$ can be defined as

$$TFD(t,\omega) = \int_{-\infty}^{\infty}\int R_x'(t,\tau) e^{-j\omega\tau} d\tau \quad (2)$$

2.2. Auto-term, Crossing Term and Crossing Term Weighting Factor

In order to establish a comparison criterion between the different time-frequency distributions considered in this paper, it is necessary to develop a method that may determine the degree in which the various components of a signal interact when the time frequency distribution is calculated (Cardoso, et al., 1996).

Consider the following signal, which is composed of a finite number of sinusoidal signals with constant amplitude A_n, frequency ω_n and phase θ_n.

$$x(t) = \sum_{n=1}^{N} A_n e^{j(\omega_n t + \theta_n)} \quad (3)$$

The n components of the signal (3) interact between them through (1). The interactions of the components with themselves generate the so-called auto-terms of the distribution, which are always positive and constitute the spectral contents of the signal. On the other hand, interactions between different components generate the so-called crossing terms of the distribution, which can be positive or negative and are added to the spectral contents of the signal. Therefore it is desirable to minimise the crossing terms of the distributions.

Substituting equation (3) in (2) and grouping the auto-terms in the first summation and crossing terms in the second summation, a time-frequency distribution can be expressed as

$$TFD(t,\omega) = 2\pi \sum_{n=1}^{N} A_n^2 \delta(\omega-\omega_n) +$$
$$+ \sum_{n=1}^{N}\sum_{\substack{m=1\\n\neq m}}^{N} F_{TFD}(\omega) A_n A_m \cos((\omega_n-\omega_m)t + \theta_n-\theta_m) \quad (4)$$

where $F_{TFD}(\omega)$ is the crossing terms weighting factor, a quantitative measure for evaluating the different time-frequency distributions. The following sections describe the distributions according to their definition.

2.3. The Wigner Ville Distribution

According to its definition (Cardoso, et al., 1996; Fan, and Evans, 1994; Martin, and Flandrin, 1985), the Wigner Ville distribution for the continuous case is given by

$$WVD(t,\omega) = \int_{-\infty}^{\infty} x\left(t+\frac{\tau}{2}\right) x^*\left(t-\frac{\tau}{2}\right) e^{-j\omega\tau} d\tau \quad (5)$$

where t is the time and ω is the angular frequency. For the discrete case, the distribution is given by

$$DWVD(n,k) = 2 \sum_{\tau=-N+1}^{N-1} W(\tau) W^*(-\tau) e^{-j\frac{2\pi k\tau}{N}} \bullet$$
$$\bullet x(n+\tau) x^*(n-\tau) \quad (6)$$

where n represents the discrete time and k the discrete frequency. Both variables are normalised.

2.4. The Choi Williams Distribution

According to its definition (Cardoso, et al., 1996; Choi, and Williams, 1989), the Choi Williams distribution for the continuous case is given by

$$CWD(t,\omega) = \int_{-\infty}^{\infty}\sqrt{\frac{1}{4\pi\tau^2/\sigma}} \int_{-\infty}^{\infty} e^{-\frac{(t-\mu)^2}{4\tau^2/\sigma}} \bullet$$
$$\bullet x\left(\mu+\frac{\tau}{2}\right) x^*\left(\mu-\frac{\tau}{2}\right) d\mu e^{-j\omega\tau} d\tau \quad (7)$$

where $\sigma > 0$ is a scaling factor, t is the time and ω is the angular frequency. For the discrete case is given by

$$DCWD(n,k) = 2 \sum_{\tau=-N+1}^{N-1} \left(W(\tau) W^*(-\tau) e^{-j\frac{2\pi k\tau}{N}} \bullet \right.$$
$$\left. \bullet \sum_{\mu=-M}^{M} \sqrt{\frac{1}{4\pi\tau^2/\sigma}} e^{-\frac{\mu^2}{4\tau^2/\sigma}} x(\mu+n+\tau) x^*(\mu+n-\tau) \right) \quad (8)$$

where n represents the discrete time and k the discrete frequency. Both variables are normalised.

2.5. The Bessel Distribution

According to its definition (Cardoso, et al., 1996; Guo, and Durand, 1994), the Bessel distribution for the continuous case is given by

$$BD(t,\omega) = \int_{-\infty}^{\infty} \frac{2}{\pi\alpha|\tau|} \int_{-\infty}^{\infty} \sqrt{1 - \left(\frac{t-\mu}{\alpha\tau}\right)^2} U_0\left(\frac{t-\mu}{\alpha\tau}\right) \bullet$$
$$\bullet \, x\left(\mu + \frac{\tau}{2}\right) x^*\left(\mu - \frac{\tau}{2}\right) d\mu e^{-j\omega\tau} d\tau \qquad (9)$$

where $\alpha > 0$ is a scaling factor, $U_0(t)$ is a second class Chebyshev polynomial, t is the time and ω is the angular frequency. For the discrete case is given by

$$DBD(n,k) = 2 \sum_{\tau=-N+1}^{N-1} \left(W(\tau) W^*(-\tau) e^{-j\frac{2\pi k\tau}{N}} \bullet \right.$$
$$\left. \bullet \sum_{\mu=-2\alpha|\tau|}^{2\alpha|\tau|} \frac{1}{\pi\alpha|\tau|} \sqrt{1 - \left(\frac{\mu}{2\alpha\tau}\right)^2} x(\mu+n+\tau) x^*(\mu+n-\tau) \right) (10)$$

where n represents the discrete time and k the discrete frequency. Both variables are normalised.

2.6. The Born Jordan Distribution

According to its definition (Cohen, 1989), the Born Jordan distribution for the continuous case is given by:

$$BJD(t,\omega) = \frac{1}{2\alpha} \int_{-\infty}^{\infty} \frac{1}{\tau} \int_{t-\alpha\tau}^{t+\alpha\tau} x\left(\mu + \frac{\tau}{2}\right) x^*\left(\mu - \frac{\tau}{2}\right) d\mu e^{-j\omega\tau} d\tau \quad (11)$$

where $\alpha > 0$ is a scaling factor, t is the time and ω is the angular frequency. For the discrete case is given by:

$$DBJD(n,k) = 2 \sum_{\tau=-N+1}^{N-1} \left(W(\tau) W^*(-\tau) e^{-j\frac{2\pi k\tau}{N}} \bullet \right.$$
$$\left. \bullet \sum_{\mu=-2\alpha|\tau|}^{2\alpha|\tau|} \frac{1}{4\alpha|\tau|} x(\mu+n+\tau) x^*(\mu+n-\tau) \right) \qquad (12)$$

where n represents the discrete time and k the discrete frequency. Both variables are normalised.

3. EVALUATION OF THE TIME-FREQUENCY DISTRIBUTIONS BASED ON THE CROSSING TERMS WEIGHTING FACTOR.

As stated previously, an ideal crossing terms weighting factor would be one that eliminates the crossing terms. Other desirable situations outside the ideal would be that the weighting factor concentrates the crossing terms due to two different frequency components of a signal around such frequencies and not around other frequencies or spread them out over a wide range of frequencies.

Substituting the signal defined by (3) in (5), (7), (9) and (11) and arranging the crossing terms in the distributions according to (4), the crossing terms weighting factors of each distribution are obtained

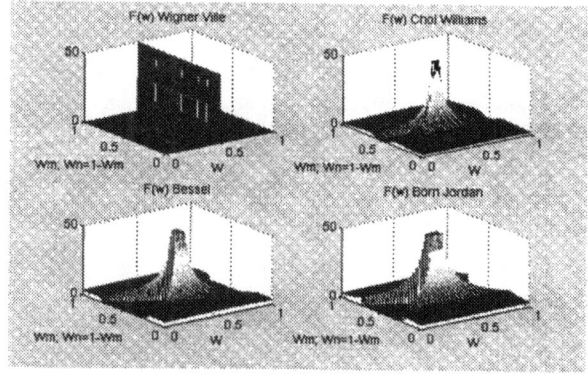

Fig. 1: Global view of the crossing terms weighting factors for the Wigner Ville, Choi Williams $(\sigma = 1)$, Bessel $(\alpha = 1)$ and Born Jordan distributions.

and defined by the following expressions.

For the Wigner Ville distribution:

$$F_{WVD}(\omega) = 2\pi\delta\left(\omega - \frac{\omega_n + \omega_m}{2}\right) \qquad (13)$$

For the Choi Williams distribution:

$$F_{CWD}(\omega) = \sqrt{\frac{\pi\sigma}{(\omega_n - \omega_m)^2}} e^{-\frac{\sigma}{4(\omega_n - \omega_m)^2}\left(\omega - \frac{\omega_n + \omega_m}{2}\right)^2} \qquad (14)$$

For the Bessel distribution:

$$F_{BD}(\omega) = \frac{4}{\alpha|\omega_n - \omega_m|} U_0\left(\frac{\omega - \frac{\omega_n + \omega_m}{2}}{\alpha(\omega_n - \omega_m)}\right) \sqrt{1 - \left(\frac{\omega - \frac{\omega_n + \omega_m}{2}}{\alpha(\omega_n - \omega_m)}\right)^2} \quad (15)$$

For the Born Jordan distribution:

$$F_{BJD}(\omega) = \frac{\pi}{\alpha|\omega_n - \omega_m|} P_{2\alpha(\omega_n - \omega_m)}\left(\omega - \frac{\omega_n + \omega_m}{2}\right) \qquad (16)$$

where $P_a(t)$ is a rectangular symmetrical pulse of duration a.

Figure 1 shows a global view of the crossing terms weighting factors for the Wigner Ville, Choi Williams, Bessel and Born Jordan distributions. The weighting factor is defined in terms of ω and contains the interacting frequencies ω_n and ω_m defined in signal (3). Such frequencies can be added or subtracted depending on the behaviour of each distribution. Consider a normalised addition, that is $\omega_n + \omega_m = 1$, then the graphs relate the weighting factor against ω_n and ω, where: $0 < \omega_n < 1$. Given the normalised addition, for each ω_n value, ω_m is given by $1 - \omega_n$.

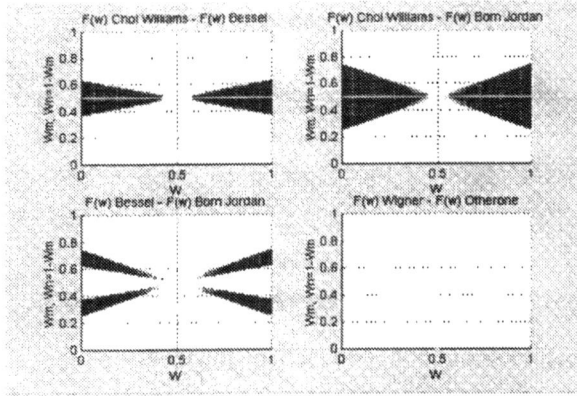

Fig. 2: Differences of the crossing terms weighting factors between Choi Williams-Bessel, Choi Williams-Born Jordan, Bessel-Born Jordan and Wigner Ville-any other distribution.

3.1. Analysis

Figure 2 depicts four graphs which show the differences of the crossing terms weighting factors between Choi Williams-Bessel, Choi Williams-Born Jordan, Bessel-Born Jordan and Wigner Ville-any other distribution.

The dark zones in the graphs correspond to points where the weighting factor of the first distribution under comparison is greater than the second one. For the purpose of this analysis scaling factors $\sigma = 1$ and $\alpha = 1$ are considered in the distributions. Its is observed that the weighting factor for the Choi Williams distribution is smaller than the Bessel's and that the Bessel's is smaller than the Born Jordan's. These results indicate that the Choi Williams distribution spreads out the crossing terms better than the Bessel's and Born Jordan's distributions and, in consequence, estimates with more precision the spectral contents of a signal in the presence of noise. In the case of the Wigner Ville distribution, it concentrates the crossing terms (due to the two frequency components of the signal) on the average of such frequencies. Therefore, the Wigner Ville distribution estimates with better precision the spectral contents of noiseless signals with small bandwidth.

It is important to point out that all the distributions are affected by a scaling factor which in turn modifies the crossing term weighting factors. As stated previously, the results have been obtained considering $\sigma = 1$ and $\alpha = 1$ but the optimum scaling factors must be found experimentally and they depend on the characteristics of the signal under study.

As an illustration, consider the Choi Williams and Bessel distributions with optimal scaling factors of $\sigma = 5$ and $\alpha = 2$ respectively. Figure 3 depicts the differences in the weighting factors between the distributions. It is observed that in general the Choi

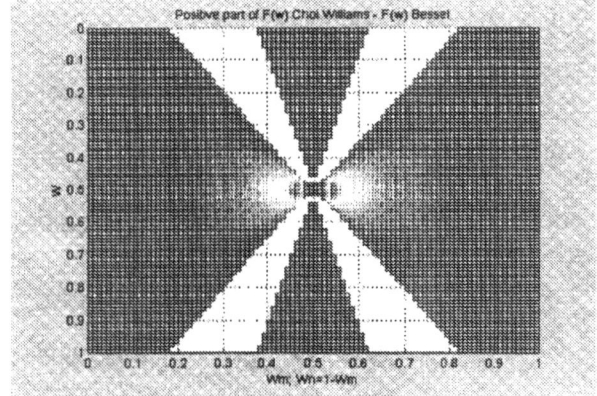

Fig. 3: Differences of the crossing terms weighting factor between the Choi Williams distribution ($\sigma = 5$) and the Bessel distribution ($\alpha = 2$).

Williams distribution presents the higher crossing terms values as shown by the differences. That explains the reason why the Bessel distribution is less sensitive to the presence of noise (Cardoso, et al., 1996).

A similar analysis can be made for the rest of the distributions, however the optimum scaling factor for each of the distributions must be obtained. This task will be ready shortly.

4. REDUCING THE COMPUTATIONAL COMPLEXITY OF THE DISCRETE DEFINITIONS

In order to evaluate the different distributions for spectral estimation, a discrete signal $x(n)$ is considered. Such a signal contains $2N - 1$ elements, where N is a power of 2 and the element range is from $-N + 1$ to $N - 1$, therefore $x(0)$ is the central element. Based on these elements, this section presents a reduction in computational terms of the number of calculations involved in the evaluation of each of the distributions considered in this paper.

4.1. The Wigner Ville distribution

Considering (6) for estimating the Wigner Ville distribution and evaluating it in n=0 (Boashash, and Black, 1987; Fan, and Evans, 1996), an equivalent simplified expression would be given by

$$DWVD(0,k) = 4\mathrm{Re}\left[\sum_{\tau=0}^{N-1} W(\tau)W^*(-\tau)e^{-j\frac{2\pi k\tau}{N}} \bullet \right.$$

$$\left. \bullet x(\tau)x^*(-\tau)\right] - 2|x(0)|^2 \qquad (17)$$

Assuming that $W(\tau)W^*(-\tau)$ is a single factor then, for each value of k in (6) evaluated in $n = 0$, there are $6N - 3$ complex multiplications, $2N - 2$ complex additions and 1 scalar multiplication, whereas in (17) there are $3N+1$ complex multiplications, N complex additions and 2 scalar multiplications.

4.2. The Choi Williams distribution

Similarly, considering (8) for estimating the Choi Williams distribution and evaluating it in $n=0$, an equivalent simplified expression would be given by

$$DCWD(0,k)= 4\operatorname{Re}\left[\sum_{\tau=0}^{N-1}\left(W(\tau)W^*(-\tau)e^{-j\frac{2\pi k\tau}{N}}\bullet\right.\right.$$
$$\left.\left.\bullet\sum_{\mu=-N+1+|\tau|}^{N-1-|\tau|}\sqrt{\frac{1}{4\pi\tau^2/\sigma}}e^{-\frac{\mu^2}{4\tau^2/\sigma}}x(\mu+\tau)x^*(\mu-\tau)\right)\right]-2|x(0)|^2 \quad (18)$$

where the summation respect to μ for $\tau=0$ is $x(0)x^*(0)$.

Assuming that $M=N-1$ and that $W(\tau)W^*(-\tau)$ and $\sqrt{\frac{1}{4\pi\tau^2/\sigma}}e^{-\frac{\mu^2}{4\tau^2/\sigma}}$ are single factors then, for each value of k in (8) evaluated in $n=0$, there are $8N^2-4N$ complex multiplications, $4N^2-6N+2$ complex additions and 1 scalar multiplication, whereas in (18) there are $2N^2-2N+1$ complex multiplications, N^2-2N complex additions and 2 scalar multiplications.

4.3. The Bessel distribution

Considering (10) for estimating the Bessel distribution and evaluating it in $n=0$, an equivalent simplified expression would be given by

$$DBD(0,k)= 4\operatorname{Re}\left[\sum_{\tau=0}^{N-1}\left(W(\tau)W^*(-\tau)e^{-j\frac{2\pi k\tau}{N}}\bullet\right.\right.$$
$$\bullet\sum_{\mu=max\{-2\alpha|\tau|,-N+1+|\tau|\}}^{min\{2\alpha|\tau|,N-1-|\tau|\}}\frac{1}{\pi\alpha|\tau|}\sqrt{1-\left(\frac{\mu}{2\alpha\tau}\right)^2}\bullet$$
$$\left.\left.\bullet x(\mu+\tau)x^*(\mu-\tau)\right)\right]-2|x(0)|^2 \quad (19)$$

where the summation respect to μ for $\tau=0$ is $x(0)x^*(0)$.

Assuming that $W(\tau)W^*(-\tau)$ and $\frac{1}{\pi\alpha|\tau|}\sqrt{1-\left(\frac{\mu}{2\alpha\tau}\right)^2}$ are single factors then, for each value of k in (10) evaluated in $n=0$, there are $8\alpha N^2-8\alpha N$ complex multiplications, $4\alpha N^2-4\alpha N-2N$ complex additions and 1 scalar multiplication, whereas in (19) there are less than $4\alpha N^2-4\alpha N+1$ complex multiplications, less than $2\alpha N^2-2\alpha N-N$ and 2 scalar multiplications.

4.4. The Born Jordan distribution

Considering (12) for estimating the Born Jordan distribution and evaluating it in $n=0$, an equivalent simplified expression would be given by

$$DBJD(0,k)= 4\operatorname{Re}\left[\sum_{\tau=0}^{N-1}\left(W(\tau)W^*(-\tau)e^{-j\frac{2\pi k\tau}{N}}\bullet\right.\right.$$
$$\left.\left.\bullet\sum_{\mu=max\{-2\alpha|\tau|,-N+1+|\tau|\}}^{min\{2\alpha|\tau|,N-1-|\tau|\}}\frac{1}{2\alpha|\tau|}x(\mu+\tau)x^*(\mu-\tau)\right)\right]-2|x(0)|^2 \quad (20)$$

where the summation respect to μ when $\tau=0$ is $x(0)x^*(0)$.

Assuming that $W(\tau)W^*(-\tau)$ and $\frac{1}{2\alpha|\tau|}$ are single factors then, for each value of k in (12) evaluated in $n=0$, there are $8\alpha N^2-8\alpha N$ complex multiplications, $4\alpha N^2-4\alpha N-2N$ complex additions and 1 scalar multiplication, whereas in (20) there are less than $4\alpha N^2-4\alpha N+1$, less than $2\alpha N^2-2\alpha N-N$ complex additions and 2 scalar multiplications.

5. PARALLEL PROCESSING OF THE TIME-FREQUENCY DISTRIBUTIONS.

As stated previously, the use of time-frequency distributions for the spectral estimation of signals opens the possibility of analysing signals that could be non-stationary. However, the computational cost is high.

In view of this, this paper has proposed a reduction in the amount of calculations involved for evaluating the original definitions of each distribution, as developed in the previous section.

In addition it is proposed the use of parallel processing techniques to further reduce the time required to perform the evaluations. In particular, a pipeline scheme is used with three stages. The first stage calculates the analytic signal $x_a(t)$ of the real signal given by $x(t)$, where $x_a(t)$ is the inverse Fourier transform of $X_a(f)$, which is given by:

$$X_a(f)=\begin{cases} 2X(f) & f>0 \\ X(f) & f=0 \\ 0 & f<0 \end{cases} \quad (21)$$

Similarly, $X(f)$ is the Fourier transform of $x(t)$.

The second stage calculates the generalised time-indexed auto-correlation function $R_x'(t,\omega)$ for $t=0$ of $x_a(t)$.

Finally, the third stage calculates the Fourier transform of $R_x'(0,\omega)$, which is the time-frequency

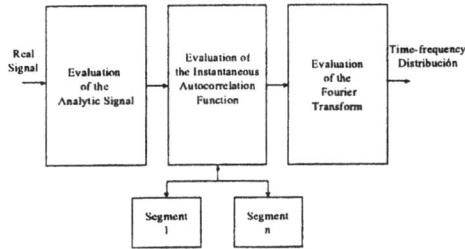

Fig. 4: General Parallel Processing scheme for the evaluation of the time-frequency distributions.

distribution $TFD(t,\omega)$ for $t = 0$ of the real signal $x(t)$.

Figure 4 shows the pipeline structure of the process.

For the first and third stages the calculations are relatively simple and a Fast Fourier Transform (FFT) algorithm is used.

However, the second stage requires of a more complex process, therefore this stage is further exploited using for this purpose a parallel farm computational model in a star topology. Here, each node calculates a set of operations of the generalised time-indexed auto-correlation function.

Although the expressions for the evaluation of each of the time-frequency distributions are different, this second stage can be adapted easily adding or subtracting processors according to the needs. A further analysis of the parallel implementation is a subject of a future work.

6. CONCLUSIONS

Conventional methods for spectral estimation are limited to the analysis of stationary signals to produce a good estimate. However, these methods offer poor resolution when dealing with non-stationary signals.

This paper has presented some alternative methods based on the so-called time-frequency distributions for spectral analysis. Four methods based on the Cohen's class have been analysed, namely the Wigner Ville, the Choi Williams, the Bessel and the Born Jordan distributions. A comparison criterion based on the crossing terms weighting factors has been proposed showing that the Choi Williams distribution spreads out the crossing terms better than the Bessel's and Born Jordan's distributions and, in consequence, estimates with more precision the spectral contents of a signal in the presence of noise, whereas the Wigner Ville distribution estimates with better precision the spectral contents of noiseless signals with small bandwidth. A further analysis has

to be conducted taking into account the optimum scaling factors which must be obtained experimentally (e.g. $\sigma = 5$ and $\alpha = 2$ for the Choi Williams and Bessel distributions, respectively). Note that for this case, the Bessel distribution spreads out the crossing terms better than the Choi William distribution.

This work also has proposed a simplification in complexity of the expressions utilised for calculating the time-frequency distributions giving as a result a reduction of at least half the operations involved in the original definition.

Finally, this paper has proposed a parallel processing scheme for the computation of the time-frequency distribution methods. Here, a pipeline scheme with three stages is utilised, corresponding to the second stage to deal with the more expensive computational process (evaluation of the generalised time-indexed auto-correlation function). A generalised scheme has been described which can adapt easily its topology according to the time-frequency distribution under consideration. Further analysis of the time performance of the system implementation will follow shortly.

7. ACKNOWLEDGEMENTS

The authors acknowledge the Universidad Nacional Autónoma de México, UNAM (PAPIIT IN106796), CONACYT-REDII(7350-858), CONACYT-México (2146P-A9507) and CONACYT-México (27982-A) for the financial support.

REFERENCES

Boashash, B. and P. Black (1987). An Efficient Real-Time Implementation of the Wigner-Ville Distribution. *IEEE Transactions on Acoustics, Speech, and Signal Processing.* **ASSP-35**. 1611-1618.

Cardoso, J. G. Ruano and P. Fish (1996). Nonstationary Broadening Reduction in Pulsed Doppler Spectrum Measurements Using Time-Frequency Estimators. *IEEE Transactions on Biomedical Engineering.* **43**. 1176-1186.

Choi, H. and W. Williams (1989). Improved Time-Frequency Representation of Multicomponent Signals Using Exponential Kernels. *IEEE Transactions on Acoustics, Speech and Signal Processing.* 37. 862-871.

Cohen, L. (1989). Time-Frequency Distributions -A Review. *Proceedings of the IEEE.* **77**. 941-981.

Fan, L. and D. Evans (1994). Extracting Instantaneous Mean Frequency Information from Doppler Signals Using the Wigner Distribution Function. *Ultrasound in Med. & Biol.* **20**. 429-443.

Guo, Z., L. Durand and H. Lee (1994). The Time-Frequency Distributions of Nonstationary Signals Based on a Bessel Kernel. *IEEE Transactions on Signal Processing.* **42**. 1700-1707.

Kay, S. (1988). *Modern Spectral Estimation. Theory & Application.* Prentice Hall. New Jersey.

Martin, W. and P. Flandrin (1985). Wigner-Ville Spectral Analysis of Nonstationary Processes. *IEEE Transactions on Acoustics, Speech, and Signal Processing.* **ASSP-33**. 1461-1470.

A CONTROL SYSTEM PROCESSOR ARCHITECTURE FOR COMPLEX LTI CONTROLLERS

**Roger Goodall[1], Simon Jones[1], Rene Cumplido-Parra[1],
Fiona Mitchell[1] and Stephen Bateman[2]**

[1]*Department of Electronic & Electrical Engineering
Loughborough University, Loughborough, Leics. LE11 3TU. UK.
Email: R.M.Goodall@lboro.ac.uk*
[2] *Gatefield Corporation, USA*

Abstract: The paper describes the use of FPGA-type devices for implementing complex, high sample rate, linear time invariant controllers. The overall methodology involves re-formulating the controller into a particular discrete state-space representation, which is optimised for numerical efficiency, then programming this into a specially-designed Control System Processor (CSP) implemented using a "programmable ASIC" device.

Copyright © 2000 IFAC

Keywords: Control algorithms, Architectures, Arithmetic and logic units, Control applications, Hardware.

1. INTRODUCTION

1.1 *Trends in control*

There is now a variety of control design methods by which appropriate control laws can be created for complex multi-variable systems, but the actual implementation of control laws is a part of the design process which most control engineers want to achieve as straightforwardly and transparently as possible. One approach is to programme a fixed point microprocessor (µP) device in a high level language, using floating point variables so that numerical issues are not a concern, but the computational overhead is large and surprising restrictions in sample rate are found. A second approach is to use a fixed point digital signal processor (DSP) which has an architecture better targeted for computationally-intensive applications, and these offer some speed advantage over a µP. Other options are to use a number of parallel processors or sophisticated floating-point DSPs, but this doesn't result in a cost-effective solution, especially for high-volume embedded control applications.

The difficulty is that there are particular numerical requirements in control system processing for which standard processor devices are not well suited, in particular arising from the high sample rates which are need to avoid adverse effects of sample delays upon stability. These could be satisfied in either µP or DSP devices using "hand-crafted" numerical

routines, probably written in Assembler language, but as mentioned above control engineers generally have neither the will nor the skill to do this. There is therefore a clear need to understand the numerical requirements properly, to identity optimised forms for implementing control laws, and to translate these into efficient processor architectures.

1.2 *Trends in silicon*

Continuing improvements in microelectronic technology has made feasible large reprogrammable silicon chips (Field Programmable Gate Arrays – FPGAs) which can be configured to realise complex computational systems without incurring either the delay or the costs associated with full-custom silicon. As a result electronic designers and control engineers are looking once again at the potential of designing low-cost, high-performance special-purpose hardware for embedded real-time control. With current chip complexities of up to 2 million designable gates and circuit density growth of 100% every 18 months it appears that the complexity of the algorithm is limited only by design capability and not by silicon complexity.

1.3 *The use of targetted architectures*

It is well accepted that customisation of silicon offers cost and performance advantages over standard components. Furthermore FPGAs open up this market to a much smaller product volume.

Contemporary microprocessors offer high-performance at the price of increased cost and power consumption. Furthermore, there is good evidence that the extensive use of floating-point numbers in calculation results in large chips and slow operation. Our view is that by taking a considered view of the numerical and calculation requirements of the algorithm allows special purpose processors to be considered which provide well-targeted support of control laws. Such systems are likely to be smaller, cheaper, faster and lower power than conventional signal processors. In the past there has been much work on special purpose processors for control (Jaswa et al, 1985; Spary and Jones, 1991) but while they are intriguing ideas the cost of producing custom silicon proved prohibitive for initial exploitation and restricted experimentation with different architectural constructs. With High-level design tools such as VHDL and logic synthesis CAD suites allied to large low-cost reprogrammable FPGAs, the constraints no longer apply and we can now develop this area with full enthusiasm.

2. CONTROLLER FORMULATIONS

2.1 z and δ operators

The discrete state-space form is a natural way of expressing the implementation equations for complex controllers, and for this reason is used as a basis in this paper. Transforming other expressions, e.g. discrete transfer functions or continuous expressions, into this form is relatively straightforward.

$$zX = A_z X + B_z U$$
$$Y = C_z X + D_z U \tag{1}$$

The numerical problems associated with discrete-time control, in which the sample frequency will typically be two orders of magnitude higher than the controller bandwidth, are well known when the z operator is used (Liu, 1971). Similarly it is recognised that the use of the δ operator overcomes a number of these problems, in which case the state equation becomes

$$\delta X = A_\delta X + B_\delta U \tag{2}$$

This is sometimes defined as $\delta = (z - 1)T$ (where T is the sample period), in which case there is a unification between discrete and continuous time since $\delta \rightarrow s$ (the Laplace operator) as $T \rightarrow 0$ (Middleton and Goodwin, 1990). In fact for the relatively high sample frequencies found for practical controllers $\delta \approx s$ is quite a realistic approximation and the coefficients in A_δ and B_δ become almost independent of the sample period. The effect of sample period must of course be taken into account when implementing δ, and an alternative simpler definition is to use $\delta = z-1$ (Goodall and Brown, 1985), in which case the

correspondence between δ and s is lost but implementation is more direct.

2.2 Implementation equations

The general form of the actual control equations which will be implemented is

$$X(n+1)T = AX(nT) + BU(nT)$$
$$Y(nT) = CX(nT) + DU(nT) \tag{3}$$

This corresponds directly to the z formulation given in Eqn (1), but its straightforward to show that it can also be used to represent the δ form with a different choice of controller states, and of course with corresponding changes in A, B and C. The comparison between the use of z and δ operators is valuable because it highlights the numerical problems which can arise and offers a practical solution, but in fact the choice of controller states goes beyond just the "z or δ?" issue. It provides a design freedom which the implementer can use: for a given control law the most appropriate set of states can be freely chosen to optimise the controller formulation from the numerical point of view. Some researchers have studied the exploitation of this design freedom in a rigorous manner to create so-called "minimal realisations" (Gevers and Li, 1993). However it's important to appreciate that these are only optimal in the strict mathematical sense because generally the controller A matrix will be fully populated with elements, for each of which a multiplication is required. Other formulations may be strictly sub-optimal by comparison, but many of the A matrix elements will be 0 or 1. If the full matrix equation is calculated this makes no difference, but if the "structure" of the A matrix is recognised it is possible to extract the essential equations from the full matrix formulation and thereby reduce considerably the number of computations which are needed.

2.3 Formulation used for the CSP

In this research study we use a "modified canonic" form based upon the δ operator. This is illustrated diagrammatically in Figure 1 for a fourth-order SISO controller.

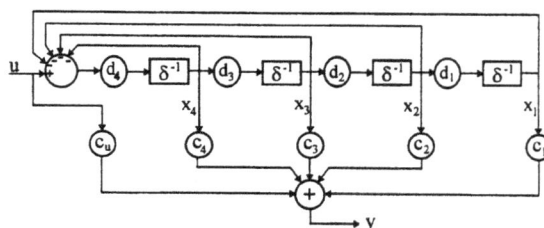

Fig. 1. Modified canonic δ formulation.

The corresponding state equations are:

$$\begin{bmatrix} x_1 \\ x_2 \\ x_3 \\ x_4 \end{bmatrix}_{n+1} = \begin{bmatrix} 1 & d_1 & 0 & 0 \\ 0 & 1 & d_2 & 0 \\ 0 & 0 & 1 & d_3 \\ 1-d_4 & -d_4 & -d_4 & -d_4 \end{bmatrix} \begin{bmatrix} x_1 \\ x_2 \\ x_3 \\ x_4 \end{bmatrix}_n + \begin{bmatrix} 0 \\ 0 \\ 0 \\ d_4 \end{bmatrix} u_n \quad (4)$$

$$y_n = \begin{bmatrix} c_1 & c_2 & c_3 & c_4 \end{bmatrix} \begin{bmatrix} x_1 \\ x_2 \\ x_3 \\ x_4 \end{bmatrix}_n + \begin{bmatrix} c_u \end{bmatrix} u_n$$

and the actual equations used for real-time implementation are as below; firstly the calculation of the output, then an update of the states ready for the next sample:

$$y := c_1 x_1 + c_2 x_2 + c_3 x_3 + c_4 x_4 + c_u u$$
$$x_{temp} := x_1 + x_2 + x_3 + x_4$$
$$x_1 := x_1 + d_1 x_2$$
$$x_2 := x_2 + d_2 x_3 \quad (5)$$
$$x_3 := x_3 + d_3 x_4$$
$$x_4 := x_4 - d_4 x_{temp} + d_4 u$$

Notice that x_{temp} is used to store the sum of the old values of x_1 to x_4, which thereby avoids having to retain old values for the states while the new values are calculated – the state variables are then simply overwritten at each calculation.

This particular formulation, although its state equation appears a little more complex, in fact yields straightforward real-time equations, and the particular features are as follows:

- uses the δ operator to minimise coefficient sensitivity, and to reduce wordlength overall

- preserves a structured A matrix such that a full matrix calculation is unnecessary

- uses "forward path" multipliers (see Figure 1) with the δ^{-1} operations to create controller variables which are all well-scaled, i.e. they all have the same nominal maximum values

- avoids the need to convert the controller into cascaded first/second order sections, something which is almost essential for high-order controllers using formulations based upon the z operator.

The CSP architecture has also been designed to allow for "block-structured" controllers, in which a number of controller blocks, each implemented by the formulation described in section 2.3, can be arbitrarily interconnected from inputs to outputs. This approach can be used to reduce controller complexity if it is possible to identify appropriate sub-structures. More importantly it enables complex classically-designed controllers to be implemented readily, because such controllers are almost invariably built up through intuitive understanding and experience using an application-specific set of loops, compensators and filters, i.e. the block-structured form.

2.4 Numerical requirements

Using this standard controller formulation the requirements for coefficients and controller state variables become relatively standardised across a wide range of applications, and these are illustrated in Figure 2.

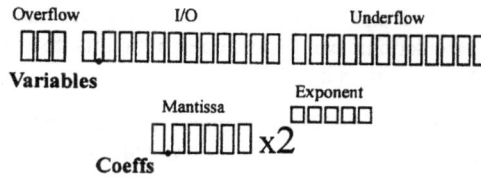

Fig. 2. Numerical specification.

The variables are 27 bit fixed point, with the input values brought in as signed fractional form, a small 3-bit allowance for overflow (although this is a nominal requirement because of the good scaling properties mentioned previously), and 12 fractional bits for underflow.

The coefficients are held in a simple low-precision floating point form, with 6 bits for the mantissa and 5 bits for the exponent. In general the coefficients are fractional, with values which become progressively smaller as the sample frequency is increased, but a positive exponent is provided to implement greater than unity gains, a small number of which are associated with most controllers.

This numerical specification will implement successfully the vast majority of LTI controller examples, and allows for the sample frequency being at least three orders of magnitude higher than the lowest pole in the controller. Of course if there are exceptional requirements it is always possible to reprogramme the CSP hardware in the FPGA, maintaining the essential principles but extending the hardware precision as required. An example of implementing extremely high sample frequency digital filters using the modified canonic δ approach can be found in Goodall and Donaghue (1993).

3. ARCHITECTURE OF PROCESSOR

3.1 Overall scheme

The processor design is extremely simple. It comprises a 4-port register bank (3 read, 1 write) allied to a special-purpose multiply accumulator (MAC). This MAC calculates D = A*B +C in a single cycle and writes the result back to the register bank. The A input is in coefficient format (11 bits) and the B and C values are in variable format (27 bits). A detailed low level design has been used to speed up the MAC operation. The system is pipelined such that there is a latency of 4 clock cycles between instruction issues and the result being written back to the register bank. The Java-based compiler ensures that instruction dependencies are observed through an appropriate series of instruction issues.

177

The processor only has four instructions (see Table 1). The MAC instruction executes a multiply-accumulation operation on the operands indicated by the source addresses and stores the result in the destination address. This instruction performs the matrix multiplication, according to the state-space representation of the control system. Because the constants 0, 1 and -1 are stored in the register bank, other calculations can be mapped into this format (e.g. D=C is achieved by setting A to 0).

There are no conditional jumps in the system. Unconditional jumps are supported for user programming. However, the code generator flattens all but the exterior loop. The program counter starts at zero and increments until it reaches the value stored in the 'jump1 register'. It is then reset to the value in the 'jump2 register'. This permits an initial start up sequence to be followed and then a main loop to be repeatedly executed.

The READ instruction loads initial values from the data ROM into the register bank during the initialisation process. Also, when the algorithm loop has begun, this instruction is used to read sampled input data from the input bus and to indicate the conversion time for the ADC's. Finally, the WRITE instruction is used to transfer the output values to the output bus.

Table 1. CSP instructions.

Instruction	Function	Description
MAC d, s1, s2, s3	(s1*s2) + s3 -> d	Multiply accumulation operation
WRITEPC sel, s	s -> pc[sel]	Write to program counter registers (PC,jump1,jump2)
READ d, in_pt, s	If in_pt = 0 DataRom[s] -> d else input[in_pt] -> d	Read from data ROM or input ADC
WRITE out_pt, s	s -> output[out_pt]	Read from register bank and write to output DAC

The program is stored in the same chip as the processor together with simple Boolean functions for interlocking etc. A/D and D/A interfaces are also provided.

The processor is implemented together with the appropriate programme in an Actel ProAsic FPGA which is a flash-programmable device (non-volatile), offering a one-chip solution. Figure 3 shows a block diagram of the CSP architecture and figure 4 shows the CSP I/O interface.

3.2 Complexity and clock speed.

Table 2 shows the CSP complexity in terms of ProAsic tiles and equivalent gates. Everything except the program and data ROM are fixed in size; these are hardwired, and their size and speed depends upon the control algorithm being implemented. The figures shown are for the 4th order example controller specified in section 2.3.

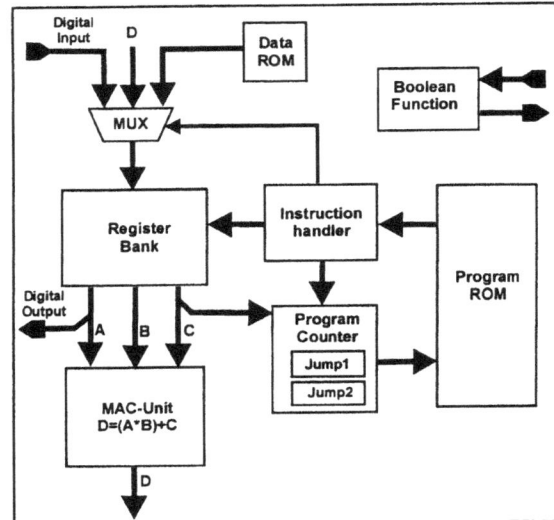

Fig. 3. Processor architecture.

The synthesis of the CSP results in an overall gate count of fewer than 21,000 gates and a delay of 20ns. The device has been successfully operated at a clock frequency of 50 MHz. As such the CSP is a compact low cost core capable of implementing the most demanding real-time systems. The relatively small size of the processor core leaves much of the FPGA free such that it can be used to carry out other functions typically associated with real-time control – logical interlocking functions, background tasks such as gain-scheduling, etc.

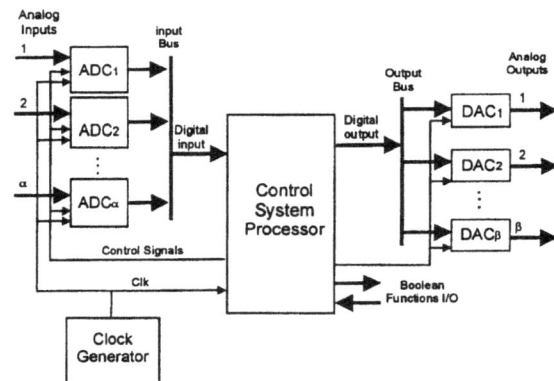

Fig. 4. Processor interface.

Table 2. CSP complexity.

Block	Tiles (ProAsic)	Equivalent gates
Instruction Handler	101	808
MAC Unit	1105	8840
Program counter	175	1400
I/O Block	60	480
Pipeline registers	120	960
Program ROM	900	7200
Data ROM	80	640
Total	**2541**	**20328**

4. IMPLEMENTATION PROCESS

4.1 Controller conversion into standard form

For the CSP to be used it is necessary that any controller can be converted into the modified canonic form, and there is a number of ways in which this can be achieved. The detail is not appropriate here, but the following paragraphs give indications of the kind of processes required, any of which can be readily expressed in Matlab, for example.

If the controller already exists in controller canonical form in which row 1 of the state matrix A has the "feedback" coefficients, a two-stage transformation process can be employed. The first stage involves the expression $\delta = z - 1$, which to convert from z to δ implies a transformation matrix involving binomial expansions of this expression. For example, for a fourth order system this gives

$$T_1 = \begin{bmatrix} 1 & -3 & 3 & -1 \\ 0 & 1 & -2 & 1 \\ 0 & 0 & 1 & -1 \\ 0 & 0 & 0 & 1 \end{bmatrix} \quad giving \; A_1 = T_1 A T_1^{-1} \quad (6)$$

The first row of the transformed state matrix A_1 now contains the "feedback" coefficients for the δ form. If this first row is $[r_1 \; r_2 \; r_3 \; r_4]$, then a second transformation can be used to move these coefficients onto the forward path, as required for the modified form:

$$T_2 = diag(r_1 \; r_2 \; r_3 \; r_4) \quad (7)$$

Writing the overall transformation $T = T_2 T_1$, the new discrete state-space representation is given by $A' = T A T^{-1}$; $B' = TB$; $C' = CT^{-1}$; $D' = D$.

Alternatively if the required controller is expressed in an unstructured form (i.e. such that there are few or no zero and unity coefficients), the easiest approach is to use T_1 to convert from z to δ, then determine the eigenvalues of the δ state matrix. From these it is straightforward to identify what should be the appropriate state matrix for the modified form, and a similarity transform can be used to determine the corresponding transformation matrix, after which the matrix conversion given in the previous paragraph completes the process. (Note that for complex high sample rate controllers many of the eigenvalues will be close to $z = 1$, but the conversion to δ form moves these close to zero and hence facilitates the transformation process numerically.)

Sometimes the controller may be expressed in continuous form, either as transfer functions in s or as a continuous state-space system, in which case the most appropriate approach is to convert directly from s to δ, for example using the algebraic process involved in the binomial transform, and translating this into modified canonic form follows straightforwardly.

4.2 Programming the CSP

A software development system has been constructed in Java. This permits the user to write a program in assembly language using the CSP instructions and to simulate its operation. When the program is correct, code can be generated. This code is then converted into VHDL (as a ROM element) and added to the VHDL code. This is then synthesised and placed and routed. Note that this implies that system clock speed can be a function of program complexity! The chip is then placed in the system board and operation commences on an asserted start signal.

5. RESULTS

Results of some simulations of the CSP are shown. These results have been cross-referenced against the results obtained with a Matlab program. Consider a fourth order 1Hz Butterworth filter as that of section 2.3 with a sample frequency of 100Hz. The coefficient values are:

$$d_1 = 0.022, \; d_2 = 0.0446, \; d_3 = 0.088, \; d_4 = 0.178$$
$$c_1 = c_2 = c_3 = c_u = 0 \; and \; c_4 = 1$$

A normalised input step data stream was used in the simulations (i.e. $VE_{min} = -1$ and $VE_{max} = 1$).

$$u(t) = \begin{cases} 0.5 & t > 0 \\ 0 & otherwise \end{cases} \quad (8)$$

Figures 5 and 6 show the output and state variable values of the CSP VHDL Model and Matlab program with the step input. Clearly, the values obtained with the two approaches are very similar. This demonstrates that the CSP specification is correct and can be implemented in practical applications.

The same example has been used to make benchmark comparisons with other processors, and the results are summarised in Table 3. This includes results for Texas Instruments TMS320C31X floating-point DSP's and TMS320C54X fixed-point DSP's. The real-time code for these processors has been carefully assessed to ensure that as fair a comparison as possible is presented. To perform the calculations, compiled code with 32-bit IEEE standard floating-point variables has been used in the 'C31 devices and with 32-bit signed integer for the 'C54's. The right hand column presents the computation time normalised to that of the CSP; the closest in performance is the 60MHz 'C31-60 device, which still takes 3.48 times as long to compute. The fastest fixed-point device is the 160MHz 'C5416 with 32 x 16 bit multiplications which takes 9.1 times as long as the CSP. Another example uses a 7th order two-input single-output controller. In the case, the 'C31-60 takes 4.65 times as long to compute and the 'C5416 takes 8.3 times.

Table 3. Benchmark results.

Processor	Frequency (MHz)	Clk Period (ns)	Average clock cycles per instruction	Number of instructions	Computation time (ns)	Maximum sample frequency (kS/s)	Normalised computation time
CSP	50	20	1	23	460	2173	1
TMS320C31-27	27	37	2	48	3552	281	7.73
TMS320C31-60	60	16.7	2	48	1603	623	3.48
TMS320C541	40	25	1.49	450	16762	59	36.40
TMS320VC5410	100	10	1.49	450	6705	149	14.57
TMS320VC5416	160	6.25	1.49	450	4190	238	9.10

6. CONCLUSIONS

The CSP is a compact, high-speed special purpose processor, which enables a low-cost solution to a wide range of LTI control problems. It's important to appreciate that, although the CSP outperforms even the fastest of the other processors by a significant margin, it is very much simpler. Indeed its modest gate count confers a number of advantages, namely reduced cost due to small die size and simpler packaging, low power permitting considerable periods of operation using battery power. There is also the possibility of being integrated into a silicon "system-on-a-chip" to provide high-performance computation as part of a more complex system solution.

The CSP concept is supported by a software development environment and is currently being tested with controllers drawn from a range of applications, including railway suspension control and aircraft flight control. Further opportunities for its use are as an IP core to enable systems integrators to utilise its capabilities. The CSP's dedicated architecture and careful numerical formulation ensure that it will perform deterministically in a real-time embedded control environment, although it is recognised that other functions are necessary in such applications for which the CSP is not well suited. Research is continuing to address ways in which the variety of functions required for high-performance real-time control can be most effectively achieved.

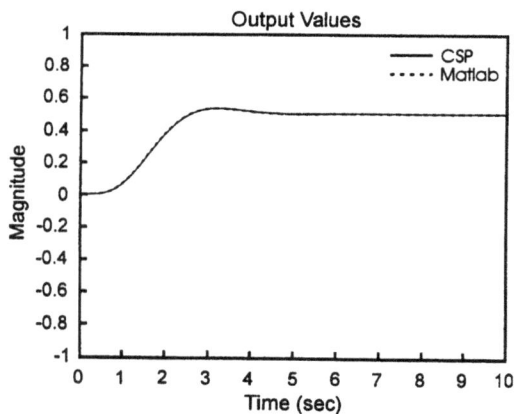

Fig. 6. CSP and Matlab state variables.

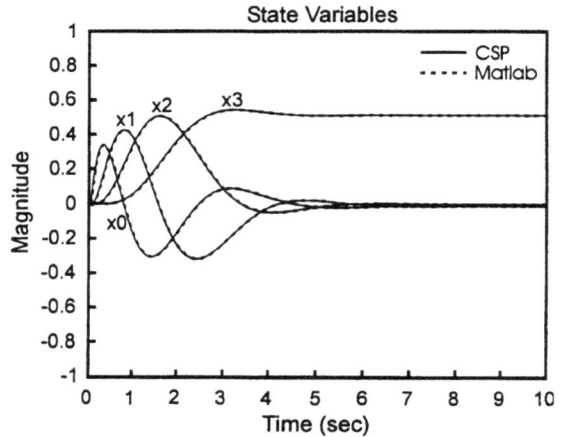

Fig. 5. CSP and Matlab output values.

REFERENCES

Gevers, M. and G. Li (1993). Parameterization in Control, estimation and filtering. *Springer-Verlag.*

Goodall, R. M. and D. S. Brown (1985). High speed digital controllers using an 8-bit microprocessor. *Software and Microsystems*, **4**, pp. 109-116.

Goodall, R. M. and B. Donaghue (1993). Very high sample rate digital filters using the δ operator. *Proceedings IEE-G*, Pt G, **140**, pp. 199-206.

Jaswa, V.C., C. E. Thomas and J. Pedicone, J. (1985). CPAC: Concurrent processor architecture for control. *IEEE Transactions on computers*, **34**, pp. 163-169.

Liu, B (1971). Effect of finite wordlength on the accuracy of digital filters – a review. *IEEE Trans Circuit Theory*, **18**, pp. 670-677.

Middleton, R. H. and G. C. Goodwin (1990). Digital control and estimation – a unified approach. *Prentice Hall.*

Spray, A. and S. Jones (1991). PACE: A regular array for implementing regularly and irregularly structured algorithms. *IEE Proceedings-G*, **138**, pp. 613-619.

EMPLOYING A MICROCONTROLLER ARCHITECTURE FOR THE IMPLEMENTATION OF AN ADVANCED PID AUTOTUNER

Alberto Leva, Roberto Manenti*

Dipartimento di Elettronica e Informazione, Politecnico di Milano
Via Ponzio, 34/5 - 20133 Milano (Italy)
*Former student at Dipartimento di Elettronica e Informazione

Abstract: The paper describes the implementation of a fast, IMC-based autotuning PID on a low cost, microcontroller-based architecture. After a brief overview of the presented device, the most important aspects of the control law and of the tuning procedure, namely the use of integer computations and the problems encountered due to the smalla amount of memory available, are pointed out. A test performed with the device are also reported. *Copyright © 2000 IFAC*

Keywords: Autotuners, PID control, Process control.

1. INTRODUCTION

The purpose of this paper is to describe the implementation of a fast autotuning PID regulator on an extremely low cost architecture, namely two PIC16C77 microcontrollers. This work has been done along a research path with the main motivation of exploring the difficulties to be faced when such an architecture has to host a rather sophisticated synthesis method.

Microcontrollers are nowadays widely employed for small control applications, both in the industrial and in the consumer domain, due to their ease of use: they are cheap, easy to interface and require just a minimum of external hardware, which makes them almost ideal e.g. for a variety of embedded systems.

However, at present their utilization is significantly confined to "low-end" applications, since their technical limitations (particularly the small amount of memory available and sometimes the lack of development tools with sufficient calculus capabilities) normally prevent them from being considered for implementing more sophisticated control tools.

As a consequence, numerous control techniques are not applied where - for technical and sometimes also "historical" reasons - the device employed for

accomplishing the task is a microcontroller: in particular, small microcontroller-based embedded regulators normally do not have autotuning capabilities.

This is in some sense a pity, because the results available nowadays in the field of automatic tuning are numerous and powerful (Doyle *et al.*, 1992; Isermann and Lachmann, 1992; Åström and Hägglund., 1995). Employing these results in those loops which are normally controlled by stand-alone devices, quite often based on microcontroller architectures, would surely be a significant aid not only in small applications, but also in larger ones where these devices and loops form the "lower" layer of the overall control system.

This is why the research mentioned previously has been started, and the results obtained so far appear to indicate that the limits of microcontroller-based architectures are significantly broader than expected and often assumed: the device presented herein is the first of these results. The paper is organized as follows: section 2 gives an overview of the device, while sections 3 and 4 concentrate on the digital control law and on the tuning procedure respectively. Section 5 reports a test to illustrate the device operation, and finally in section 6 some conclusions are drawn.

2. OVERVIEW OF THE DEVICE

The device presented in this paper implements a PID regulator in the ISA output derivation form, which in the frequency domain is expressed as

$$u(s) = K\left[br(s) - y(s) + \frac{1}{sT_i}(r(s) - y(s)) - \frac{sT_d}{1 + sT_d/N}y(s)\right]$$

where $r(s)$, $y(s)$ and $u(s)$ are the Laplace transforms of the set point, of the process output and of the control signal respectively. Moreover, it performs the automatic PID tuning by means of a modified version of the IMC method as introduced in (Morari and Zafiriou, 1989), capable of synthesizing also parameter N.

The method is applied to a classical first order plus delay process model, identified by means of an open loop step experiment and by a revised version of the well known method of areas capable of coping also with dominant delay and (moderately) underdamped processes. Finally, a simple empirical rule is employed for synthesizing parameter b.

Under a functional point of view, the main characteristics of the presented autotuner can be summarized as follows. First, the PID control law is implemented by exclusive use of *integer* computations, so as to achieve small sampling times and to reduce memory occupation; internal computations employ 16 bits. Conversely, the tuning procedure involves floating point computations. A specific scaling method is then adopted to compute the integer parameters for the PID control law, so as to optimize the precision of integer computations and to achieve the best use of the 8 bit A/D and D/A converters.

Moreover, the duration of the open loop step experiment being unpredictable without prior process knowledge, an on line algorithm is employed for sub-sampling the recorded process response *as it is collected*. This allows efficient memory usage (only 128 bytes are required for the response storage, regardless of its duration and of the sampling time in use) and also a consistent choice of the sampling time to be employed by the controller after the tuning phase. The ability of selecting the sampling time automatically, coupled with the rather high speed achievable thanks to integer computations, greatly improves the device flexibility.

The user interface has been designed so as to resemble closely that of industrial PIDs. It allows 3 modes of operation: manual, automatic and tuning. The first two are absolutely standard, while the third is a one-key operation requiring no design parameters. The only parameter required by the IMC method related, basically to the required closed-loop settling time, is selected by the tuning procedure automatically: the rule employed is empirical, yet proven effective by experience. As a result, for tuning the PID the user is only requested to start (and possibly abort) the operation. Finally, particular care

has been taken in reducing the process upset during the tuning phase and in rejecting the effects of noise effectively.

The device employs two PIC microcontrollers and just a minimum of additional hardware. One of the two microcontrollers is devoted to implementing the PID, performing the open loop experiment on the process and identifying its model (operations that involve mostly integer computations). The other one implements the autotuner and takes care of the user interface (a four-digit led display and a small keyboard). The A/D converter resides in the first microcontroller, while the D/A is external. Communication between the devices is asynchronous for better performance and has required the definition of a convenient protocol (not described here for brevity). Clearly, all the normal functionalities of industrial PIDs are present, such as anti-windup, bumpless A/M transfer, and so forth. The maximum sampling rate achievable is approximately 100 Hz. The prototype is depicted in figure 1.

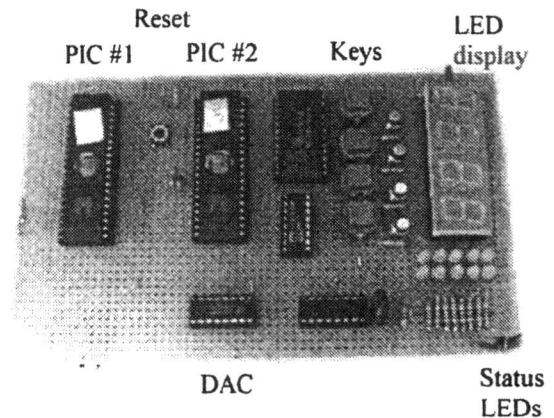

Fig. 1. The prototype.

3. THE CONTROL LAW

The output derivation ISA-PID, when discretized by means of the implicit Euler method and written in incremental form except for the derivative term (as normally done in industrial devices for reasons too long to explain here and not relevant for this paper), results in the digital law

$$P(k) = P(k-1) + K\left[b\left(y^\circ(k) - y^\circ(k-1)\right) - \left(y(k) - y(k-1)\right)\right]$$

$$I(k) = I(k-1) + K\frac{T_s}{T_i}\left[y^\circ(k) - y(k)\right]$$

$$D(k) = \frac{T_d}{T_d + NT_s}D(k-1) - \frac{KNT_d}{T_d + NT_s}\left(y(k) - y(k-1)\right)$$

$$U(k) = P(k) + I(k) + D(k)$$

where T_s is the sampling time.

In this work, integer computations have been used for improving speed. However the PID parameters are naturally computed by the tuning procedure (described later) as floating point numbers, so two problems exist: first an inefficient use of the limited precision provided by the 8-bit architecture can arise, and then the integral gain (having T_s in the numerator) can be zero in integer numbers, thus making the integral action vanish. This is one of the reasons why most microprocessor-based autotuning PIDs use floating point computations, the range of parameters computed by the tuner in practical applications being unpredictable.

The alternative used here is an *ad-hoc* parameter scaling and possibly a convenient, automatic choice of the sampling time. The autotuner operates on the basis of a record of the process step response, which has been obtained during an identification phase (described later) and is limited to 128 samples. As such, after this phase both the necessary record of the step response and the sampling time T_{sid} it has been collected with are available. Then the tuning procedure computes the PID parameters K, T_i, T_d, N and b in the continuous time domain.

Finally, the digital control law is implemented as follows:

1. The sampling time is chosen as $T_s = 2T_{sid}$ so that the process settling time correspond to approximately 64 control steps, which is a quite reasonable practical compromise.

2. The scaling floating point quantities

$$A = \max\left(1, K, Kb, \frac{KT_s}{T_i}\right)$$

$$B = \max\left(\frac{T_d}{T_d + NT_s}, \frac{KNT_d}{T_d + NT_s}\right)$$

are computed.

3. The integer control law parameters

```
IKPSP = round(K*b/A*255)/2
IKPPV = round(K/A*255)/2
IKI   = round(K*Tsc/Ti/A*255)/2
IKDD  = round(Td/(Td+N*Tsc)
        /B*255)/2
IKDPV = round(K*N*Td/(Td+N*Tsc)
        /B*255)/2
```

are obtained by floating point computations and then converted to integers. Trivial details are omitted, suffice to say the 255 comes from the 8-bit range and the division by 2 (actually implemented in the PIC microcontroller as a right shift by 1 bit) from the necessity of employing the same bit range for signed and unsigned quantities.

3. The integer scaling quantities

```
IA = ceil(A)
IB = ceil(B)
```

are calculated.

4. The integer control law is computed as described by the following pseudocode:

```
IDY°    = IY° - IY°old
IDY     = IY - Iyold
IY°old  = IY°
IYold   = IY
IDP     = IKPSP*IDY°-IKPPV*IDY
IDI     = IKI*(IY°-IY)
ID      = (IKDD*Idold
          -IKDPV*IDY)/128
IDD     = ID - Idold
IDold   = ID
IDU     = (IDP+IDI)*IA+IDD*IB
IU      = IUold+IDU
if IU<0
    ICS = 0
if IU>32767
    IU = 32767
IUold = IU
IUout = IU/128
```

Here too details are omitted for brevity: suffice to say that the prefix "I" denotes integer quantities, the prefix "D" the variation from the previous step, the suffix "old" denotes the values at the previous step. Internal quantities and parameters are represented with 16 bits, while inputs and outputs with 8 bits (this explains the last shift).

The rationale behind the choice of A and B is to scale the PID parameters so that at least one of the gains involved in the three actions use all the available bits, while the integral gain in the integer control law cannot vanish due to an incorrect choice of the sampling time (recall that in autotuners it is highly desirable that this choice be automated, as in the one presented herein).

Though very simple, this way of implementing the PID has proven to be very effective. The IMC-based autotuner described in the next section can be used without concerns on the range of the PID parameters it will output: in any case, it is ensured that the available precision will be used efficiently.

Moreover, floating point computations are confined in the initial phase, i.e. when the parameters of the integer control law are computed after the tuning operation has been completed. The rest of the algorithm contains only sums, products and shifts, thus it can be implemented on any microcontroller (not only the PIC) very easily and the execution is quite fast. This is witnessed by the obtained maximum sampling rate of 100 Hz, which is a fairly good result for such a low cost device.

4. THE TUNING PROCEDURE

As previously said, the PID is synthesized by the IMC-based procedure described in (Leva and Colombo, 1998). The only peculiarity with respect to the standard IMC-PID tuning method reported in (Morari and Zafiriou, 1989) is that parameter N is taken into account *explicitly* so as to avoid the error in the position of the regulator zeros induced by a subsequent introduction of N done by assuming for it a prespecified value.

Assuming an asymptotically stable, type 0 process model in the form of a first order transfer function with delay, i.e.

$$P_m(s) = \mu \frac{e^{-s\tau}}{1 + sT}$$

the formulae for computing parameters K, T_i, T_d and N are

$$T_i = T + \frac{\tau^2}{2(T_f + \tau)}$$

$$K_p = \frac{T_i}{\mu(T_f + \tau)}$$

$$N = \frac{T(T_f + \tau)}{T_i T_f} - 1$$

$$T_d = \frac{N T_f \tau}{2(T_f + \tau)}$$

where T_f is a design parameter relative to the desired closed-loop bandwith, which in this work is computed automatically as $2T$. Parameter b, not computed by the IMC formulae, is obtained by a simple, empirical rule aimed basically at improving the set point tracking speed without inducing overshoots. To avoid details, this rule can be summarized as follows.

1. A first value of b is computed as the inverse of the maximum overshoot during the step test, i.e. as the ratio between the amplitude of the input step and the difference between the maximum value reached by the control variable and the value it had before the step. Taking this vaue (and not the minimum reached during the experiment) prevents possible undershoots of the control variable, e.g. when dealing with a non-minimum phase process, to result in an incorrect choice of b.

2. Since introducing the set point weight b means that the regulator is described by the block diagram depicted in figure 2

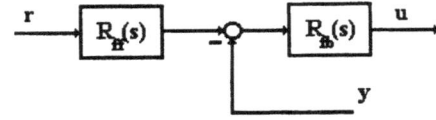

Fig. 2. Block diagram of the ISA PID.

i.e. by a two-degree of freedom regulator where the blocks $R_{ff}(s)$ and $R_{fb}(s)$ are characterized by the transfer functions

$$R_{ff}(s) = \frac{1 + s(bT_i + T_d / N) + s^2 T_i T_d b / N}{1 + s(T_i + T_d / N) + s^2 T_i T_d (1 + 1/N)}$$

$$R_{fb}(s) = K \frac{1 + s(T_i + T_d / N) + s^2 T_i T_d (1 + 1/N)}{sT_i + s^2 T_i T_d / N}$$

respectively, it is checked that the zeros of $R_{ff}(s)$ are "not too far" (say not more than one decade) from the dominant dynamics of the closed loop system, a measure of which is assumed to be given by Ti. In this case, the value of b is conveniently limited.

Before the tuning phase, which resorts to the previous formulae, it is necessary to estimate the process model $P_m(s)$. i.e. parameters μ, T and τ. This is done by a slightly modified version of the well known method of areas, described in (Leva and Colombo, 1998) and capable of coping also with (moderately) oscillating dynamics. This identification phase, in turn, requires a record of the process step response, which is obtained as follows:

1. The PID is set to tracking mode, the control signal remains constant and the system records the process output for a prespecified number of samples (denoted by N_l and chosen by the operator when configuring the device with a default of 20) so as to estimate the noise band.

2. A step is applied to the control input: also the amplitude of this step can be chosen by the operator when configuring the device, and of course care is taken not to exceed the control saturation limits.

3. The autotuner starts recording the process output at thr maximum speed, i.e. with a sampling time of 0.01s, and waits for a steady state is reached. This is detected when the last N_l samples of the process output differ less than the noise band, while their average differs from the average of the process output in the N_l samples before the control step more than twice the noise band.

4. If the number of 128 recorded samples is reached before steady state is detected, the

system puts every second recorded sample in the first 64 positions of the record vector, and then continues sampling at twice the previous sampling time.

This method too is very simple, easy to implement and rapid, yet it is effective for the required purpose. There is virtually no limit to the duration of the process step response, and at the end of the procedure a record of it is available with a number of samples ranging from 64 to 128, which is surely a reasonable description of any process response that can be encountered in practical applications where a PID can be employed.

5. AN OPERATION EXAMPLE

A prototype version of the presented device (the one depicted, at an even more preliminary stage, in figure 1) has been completed and tested. At present, the code is being prepared for porting to a new, more powerful family of PIC microcontrollers and further optimized, reducing the use of the C language (which in the first version was quite extensive so as to simplify the unavoidable algorithmical modifications and improvements) and exploiting the PIC assembler language as much as possible; this should lead to put all the required software (PID, autotuner and user interface) in a single PIC instead of two, thus eliminating all the communication problems, improving speed (as an obvious consequence) and simplifying the board layout significantly.

As an example of the device operation in its present state, however, a test run of the entire tuning algorithm (i.e. comprehensive of data collection, sub-sampling, IMC-based floating point tuning, choice of the sampling time, parameter scaling, conversion to integer parameters and integer control law) is reported.

The process employed in this test (simulated by one of the two PICs) has unity gain, two poles with a natural frequency of 0.2 r/s and a damping factor of 0.6, and a delay of 4s; moreover, a 2% f.s. additive output noise is present.

The ISA-PID parameters computed by the autotuner in the continuous time domain are

$$K = 0.26091$$

$$T_i = 6.1459s$$

$$T_d = 0.45265s$$

$$N = 0.18604$$

$$b = 0.92362$$

The parameters of the integer control law are not reported here, because they would be practically meaningless to the reader. The initial sampling time was set to 0.01s, while after the identification phase the autotuner has chosen for it a value of 0.64s,

leading to a step response record composed of 94 samples.

Figure 3 reports the results of the step test and the sub-sampled step response record obtained from it, while in figure 4 the closed-loop response of the control system with the tuned regulator to a set point step variation of amplitude 50 is depicted. All the variables shown are in the native range 0-255, and the process output samples in the closed-loop test of figure 4 have been interpolated to better distinguish the output signal from the control.

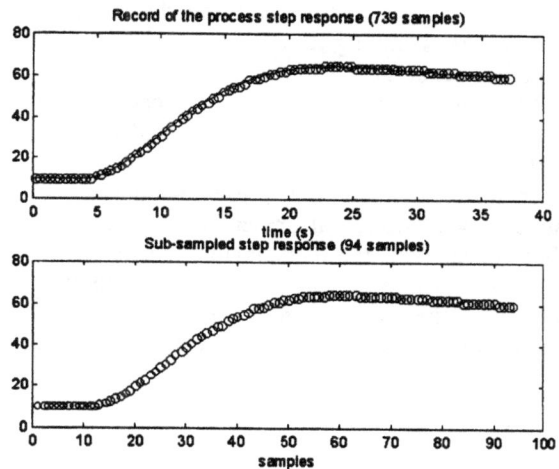

Fig. 3. Identification phase.

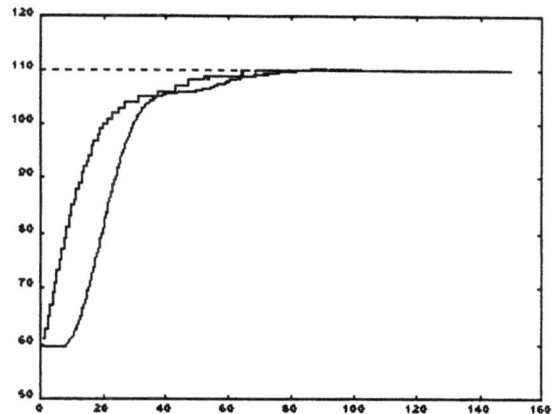

Fig. 4. Tuning results.

It can be noticed that the behaviour of the closed-loop system is definitely satisfactory. Note also that the sampling time chosen by the autotuner is 64 times the initial value. This shows the effectiveness of the method proposed, and allows to state that the resulting autotuner can be safely used with process characterized by very different (and unknown *a priori*) response times. Also the implementation of the PID regulator with integer computations behaves fine, which shows that the scaling method employed guarantees a consistent regulator (and, in particular, that the integral action cannot vanish) when, again, applying the method to processes with different dynamics.

6. CONCLUSIONS

An IMC-based autotuning PID implemented with two PIC microcontrollers has been presented. The work is aimed at showing that, with a convenient design of the algorithm (i.e. both of the regulator and of the correesponding autotuner) and with an efficient implementation, also a very low-cost architecture can successfully host quite complex and sophisticated tuning methods. The presented device, for example, encompasses some capabilities (such as the quite high speed due to the use of integer computations in the control law and the automatic choice of the sampling time on the basis of the information concerning the process gathered during the identification phase) that are typical of more complex and sophisticated devices, based on more powerful architectures.

The characteristics of the presented device are definitely good, though at present only a prototype has been developed, and its behaviour is satisfactory. Further work is in progress to improve its operation, to optimize the code, and to repeat the same implementation and engineering experience with other automatic synthesis techniques.

ACKNOWLEDGEMENTS

This work has been partially supported by MURST (Ministero dell'Università e della Ricerca Scientifica e Tecnologica) and by CNR (Consiglio Nazionale delle Ricerche).

REFERENCES

Åström K.J. and T. Hägglund (1995), "PID Controllers: Theory, Design and Tuning – Second Edition", *Instrument Society of America*.

Doyle J.C., B.A. Francis and A.R. Tannenbaum (1992), "Feedback Control Theory", *MacMillan*.

Isermann R. and K.H. Lachmann (1992), "Adaptive Control Systems", *Prentice-Hall*.

Leva A. and A.M. Colombo (1998), "IMC-Based Synthesis of the Feedback Block of ISA-PID Regulators", Int. Rep. 98-074, Dipartimento di Elettronica e Informazione, Politecnico di Milano.

Morari M. and E. Zafiriou (1989), "Robust Process Control", *Prentice-Hall*.

INTERPRETED PETRI NET APPROACH FOR DESIGN OF DEDICATED REACTIVE SYSTEMS

Marian Adamski, Marek Wegrzyn

Technical University of Zielona Gora
Computer Engineering and Electronics Department
ul. Podgorna 50, 65-246 Zielona Gora, POLAND
<M.Adamski, M.Wegrzyn @iie.pz.zgora.pl>.

Abstract: In the paper a structured synthesis method based on the direct mapping of hierarchical interpreted Petri nets into field programmable logic is presented. The described methodology is especially useful for designing of Industrial Application Specific Logic Controllers (ASLCs) with FPGA. The specification of the reactive system in the form of symbolic, *if-then* or *if-then-else* conditional decision rules is transformed into a HDL format that is accepted by standard FPGA simulators and synthesis tools. The Concurrent State Machine model of Logic Controller is verified using the well-developed Petri net theory, and then it is translated through automated processes into selected FPGA specification format. *Copyright © 2000 IFAC*

Keywords: Process Control, Programmable logic controllers, Petri-nets, Rule-based systems, Sequential control algorithms, Digital systems, Logic arrays.

1. INTRODUCTION

The main aim of this paper is a presentation of new methods of direct mapping of Logic Controller (LC) programs into Field Programmable Logic (FPL). The behavioural specification of Logic Controller programs (Halang & Jung, 1994), related with subset of IEC 1131-3 standard, is transformed into a Hierarchical Interpreted Petri Net model. The textual, rule-based description of Petri net, called Petri Net Specification Format - version 2 (PNSF2), is used as an entry format for dedicated CAD tool. The experimental results show that proposed methods produces economical, structured and flexible FPGA implementations of dedicated Reprogrammable Logic Controllers (Mandado, *et al.*, 1996).

Hierarchical, safe (1-bounded) Petri nets provide a mechanism, which is suited to represent both parallelism (concurrency) and hierarchy in digital processes. The control part of the designed discrete-event system may be described and verified using the well-developed Petri net theory (Lee & Favrel 1985; David & Alla, 1992; Zakrevskij, 1990; Zhou 1992).

In the first of presented methodologies, a rule-based behavioural specification of Petri net model is directly transformed into hardware library (on the netlist level) by means of Xilinx Foundation system (Wegrzyn, Adamski, *et al.*, 1997, 1998). The *if-then* decision rules (non-procedural conditional statements) are mapped into Xilinx XNF format after some simple additional symbolic transformations. From the conceptual point of view, a set of event-oriented (Petri net transition-oriented) *if-then* conditionals (Adamski, 1991; Sagoo & Holding, 1991) is easily replaced by the equivalent local state-oriented (Petri net place-oriented) *if-then-else* decision rules. In such a way, the Petri net image in FPGA structure is straightforward and simple. Standard design tools perform the final multi-level, combinational optimisation, placement and routing.

Petri nets can be also behaviourally specified by means of using hardware description languages (HDLs). They are easily described in traditional simple languages like Palasm (Adamski, 1991). Petri net specification, modelling, validation and FPL synthesis can be directed towards modern, well-accepted VHDL or Verilog based CAD tools. The second proposed method of structured implementation of Petri net is based on self-evident mapping of decision rules into VHDL statements (Wegrzyn, Wolanski, *et al.*, 1997; Adamski, 1999). The *if-then-else* conditionals may serve both as flexible and formal intermediate forms.

2. BACKGROUND AND MOTIVATION

Control oriented, 1-bounded Petri net is considered as a formal model for the set of decision rules. It is used also as a distributed, local state based model for the generated HDL descriptions. The behavioural textual descriptions of Control Interpreted Petri Nets (CIPN) (David & Alla, 1992) may be formally transformed into a format accepted by comfortable and effective CAD tools (Adamski, 1998).

In the considered interpreted Petri net model, a particular combination of Boolean external inputs, and eventually some internal conditions, have to be true, for the transition to be fired (to be occurred).

Conditional logic is treated as an intermediate behavioural level language for the description of concurrent digital systems. Implicitly, it gives the *procedural* specification of distributed, concurrent digital system, implemented in FPL as a single State Machine with Data Path. The global states (Petri net markings) of Concurrent State Machine (direct hardware implementation of Petri net) are formed from local states (Petri net places). The places are encoded in such a way, that the global state code is implicitly obtained by superposition of the local state codes. A rule-based specification, considered in the paper, is composed from discrete local state symbols (places P), input signal symbols (X) and output signal symbols (Y) of the controller. The name of transition (t) is treated only as a rule label.

The textual description of developed implementation in Field Programmable Logic is not procedural, as Petri net evidently is, but rather *declarative*. The designed control part of a digital system is treated as a reasoning system implemented in the regular structure (for example - array) that is built from logic cells. The decision rule specification, as a textual equivalent of Petri net, is easily represented by using HDLs, and then mapped into hardware. The manual or automatic translation is straightforward.

The simplest technique for Petri net place encoding is to use one-to-one mapping of places onto flip-flops in the style of a one-hot state assignment (Wegrzyn, Adamski, *et al.*, 1997; Wegrzyn, *et al.*, 1998). If the programmable device, such like CPLD, contains a limited number of cells (flip-flops) and pins, some more efficient heuristic strategies should be employed (Adamski, 1991; Bilinski, *et al.*, 1994).

3. PETRI NET AS A SPECIFICATION OF LOGIC CONTROLLER

A. COLOURED CONTROL INTERPRETED PETRI NET

Petri nets, as a graphical and mathematical tool, provides a unified structured methodology for effective design of discrete-event systems. They can be applied in various stages of the implementation, starting from discrete event system description and finishing with its FPL realisation. Several modelling techniques with Petri nets have been proposed. They are based on software tools that help designers to specify, develop and simulate (animate) the system, on a conceptual and abstract level. Universal Petri net tools are usually oriented mostly towards the verification (formal analysis) of pure, mathematical, non-interpreted models. One of the most powerful and well-known systems is Design/CPN that is dedicated to Coloured Petri Nets (CPN) (Jensen, 1992).

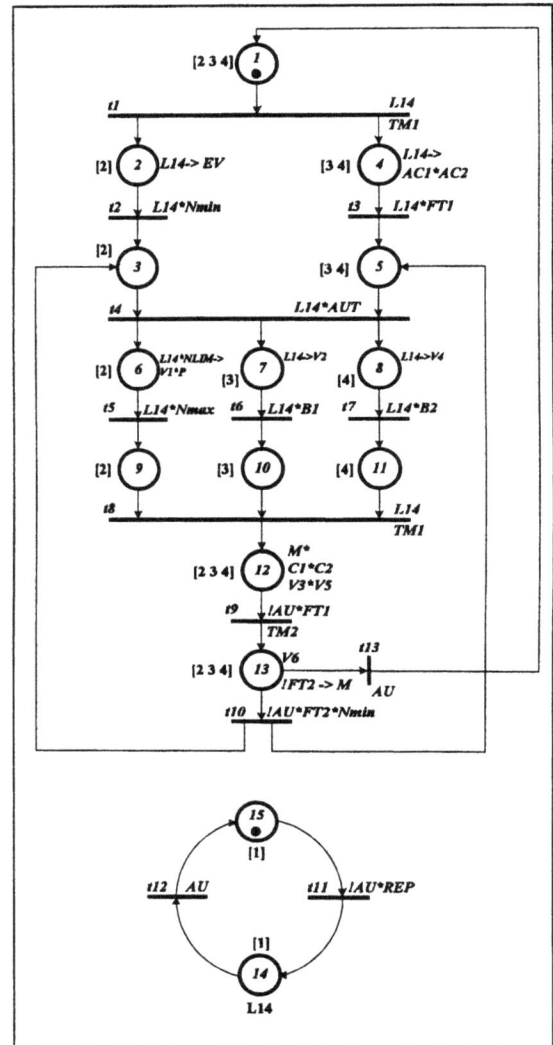

Fig. 1. CCIPN model of the controller

188

A *coloured control-oriented interpreted Petri net (CCIPN)* (Wegrzyn, Wolanski, *et al.*, 1997) is a modification of Control Interpreted Petri Net - CIPN (David & Alla, 1992). Introducing coloured tokens and places into bounded Place/Transition net extends the original CIPN concept. On the other hand, proposed kind of net could be also treated as a very restricted, control interpreted version of Jensen's Coloured Petri Net. Since no more than one token with particular colour can ever be in any place, the net is *1-bounded (safe)* in respect with any colour. It is assumed that the colour of the place and the colour of the token contained in it should be always consistent. The interpretation could be expanded into coloured transitions, coloured arcs, and coloured external inputs and outputs.

Colours can be used to determine and explicitly represent the partial relation of concurrency among places (Fig.1). The set of rules that have to be satisfied during such kind of net colouring is given in paper (Adamski & Wegrzyn, 1994). Colours *[1],[2],[3]* and *[4]* are explicitly affiliated to places *{p1 ,..., p15}*, and implicitly attached to transitions *{t1, ..., t13}*. The colour subset of any transition is a union of colour subsets of its input places. It is exactly the same, as the union of colours of its output places. For example, the colour subset of transition *t8* is {2,3,4}.

P-coloured, safe Petri nets have been successfully used in digital system specification, replacing and integrating into one compact model several linked sequential state diagrams (graphs), which are previously transformed into interpreted subnets. From that point of view, colours can be treated as distinguished labels of particular sequential state machine subnets. The Petri net from Fig. 1 consists of four, partially overlapping state machine subnets, distinguished by colour [1], [2], [3] and [4].

B. *HIERARCHICAL COLOURED CONTROL INTERPRETED PETRI NET*

Hierarchical Petri Net is based on a concept of a safe, (1-bounded) control interpreted Petri Net, where abstraction and modularization is applied by means of introducing or refining the structured macroplaces. Such macroplaces represent the hierarchically organised, well-structured subnets (Fig. 2).

The *sequentially related macroplaces*, for example *{M1,M3,M6}* or *{M6,M7,M8}*, describe the depended, but isolated parts of specification. *The concurrently related macroplaces*, for example *{M3,M4,M5}* describe the distinguished, independent parts of specification. Some macroplaces, like *{M3,M4,M5}* or *{M1,M2}*, are evidently *parallel*. The hierarchy (depth) of structured macroplaces is obtained by allowing the refinement of macroplace

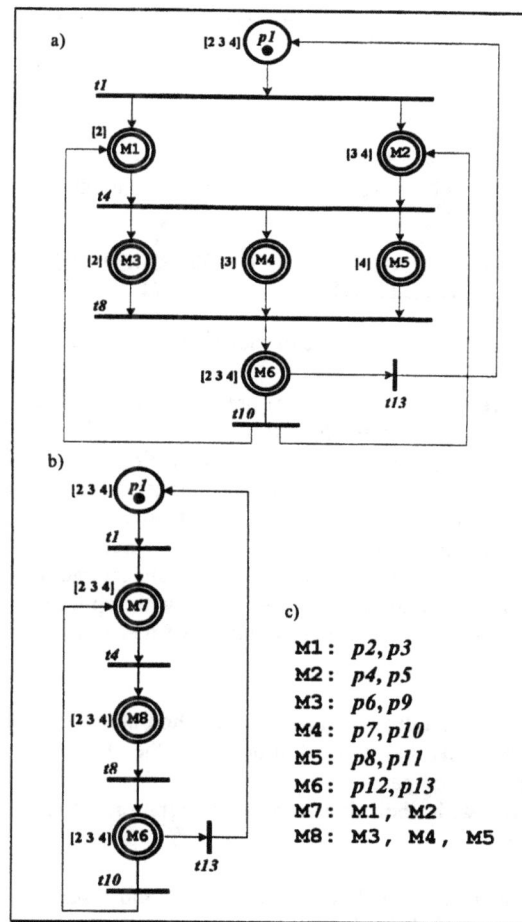

Fig. 2. Macronet in the first level (a) and the second level (b) of hierarchy; (c) description of the hierarchically structured macroplaces {M1, ..., M8}

by subnets. Each macroplace has a particular subset of associated colours. The subset of all colours related with the input arcs of the structured macroplace must be the same, as the subset of colours attached to the output arcs of considered macroplace. All tokens in concurrently related macroplaces have different colours.

C. *BEHAVIOURAL SPECIFICATION WITH PETRI NETS*

During the hierarchical specification Petri net consisting of the finite set of partially ordered subnets is created. The recommended stepwise methodology is based on top-down or bottom-up approaches.

Top-down construction of hierarchical Petri net starts from a very abstract model of the concurrent system, which is refined up to the required level of detail. The macroplaces of the macronet are gradually replaced by appropriate subnets by means of the repeated substitutions, until the final subnets, only with ordinary places and transitions, are obtained.

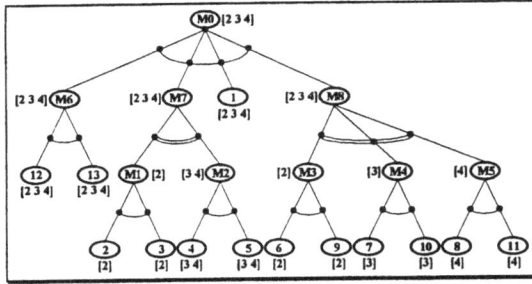

Fig. 3. Hierarchy and concurrency between subnets

Table 1. Description of coloured macroplaces

Macroplaces	Internal places
M1 [2]	p2[2] p3[2]
M2 [3 4]	p4[3 4] p5[3 4]
M3 [2]	p6[2] p9[2]
M4 [3]	p7[3] p10[3]
M5 [4]	p8[4] p11[4]
M6 [2 3 4]	p12[2 3 4] p13[2 3 4]
M7 [2 3 4]	M1 [2] M2 [3 4]
M8 [2 3 4]	M3 [2] M4 [3] M5 [4]

The reverse way begins from previously given 'flat' Petri net, for that some formal reduction rules (Lee & Favrel, 1985) should be applied. Application of AND composition for evidently parallel subnets, and OR composition for evidently sequential subnets abstract conventional one-level (flat) Petri nets into hierarchical macronet. The bottom-up understanding of behavioural specification is related with top-down structured view and implementation of the designed system.

Bottom-up modular composition method starts from the Petri net behavioural models of the different, distinguished parts of control specification and follows with their hierarchical integration (Zhou, 1992).

The subnets are implemented separately and assumed to operate independently and concurrently, unless their activities are synchronised. The *shared inter-level transitions* are used to compose together the hierarchically ordered subnets.

The hierarchically structured Petri net (Fig. 2) consists of subnets, which, except possibly *Base Net*, are well-formed blocks. The concurrency relation between subnets is depicted by means of colours that are attached explicitly to the places and tokens and implicitly to the transitions and arcs. The set of subnets is partially ordered. The coloured relation tree (Fig. 3) represents graphically both the hierarchy and concurrency relations among subnets. The Base Net is on the root of the tree. It contains the double macroplaces, which stand for the hierarchically structured subnets at the lower level of hierarchy. The colours are used to distinguish particular sequential processes and to validate the design by continuously controlling the place invariants (P-colours) during the composition of the net. Macroplaces could be roughly interpreted as *superstates*, in the sense introduced in the Harel's *statecharts* (Harrel, 1987).

Fig. 2 and Fig. 3 together show a considered hierarchical Petri net specification. The root macroplace M0 represents the OR superstate with four sequentially related components, which are described by means of macroplaces M6, M7, M8 and

place *p1*. Place *p1* initializes the controller. The AND superstate, which is depicted as macroplace *M7*, contains concurrent subnets abstracted as macroplaces *M1* and *M2*. Macroplace *M6* denotes the OR superstate with components *p12*, *p13*. In the lower level of hierarchy *M2* is treated as an OR superstate, with sequentialy related components *p4* and *p5*. Detailed description of macroplaces with attached colours is presented in Table 1.

4. FIELD PROGRAMMABLE LOGIC BASED DESIGN WITH PETRI NET

In this section, the specification of industrial controller (Toulotte, 1978) is developed, to be later related with coloured control interpreted Petri net (CCIPN) Fig. 4 depicts the controlled part of a designed simple reactive system.

In the considered Field Programmable Logic implementation, transitions between local states occur simultaneously, on all levels of hierarchy. It is realised by means of using the same clock for all logic blocks. To implement concurrent state machine, a global state register can be used to reflect all current, concurrently related, local states. As an opposite option it is possible to distribute the local state register into several single logic blocks. In such a case, each Petri net place may be implemented separately as two local states: active - if place is marked, passive - if place is empty.

The direct mapping of Petri net into FPL is based on correspondence between a transition and a simple combinational circuit and correspondence between a place and a clearly defined subset of state register. In dealing with concurrency the designer is confronted with some problems that will not arise in the logic synthesis of sequential systems. One of them is the concurrent local state encoding.

The one-hot-like method produces fast designs with a simple combinational part, especially for implementations in FPGA (Wegrzyn, Adamski, *et al.*, 1997). It is not possible to assume that all flip-flops attached to places, are set to 0 since several

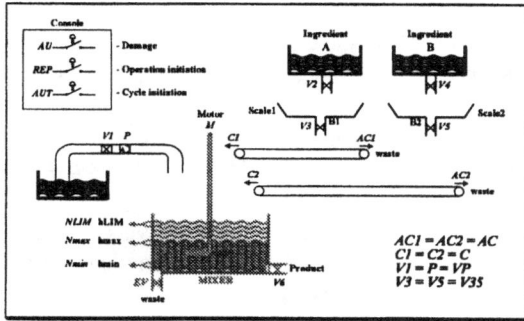

Fig. 4. The controlled part of reactive system

places can be marked simultaneously. Some efficient state assignment techniques are referred to or presented in (Adamski, 1991; Bilinski, et al., 1994). They are both suited to CPLD and FPGAs. The encoding procedure assigns to each place or macroplace a Boolean term. This is combined from all Boolean terms, which are assigned to higher-level macroplaces it belongs to, and a particular 'private' part. The state variables could be eventually shared among subnets but all concurrent subnets should not have orthogonal codes (Adamski, 1991).

The use of concurrent state machine techniques in FPL designs has several advantages. They create near-optimal controller implementations (in terms of performance), tailored to the exact requirements. Practical designs often fit in a single FPGA package since there are no redundant elements. Any particular function, which is not included in the library, can be effectively implemented. Controller outputs respond

```
.clock CLK
.inputs AU AUT REP
.comb_outputs M14

.part console
.places P14 P15
.transitions t11 t12
.net
 t11: P15 * !AU * REP |- P14;
 t12: P14 * AU |- P15;
.MooreOutputs
 P14 |- M14;
.marking P15
.endp

.part mixer
.inputs Nmin Nmax NLIM B1 B2 FT1 FT2
.comb_outputs V1 V2 V3 V4 V5 V6 P M
                C1 C2 AC1 AC2 EV
.reg_outputs TM1 TM2 //for T1 (FT1), T2 (FT2)
.places P1 P2 P3 P4 P5 P6 P7 P8 P9 P10
         P11 P12 P13
.transitions t1 t2 t3 t4 t5 t6 t7 t8 t9
             t10 t13
.net
 t1: P1 * M14 |- P2 * P4 * TM1;
 t2: P2 * M14 * Nmin |- P3;
 ...
.MooreOutputs
 P12 |- M * C1 * C2 * V3 * V5;
 P13 |- V6;
.MealyOutputs
 P2 * M14 |- EV;
 P4 * M14 |- AC1 * AC2;
 ...
.endp
```

Fig. 5. A part of PNSF2 specification

to inputs with high speed, in predictable and repeatable way, without glitches. The behavioural specification, which is mapped in the programmable logic can be formally verified or validated by means of advanced CAD tools.

The standard CAD tools have been combined with dedicated design environment PeNCAD (Wegrzyn, Adamski, et al., 1997). As an intermediate format the rule-based textual specification, called PNSF2, is used (Fig. 5). The design experiments have been developed by means of using Aldec Active-HDL, Model Technology ModelSIM, Synopsys FPGA Express and Xilinx Foundation Series, and tested on Xilinx FPGA demoboard with XC4005E device.

5. SPECIFICATION AND IMPLEMENTATION OF REAL-TIME REACTIVE SYSTEMS

Application Specific Logic Controllers are dedicated devices embedded within an application and implemented with programmable logic (Mandado, et al., 1996; Wegrzyn, Adamski, et al., 1997). To define the behaviour of reactive system, it is necessary to specify the set of allowed sequences of conditions, actions and events (Zakrevskij, 1990). The papers (Adamski, 1998; 1999) present a VHDL package (front-end) that allows RTL level descriptions and simulations of Application Specific Logic Controllers. The importance of this approach lies on its portability to any VHDL simulator, for example Aldec Active-CAD. It makes possible to validate of the indented behaviour of the designed microsystem in an user-friendly environment. For example, VHDL description of the logic controller is validated using "test-benches", which models roughly the controlled subsystem behaviour.

The Reprogrammable Logic Controller is considered as an abstract reasoning system (rule based system) implemented in reconfigurable hardware (FPGA or CPLD). The mapping between inputs, outputs and local internal states of the system is described in a formal manner by means of logic rules (represented as sequents) with some temporal operators, especially with operator 'next' @ (Adamski, 1991; Sagoo & Holding, 1991). The correctness preserving synthesis, based on Gentzen calculus, is treated as a formal transformation of the initial set of compound rules (*Specification*) into another set of compound rules (*Implementation*).

Transition - based rules (sequents) are usually described as labelled (by the name of transition) *if - then* non-procedural conditionals:

Transition 1: if P1 and L14 then next P2 and next P4
T1: P1 * L14 |- @P2 * @P4

Place based rules (sequents) are labelled (by the name of place) *if then else* nested conditionals:

Place P1:   *if P1 then (if L14 then next P2 and next*
P4) else next P1
P1: P1 |-   (L14 -> @P2 * @P4) | @P1

VHDL supports conditional-statement constructs, which can be used to describe Petri net-based Concurrent State Machine (CSM).

6. CONCLUSIONS

This paper has been presenting a hierarchical approach for modelling and design of discrete logic controllers, which is based on Petri nets and Concurrent State Machine principles. The proposed methodology combines the principles of Petri net refinement and Petri net composition, with hierarchical mapping of the net into reprogrammable logic devices.

The Petri net is directly mapped into FPL without explicit enumeration of all possible global states. The specification is given in terms of local state changes.

The proposed method of Petri net-based realisation of Logic Controllers considers explicit hierarchy included in netlist formats, or hierarchical constructs of VHDL or Verilog. The syntactic and semantic compatibility of Petri net descriptions and decision rules with structured netlist or HDLs statements are as close as possible.

REFERENCES

Adamski, M. (1991). "Parallel Controller Implementation using Standard PLD Software", in: W.R.Moore, W.Luk (eds.), *FPGAs*, Abingdon EE&CS Books, Abingdon, England, pp.296-304.

Adamski, M. (1998). SFC, Petri Nets and Application Specific Logic Controllers. In: *Proc. Of the IEEE Int. Conf. on Systems, Man, and Cybernetics*. San Diego, USA, 11-14.10.1998, pp.728-733.

Adamski, M. (1999). Application Specific Logic Controllers for Safety Critical Systems. In: *Proc. of the 1999 IFAC Triennial World Congress*, Beijing, China, **Vol.Q**, pp.519-524.

Adamski, M. & Wegrzyn, M. (1994). Hierarchically Structured Coloured Petri Net Specification and Validation of Concurrent Controllers. In: *Proc. of the 39th International Scientific Colloquium IWK'94*. Ilmenau, Germany, **Band 1**, pp.517-522.

Bilinski, K., Adamski, M., Saul, J.M. & Dagless, E.L. (1994). Petri net based algorithms for parallel controller synthesis. In: *IEE Proceedings, Computers and Digital Technique*, **Vol.141**, No.6, Nov.1994, pp.405-412.

David, R. & Alla, H. (1992). Petri Nets & Grafcet. Tools for modelling discrete event systems. Prentice Hall, New York.

Halang, W.A. & Jung, S.-K. (1994), A Programmable Logic Controller for Safety Critical Systems. In: *High Integrity Systems, 1, 2*, pp.179-193

Harrel, D. (1987). Statecharts: A Visual Formalism for Complex Systems. In: *Science of Computer Programming*, **Vol.8**, pp.231-274.

Jensen, K. (1992). Coloured Petri Nets. Basic Concept, Analysis Methods and Practical Use, **Vol.1**, Basic Concepts. Springer-Verlag, Berlin.

Lee, K.H. & Favrel, J. (1985). Hierarchical Reduction Method for Analysis and Decomposition of Petri Nets, In: *IEEE Trans. System, Man Cybernetics*, **Vol.17**, No.2, pp.297-303.

Mandado, E., Marcos, J. & Perez, S.A. (1996). *Programmable Logic Devices and Logic Controllers*. Prentice Hall, London.

Sagoo J.S. & Holding D.J. (1991). A comparison of temporal Petri net based techniques in the specification and design of hard real-time systems, In: *Microprocessing and Microprogramming*, **Vol.32**, No.1-5, pp.111-118.

Toulotte, J.M. (1978). Reseaux de Petri et automates programmables. In: *Automatisme*, Juillet-Aout 1978, pp.200-211.

Wegrzyn, M., Wolanski, P., Adamski, M.A. & Monteiro, J.L. (1997). Coloured Petri Net Model of Application Specific Logic Controller Programs. In: *Proc. of the IEEE International Symposium on Industrial Electronics ISIE'97*, Guimaraes, Portugal, 07-11.07.1997, **Vol.I**, pp.SS158-SS163.

Wegrzyn, M. & Adamski, M. (1999). Hierarchical Approach for Design of Application Specific Logic Controller. In: *Proc. of the IEEE International Symposium on Industrial Electronics ISIE'99*, Bled, Slovenia, 12-16.07.1999, **Vol.3**, pp.1389-1394.

Wegrzyn, M., Adamski, M. & Monteiro, J.L. (1997). Reconfigurable Logic Controller with FPGA. In: *Proc. of the 4th IFAC Workshop AARTC'97*, Vilamoura, Algarve, Portugal, 9-11.04.1997, pp.247-252.

Wegrzyn, M., Adamski, M. & Monteiro, J.L. (1998). The Application of Reconfigurable Logic to Controller Design. *Control Engineering Practice*, Pergamon, **Vol.6**, 1998, pp.879-887.

Zakrevskij, A.D. (1990). On the Theory of Parallel Logical Control Algorithms. In: *Sov.Journal of Computer and System Sciences* (USA), 1990, **Vol.28**, No.5, pp.36-46.

Zhou, M.-C. (1992). A Hybrid Methodology for Synthesis of Petri Net Models For Manufacturing Systems. In: *IEEE Transactions on Robotics and Automation*, **Vol.8**, No.3, 1992, pp.350-361.

LOW COST IMPLEMENTATION OF MATHEMATICAL FUNCTIONS USING PIECEWISE INTERPOLATION

J.A. Rodríguez Mondéjar * F. de Cuadra García **
O. Nieto-Taladriz García ***

Dpto. de Electrónica y Automática ICAI UPCO, Alberto Aguilera 23, 28015 Madrid, Spain
** *Instituto de Investigación Tecnológica (IIT) ICAI UPCO, Alberto Aguilera 23, 28015 Madrid, Spain*
*** *Dpto. de Ingeniería Electrónica ETSIT UPM Ciudad Universitaria s/n, 28040 Madrid, Spain*

Abstract: This paper presents a technique for implementing non-trivial n-dimensional mathematical functions in low-cost real time control systems. The proposed technique uses recursive piecewise multilinear interpolation to approximate the original mathematical function. The paper presents the theoretical basis of the proposed method and the software tool developed for implementation. The main implementation issues are explained in detail. The developed tool includes an automatic C-code generator that implements the best software solution in terms of computational time and cost. Finally, the paper describes the application of the technique to a torque control function for a scara robot. *Copyright © 2000 IFAC*

Keywords: Embedded systems, interpolation approximation, multidimensional systems, real-time computer systems, robot control, software tool, splines.

1. INTRODUCTION

Very often the solution to practical problems, like those that appear in automation and industrial control, implies the use of non-trivial mathematical functions with a heavy computational load. The excessive computation time, or the requirement of sophisticated hardware, makes the direct use of these functions in low cost real time systems nonviable (Laplante, 1993; Motus, 1995). These functions are named *slow functions* in this article. The problem with slow functions can be due to the fact that they have a complicated mathematical expression, they need an iterative algorithm to calculate the values, or they use a lot of floating point operations.

Equation 1 shows an example of a slow function called the Franke-Little function: the main prob-

lem is the exponential operation. This function causes no problem for a Pentium microprocessor-based system, but it is very difficult to implement in an integer microcontroller such as an 8051 system.

$$f(x,y) = \frac{3}{4}e^{-\frac{1}{4}[(9x-2)^2+(9y-2)^2]} +$$
$$+\frac{3}{4}e^{-[\frac{1}{49}(9x+1)^2+\frac{1}{10}(9y+1)^2]} -$$
$$-\frac{1}{5}e^{-[(9x-4)^2+(9y-7)^2]} +$$
$$+\frac{1}{2}e^{-\frac{1}{4}[(9x-7)^2+(9y-3)^2]}$$

(1)

A solution to the problem using traditional techniques (Laplante, 1993) may be impossible. These techniques may be divided into two groups: general and custom solutions.

As general solutions, the following can be used:

- A better compiler. However, the speed-up is limited.
- One or more floating-point coprocessors. This solution is expensive for low cost systems. Moreover, only elementary functions can be speed up.
- A look-up table (Noetzel, 1989). It is a very fast solution, but the size of the memory required may be too big.
- A look-up table (Instruments, 1996) with interpolation. Although the memory size can be reduced (look-up table case), this reduction may not be enough.

In custom solutions you can:

- Try to improve the calculation algorithm, but it may be impossible to do.
- Substitute the slow function for simplified models. It demands a lengthy and sophisticated analysis. In most of the cases, the original domain of the slow function must be limited.

This paper presents an alternative approach that shows many practical advantages with respect to the previously mentioned techniques. This approach makes use of the fact that these slow functions do not necessarily imply a complicated topography. The control functions used in engineering are usually smooth and have good local behavior. This means that their values can be approximated by using interpolation or extrapolation techniques with low error (de Cuadra, 1990). For example, figure 1 shows the smoothness of equation 1. Moreover, there is usually no need for great precision in the computations due to the robustness of the (engineering) system. These two facts allow the use of interpolation techniques to replace the slow-function by another whose main characteristics are its far simpler structure, low implementation cost, fast and total predictability in terms of time and cost.

In this paper, the recursive piecewise multilinear interpolation is used to construct this approximation function, following the results of (de Cuadra, 1990). This function receives the name of abacus, according to the nomenclature of (de Cuadra, 1990).

2. RECURSIVE MULTILINEAR INTERPOLATION

The main idea behind recursive multilinear interpolation is combine the look-up table technique with the interpolation and the zoom. The main contribution of the abacus is the zoom. The original domain of the function is divided into hyperintervals according to a cartesian grid of non-homogenous step. The hyperintervals that do not

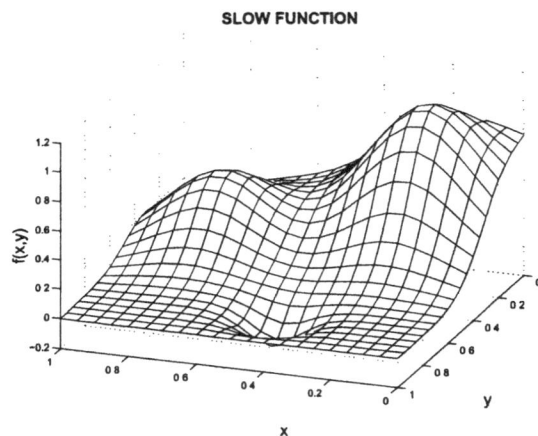

Fig. 1. Example of slow function with smooth behavior: Franke-Little function.

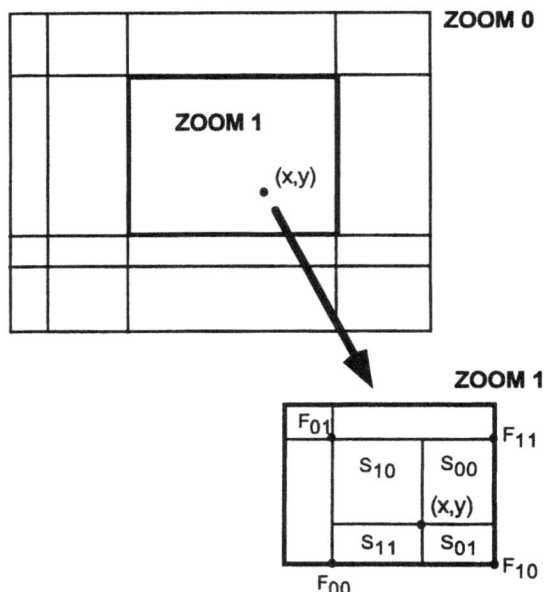

Fig. 2. Recursive multilinear interpolation.

stay within the approximation error restriction are isolated and handled as zooms (see the figure 2).

In any correct hyperinterval, the approximation function is calculated by interpolation. The result is a hierarchy of zooms that approximates the original slow function with the accuracy required. With this model, the abacus has three blocks:

(1) Memory block. It stores the constants (target function values in the nodes of the grid). The cartesian grid simplifies the organization of the data.
(2) Searching block. It searches the correct hyperinterval to interpolate the function.
(3) Interpolation block. This calculates the approximation value (abacus function) with the following formula (equation 2):

$$f^a(x) = \frac{1}{\text{Vol}_{\text{hyper}}} \sum_{\text{vertex}} \text{Vol}_{\text{vertex}}(x) f(\text{vertex}) \quad (2)$$

where Vol_{hyper} is the volume of the hyperinterval at which the point falls. Vol_{vertex} is the volume between the point and the corresponding vertex opposed to the vertex of the indicated interval. This formula is derived from the Lagrange formula (Hämmerlin and Hoffmann, 1991). For the case of 2 dimensions the formula is (equation 3):

$$f^a(x,y) = \frac{S_{00}F_{00} + S_{01}F_{01} + S_{10}F_{10} + S_{11}F_{11}}{S_{00} + S_{01} + S_{10} + S_{11}} \tag{3}$$

The *ZOOM 1* in figure 2 shows the meaning of the terms in equation 3. The abacus reduces the slow function to one search and one interpolation. The most important properties of the abacus are:

- Cost and time predictability. The searching block and the interpolator block only depend on the interface of the slow-control-function: number of inputs, input format and output format. The operations involved are comparisons, additions, multiplications and only one division (if it is necessary). Their simplicity makes the computational time totally predictable.
- Flexibility. The memory stores the topography of the slow-function. If you change the slow-function for another with the same interface, you need only change the memory values (if the memory size is large enough). The error between the slow-function and the abacus along with the slow-function topography define the minimum memory size. In the limit, with error 0, the abacus becomes the classical look-up table (Noetzel, 1989).

These properties make the abacus suitable for low cost real time systems (microprocessor or ASIC based). The main problem with the abacus construction is how to calculate the grid. Every point on the slow-function domain must satisfy the following error equation:

$$|f - f^a| \le e(x) \tag{4}$$

where f is the slow-function, f a is the abacus and $e(x)$ is the admissible error function in each point of the domain. This natural definition of the error is hard to use due to the absolute value function. With the exception of simple cases, there is not efficient algorithm to calculate the optimal grid in n-dimensional approximation problems (Dierkcx, 1995). However, reference (de Cuadra, 1990) describes a heuristic algorithm that has proved to be very efficient for calculating the grid.

The algorithm is based on the following idea: if the definition of the error of equation 4 is fulfilled

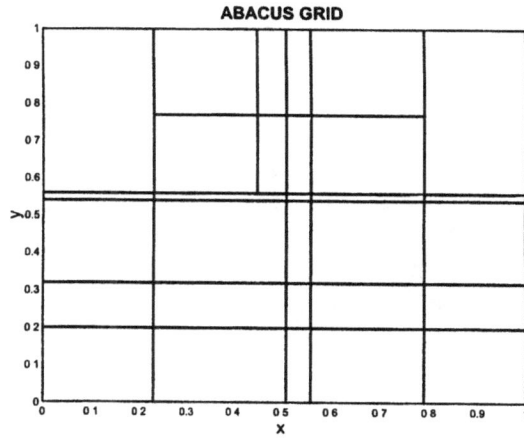

Fig. 3. Abacus of Franke-Little: division of the domain.

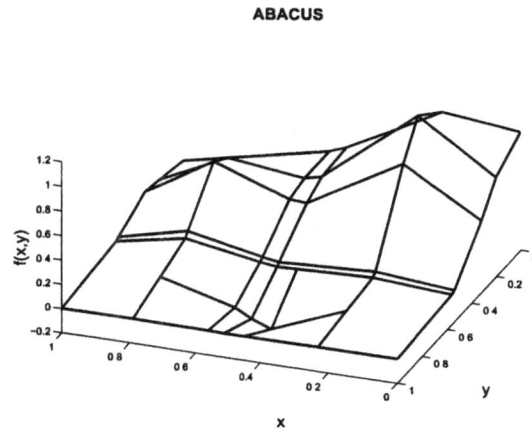

Fig. 4. Abacus of Franke-Little: splines.

in all points of the domain, it is also fulfilled on the edges that define the border of this domain. We must assume that border points belong to the slow-function domain. The n-dimensional problem has become an initial 1-dimension approximation problem in the edges of the domain. We solve it by applying the Adaptive Approximation algorithm (Hämmerlin and Hoffmann, 1991). The convergence is assured by the Jackson Theorem (de Boor, 1978). This division of edges leads to the first grid. The next step is to verify that the error is satisfied within each interval of this first grid. The Hooke&Jeeves algorithm is used to calculate the maximum error. The hyperintervals where the error restriction has been exceeded are calculated separately in the same manner (the recursive term comes from here). Figure 2 shows the mechanism. Figures 3, 4 and and 5 present the application of the technique to the Franke-Little function (equation 1) with 5% accuracy. Figure 3 shows the division of the domain in zooms and hyperintervals.

Figure 4 shows the bilinear splines (Hämmerlin and Hoffmann, 1991; de Boor, 1978) of the abacus (piecewise approximation). Figure 5 presents the full abacus function.

195

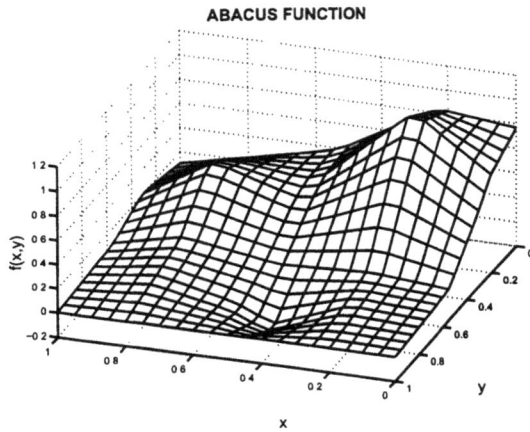

Fig. 5. Abacus of Franke-Little: function.

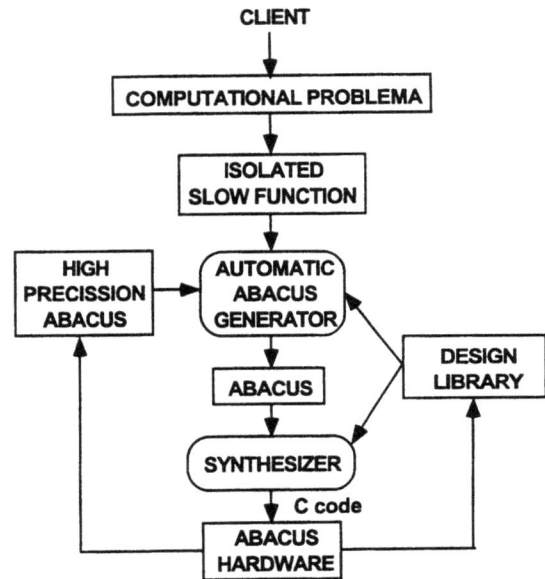

Fig. 6. Work methodology of the tool.

If you compare the original function and the abacus function in a Pentium based system, the calculation time is reduced from 12 microseconds to 3 microseconds. Nevertheless, the direct application of this method in real-time systems poses several problems:

- Fixed-point abacus. In (de Cuadra, 1990) only floating-point arithmetic is used. In low cost systems, the use of fixed-point arithmetic is very common. The Hooke&Jeeves algorithm presents serious problems of local minimums with the fixed point abacus. These have been solved by replacing the fixed-point abacus by two floating point functions which surround the fixed-point abacus.
- Excessive number of nested zooms. The search time depends on the number of nested zooms. An excessive number of nested zooms slows down the searching block. In order to prevent this slow down, the information of the inferior zoom is used to recalculate the grid of the superior zoom. This procedure is equivalent to using the samples that divide the axes of *ZOOM 1* in the *ZOOM 0* (see figure 2).

From the implementation point of view, the main contribution of this paper resides in the control of the number of nested zooms, which leads to a wide range of different feasible abacuses. The full hierarchy of zooms represents the case of minimum memory size and maximum latency, whereas the case without nested zooms represents maximum memory size and minimum latency. Nevertheless, you must build the design space to eliminate the bad abacus solutions. For example, if you decrease the number of nested zooms, the search time of the correct zoom decreases, but the search time of the correct hyperinterval in a zoom can increase, because the size of the zoom grid is bigger.

3. TOOL FOR THE IMPLEMENTATION OF ABACUS

A software tool has been developed to calculate the abacus outlined in the previous section. Figure 6 shows the work methodology of the tool. The designer provides the tool with the slow-function written in C. The rest of the information (interface of the function with the real world, maximum error allowed, maximum number of nested zooms, type of implementation, etc.) is introduced through script files or in an interactive way. The most important characteristics of the tool are:

(1) The abacus can be implemented in different formats: standard floating point, fixed point with different formats (unsigned, or two's-complement), and different data size.
(2) Automatic generator of C code. The C code of figure 7 is an example of a 2-variable abacus. The code is the same for any 2-variable function abacus. This permits the optimization of abacus implementation to be separated from the approximation of the slow function by the function abacus. This is why the tool comprises two main components: the abacus generator and the synthesizer.
(3) Study of abacus latency and size in each one of the possible solutions. The tool calculates the memory and the number of operations necessary in the worst case. With these operations and the technical sheet of the microprocessor, the response time can be estimated.
(4) Study of the error in the approximation.

Reference (Mondéjar *et al.*, 1999) outlines the ASIC implementation of the abacus.

196

```
// Searching block

while (1) {
                        // Hyperinterval
                        // searching
    i0=Search(pz->pN0,pz->Max0,x0);
    i1=Search(pz->pN1,pz->Max1,x1);
    W0=pz->Max1;
    if (!pz->pE)        // No more zooms
        break;
    pos=i1;
    pos+=i0*W0;
    e=pz->pE[pos];
    if (!e)             // Right zoom
        break;
    pz=pz->pHijos[e-1]; // Go to child zoom
}

// Interpolating block

p01=pz->pN0[i0];        // Vertex
p00=pz->pN0[i0+1];
v*=(p00-p01);           // Full volume
p01=x0-p01;             // Vertex distances
p00-=x0;
p11=pz->pN1[i1];
p10=pz->pN1[i1+1];
v*=(p10-p11);
p11=x1-p11;
p10-=x1;

W0++;                   // Interpolation
pos=i0*W0+i1;
tp=pos;
f+=pz->pC[tp]*p00*p10;
f+=pz->pC[tp+1]*p00*p11;
tp+=W0;
f+=pz->pC[tp]*p01*p10;
f+=pz->pC[tp+1]*p01*p11;

f/=v;                   // Division or shift
```

Fig. 7. Abucus C code.

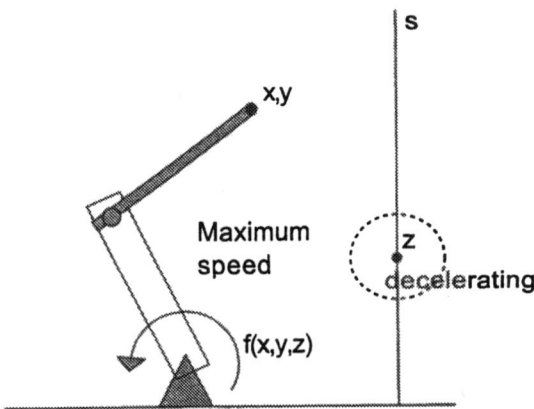

Fig. 8. Scara robot.

4. EXAMPLE OF APPLICATION

In this section, a more complex example than the Franke-Little function is presented: the application of abacus to solve the problem of controlling the first arm of a scara robot. Figure 8 shows the robot drawing.

Table 1. Different abacus implementations for a scara-robot arm control (4 levels of zooms)

Zoom architecture	Directory		Tree	
Search method	Bip.	Index	Bip.	Index
Memory access	50	34	85	67
Function calls	6	0	15	0
Multiplications	27	27	36	33
Additions	69	37	88	43
Divisions	1	1	1	1
Comparisons	39	9	58	15
Memory (bytes)	6436	11684	6532	12628
Look up table: 16MBytes, 1 memory access				

Table 2. Attribute variation vs. zoom levels

Zoom levels	Architecture	Time μs	Memory bytes
4	Directory/Index	30	11684
3	Directory/Index	15	20171
2	Tree/Index	11	37231
1	Tree/Index	7	941343

The aim is to move the end of the robot arm (position (x, y)) to position z in line s with maximum speed. It is a 3-variable function (floating point) that took 103 microseconds using a Pentium microprocessor-based system. An abacus was built for this slow function without limiting the hierarchy of zooms and with 153428 bytes of memory. The abacus latency was only 10.5 microseconds in the worst case in the same Pentium system.

Next, the experiment was repeated using a 16-bit microcontroller. The size of all the variables (inputs and output) is 8 bits. The approximation error is smaller than 1for the tool shows the results. We use four different strategies to implement the hierarchy of zooms in the memory. For each strategy, the table shows the operation numbers classified by types and the memory size necessary.

With these tables the designer chooses the most suitable design for his application. In this case and if there are no other restrictions, the second column (Directory/Index) presents the most favorable solution. If this is implanted in a Siemens 80167 20 MHz (Siemens, 1997) the estimated latency of the function is smaller than 30 microseconds. The information in the bottom row can be used to compare the look-up table with the four abacus solutions above it.

Table 2 shows the attribute variation vs. the reduction of nested zooms. This table shows that the solution using only interpolation (zoom level = 1) is unsatisfactory because of its high memory requirement.

5. CONCLUSIONS

This paper has presented a technique for the implementation of slow control functions by using a faster and cheaper function in its place. The main idea is to combine the look-up table techniques with the interpolation and the nested controlled zooms. It permits a trade-off between cost (memory size) and latency. The two examples presented show that the results obtained through this technique are very satisfactory. The designer can explore the space of possible designs quickly, thanks to the flexibility of the tool. The abacus can be effectively used in the fields of classical control, fuzzy or neuronal control and simulation, among others.

6. REFERENCES

de Boor, Carl (1978). *A Practical Guide to Splines*. Springer-Verlag. New York.

de Cuadra, F. (1990). El Problema General de la Optimización del Dise no por Ordenador: Aplicación de Técnicas de Ingeniería del Conocimiento. PhD thesis. Universidad Pontificia Comillas. Madrid.

Dierkcx, Paul (1995). *Curve and Surface Fitting with Spline*. Oxford University Press. Clarendon, Oxford.

Hämmerlin, Günter and Karl-Heinz Hoffmann (1991). *Numerical Mathematics*. Springer Verlag. New York.

Instruments, Texas (1996). *Table Look-up and Interpolation on the TMS320C2xx*. Texas Instruments.

Laplante, P.A. (1993). *Real-Time Systems Design and Analysis: an Engineer's Handbook*. IEEE Press.

Mondéjar, J. A. R., F. de Cuadra and O. Nieto-Taladriz (1999). A tool for the implementation of heavy-computational-load control functions in embedded systems. In: *Proceedings of 25th Annual Conference of the IEEE Industrial Electronics Society IECON'99*.

Motus, L. (1995). Timing problems and their handling at system integration. In: *Artificial intelligence in industrial decision making, control and automation*. Kluwer Academic Publisher.

Noetzel, A. S. (1989). An interpolating memory unit for function evaluation: analysis and design. *IEEE Transactions on Computers* **38**(3), 377–384.

Siemens (1997). *C166 Family Instruction Set Manual*. Siemens.

MATLAB-BASED REAL-TIME FRAMEWORK FOR
DISTRIBUTED CONTROL SYSTEMS

Marga M. Marcos*, J. Portillo*, J.M. Bass†

*Escuela Superior de Ingenieros de Bilbao (University of the Basque Country)
Alameda Urquijo s/n 48013 Bilbao-SPAIN
†School of Informatics, University of Wales, Bangor
e-mail: jtpmamum@bi.ehu.es, jtppobej@bi.ehu.es, J.Bass@bangor.ac.uk

Abstract: Embedded systems are computer-based systems that often must have deterministic temporal behaviour. The application software must ensure overall system performance in terms of hard real-time constraints, reliability and safety requirements. Often application constraints include distribution of input/output. In some other cases, distribution is needed to achieve demanding performance or dependability requirements. This paper presents the design and implementation of the so-called Real Time Framework (RTF) that consists of a general environment for the easy generation of distributed control software. RTF covers from the specification of the distributed system to automatic code generation. The design is based on the integration of well-known commercial software for control systems into a framework that extends its use to distributed systems using a fieldbus as communication system. The internal design includes a model of the system in each design phase and information consistency is assured through an internal database. *Copyright © 2000 IFAC*

Keywords: distributed computer control systems, real-time systems, fieldbus

1. INTRODUCTION

The design of complex real-time systems often must meet hard requirements in terms of temporal behaviour, performance and safety. Real-time applications vary in size and scope from microwave ovens to factory automation, aerospace industry or railway control. Even within industrial process control, it is possible to find applications in which several sub-systems co-operate and a sub-set of them can have hard real-time and/or safety requirements. In order to face the design of such type of applications, the availability of tools covering all design phases (from specification, simulation and design to code generation) results fundamental to succeed. In fact, software vendors are adopting a tool integration approach, following the demand from control engineering developers. The current market offers well-known software packages for system analysis and simulation that allow some exchange of information between the tools that cover the design

phase. For instance, the two packages Xmath and Statemate (Harel et al 1990) can be linked together at the code level and an interface allows the joined simulation within Statemate environment. The *Matlab / Simulink / StateFlow* environment (Mathworks 1996, 1997a, 1997b) offers graphical tools that allow joint simulation of models edited in Simulink and Stateflow. Stateflow is based on the statechart notation used in Statemate which, like Petri nets, have an underlying formalism that allows validation of aspects of the system specification. Statecharts are a form of transition diagrams that support hierarchy and concurrency.

There are many structured design methods that are targeted toward real-time systems: MASCOT, JSD, Yourdon, MOON, HOOD, DARTS, MCSE, etc. (Burns and Wellings 1995). Nevertheless, there are few CASE environments available to support the real-time systems design process. Teamwork (Cadre Technologies 1990), for example, provides an environment in which designs can be expressed using Buhr's graphical notation (Buhr 1984) or HOOD (ESA 1991). Tools that analyse the timing properties of the resulting designs, however, are not included. Other environments, which support the whole real-time system life cycle, are EPOS (Lauber (Ed) 1989) or Beacon (Spang et al, 1993), that allow code generation from hierarchical block diagrams. These diagrams can be defined graphically and from them source code in C, Fortran, Ada or assembly language, can be generated. It also allows static analysis of the code in a format suitable for the software engineer, by means of the generation of both the control and data flow diagrams. But, all these tools do not offer the possibility of analyse the control system, such as other environments familiar to the control engineer, as Xmath / SystemBuild or Matlab / Simulink / Stateflow. Thus, after the analysis in control engineer domain, it is necessary to translate manually from the block diagrams designed in Xmath or Matlab into the block diagrams suitable in the software engineer domain.

On the other hand, object-oriented modelling seems to be a natural fit for capturing the various characteristics and requirements of systems that have hard deadlines of performance (Powel 1998). UML is a third-generation modelling language that rigorously defines the semantics of the object metamodel and provides a notation for capturing and communicating object structure and behaviour. It is the natural successor to the second-generation methodologies including BOOCH, OMT, OOSE and Statecharts.

However, as far as the authors know, there is the lack of a commercial environment that integrates automatically all design phases, from specification to application code generation. Some applications generate code automatically, but they do not allow the analysis and simulation of the control system, or they do not allow the analysis of the software (as it is the case of *Matlab*). That means that it is necessary to translate manually from one model to another in order to design the complete system.

The creation of this type of integrated environment is also a focus of interest for well known research groups in universities. For instance in Bass et al (1994), Browne et al (1996), Hajji et al (1997) a Framework project developed at the University of Sheffield is described. In this case, the target system is based on transputer hardware.

In this paper, a prototype of an integrated environment for the design and development of real-time distributed control systems is presented. The basic architecture integrates different commercial tools that represent system behaviour from different points of views, from system specifications to code generation. The proposed framework assures the automatic translation between models expressed in each domain. In this form, the designer can use the power of different tools in every design phase and, once the design is completed, generate the application source code.

As mentioned above, the goal is the design of distributed control systems. In this type of applications, embedded and automation applications predictability in messages response time can be the most critical point. The currently most used fieldbuses in time critical applications have as bus assignment protocol either TDMA (Time Division Multiple Access), used for example by ARINC (AEEC 91), or priority based protocol (like CAN 1991). Regarding the MAC layer (Medium Access Control) fieldbus protocols can have destructive access (as CSMA) or non-destructive access. For this latter, the access can be controlled or not. If some type of control is used, it can be centralised, like Profibus (Decotignie et al, 1993) or WorldFip (Saba et al, 1993), or decentralised (like CAN).

The current version of RTF supports CAN bus communication. It has been selected as it is simple, low cost, decentralised, non-destructive and fixed-priority based. These characteristics make of CAN a good option for embedded and automation applications. It is also possible to apply well-known task scheduling theories for analysing the temporal behaviour (Audsley et al 1993, Tindell and Burns 1994). Future versions of RTF should include other real-time fieldbuses.

RTF integrates commercial software through the definition of view-models. However, it also allows the integration of specific tools that extend the analysis and simulation of the distributed system behaviour. Thus, in Marcos et al (1999a, 1999b) a tool for analysing and simulating the temporal behaviour is described.

The paper is laid as follows: section 2 presents the 4-view model of the distributed system in which RTF is based. In section 3 the RTF library included in *Simulink* is described, as well as the relationship between these new library blocks and the Requirements and Temporal models. Section 4 is dedicated to a detailed description of the Requirements model. Finally, an example of use is presented in section 5 and some conclusions in section 6.

2. 4 -VIEW MODEL APPROACH

The strategy uses a global database from where separate domain-specific views on the system can be obtained. The global database enables analysis of the system prior to implementation and the integration of behavioural and architectural requirements. The distributed system is described from four complimentary sub-models: Requirements, Architecture, Temporal and Process.

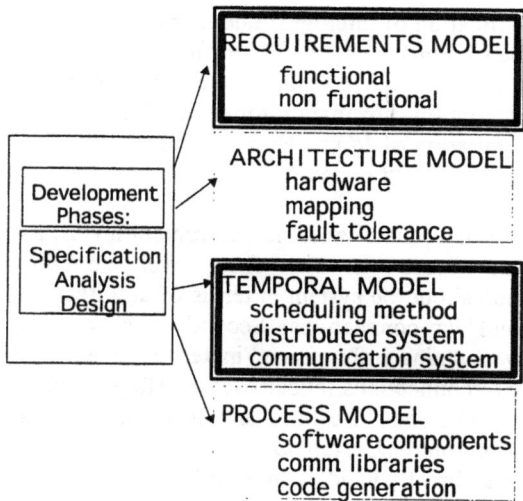

Figure 1. Specific Domain Models of the Distributed System

Automatic tools should translate the information contained in the global database into each of the 4-view models. In this sense, a Model consists of the minimum database to be used by the tools that assist the designer in the analysis, simulation and design of the proposed system.

Figure 1 represents the 4-view model of the distributed system. The proposed environment seeks to capitalise on the strengths of an iterative design approach supplemented by search/optimisation tools. Co-simulation across modelling domains is necessary to enable exploration of conflicting trade-offs between different views of the system. The impact of a modification in one model must be seen in another. For example, addition of a new functional feature may compromise performance constraints. Thus, it is very important to maintain consistency between models throughout the design process and product lifetime. RTF must assure such consistency and it has to maintain all the information needed to translate modifications from a model view to another.

The model integration software comprises four main layers: requirements model, architectural model, temporal model and process model. The requirements model is an architecture independent, behavioural, model of system functionality. The architecture model supplements the requirements model with details of the hardware architecture and processor types selected. The temporal model allows the analysis and simulation of the temporal behaviour of the system. The process model is the most detailed representation, including knowledge of software modules, their interconnection and synchronisation, as well as the communication system.

The work reported in this paper focuses on the design and implementation of a prototype of RTF based in *Matlab* and *toolboxes*. In particular, the Requirements Model and its relationship with the

Temporal Model are defined. The latter is described in detail in Marcos et al (1999b).

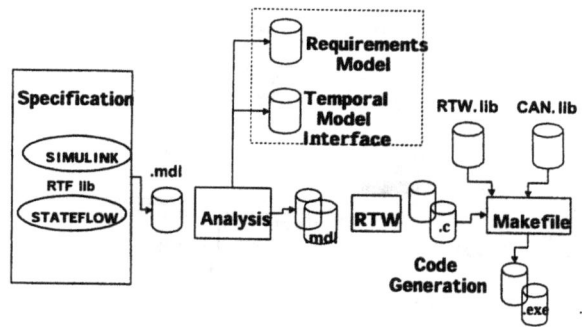

Figure 2. Prototype of matlab-based RTF

Architecture and Process Models are subject of future works. By now, the Architecture Model is defined by user in the Specification Phase and the Process Model assumes the default option of mono-processor targets under DOS operating system (one of the options in Real Time Workshop toolbox).

Thus, the current version of RTF offers an extended simulink environment for the *specification* of the distributed system. It also offers simulation capabilities inherent to both *simulink* and *stateflow* toolboxes. Processing of the model file (extension .mdl) allows the generation of the database that constitutes the Requirements Model. This model contains the information of all nodes presented in the system, its characteristics and general information is needed in the application source-code generation. To do that, several calls to the RTW coder are executed (see figure 2).

3. REAL-TIME FRAMEWORK LIBRARY BLOCKS

A new block library, called RTF_lib has been designed. It allows the user to specify the elements within the distributed system. Two type of blocks have been defined: *architecture* blocks, those that define intelligent nodes, input/output nodes and tasks. The *communication* blocks allow to specify messages sent/received between nodes. Thus, using RTF library, user can specify a distributed system like the one represented in Figure 3.

3.1 Architecture blocks

- *I/O node.* Used for the specification of distributed inputs and outputs.
- *Intelligent node.* They represent the distributed CPUs (computer, embedded system, PLC,...).
- *Action.* It allows the user to represent concurrent activities within a node (task). Parameters: period of activation (it can be inherited if it

belongs to a transaction), deadline, priority and criticality (Hard Real Time or Soft Real Time)

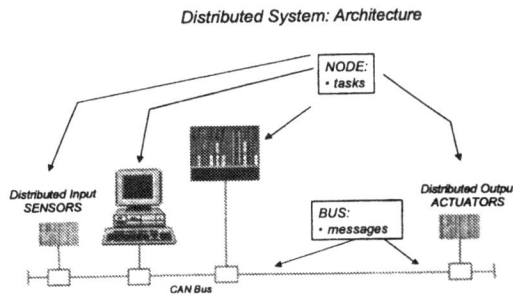

Figure 3. Elements of a distributed system

3.2 Communication blocks

- *Send*. It represents a message coming from an action of the source node. Parameters: period of activation (it can be inherited, if it belongs to a transaction), deadline, priority and criticality.
- *Receive*. It represents a message arriving to an action of the destination node. Their parameters are inherited, as it belongs to a transaction.
- *Config_Input*. Configuration of Input Nodes (analog and digital)
- *Config_Output*. Configuration of Output Nodes (analog and digital)

4. REQUIREMENTS MODEL OF THE DISTRIBUTED SYSTEM

The Requirements Model of the distributed system is built from the specification of the distributed system defined by the user during the specification phase. Initially, a graphical user interface allows the user to select the type of network, the scheduling method of intelligent nodes (if they support concurrency) and the priority assignment policy. Currently, only CAN network and Rate Monotonic Scheduling, Deadline Monotonic Scheduling and user priority assignment are supported. In second place, the general architecture is defined (using the RTF lib). Finally, the internal structure of both continuos and discrete event parts of the system are specified (using *simulink* and *stateflow* libraries).

Once the distributed system is specified, the user should launch the generation of Requirements Model procedure. The processing of the file model of the system defined in *simulink* is used to define the Requirements Model as an internal hierarchical data structure that will allow:

- Generation of a minimum temporal model of the distributed system (input to the temporal analysis and simulation tools (Marcos et al 1999b))
- Generation of the process model. In the current prototype of RTF this means:

- Identification of intelligent nodes and generation of the *simulink* file model for each of them
- Identification of I/O nodes and generation of configuration information for each of them. The configuration of I/O nodes is carried out by one of the intelligent nodes

The data structure is divided in hierarchical levels, as shown in figure 4. The first level contains information about the system in terms of architecture blocks and its connections. Second level of the hierarchy contains information, in terms of actions and its interconnections, about each intelligent node and each I/O node. The third and following levels contain the *simulink, stateflow* or user s-functions subsystems, directly handled by *simulink*.

During generation of the Requirements Model, a table of found *transactions* is built in which user defines global deadlines for each of them (see Figure 4). A transaction, in control engineering terms, is defined as a particular control loop going from sensors to actuators. A graphical user interface allows the user to define new transactions (for example, if some sensor/actuator are not distributed but belongs to the hardware of an intelligent node). This information is used by RTF to compute the priorities for actions and messages (jointly with the priority assignment policy).

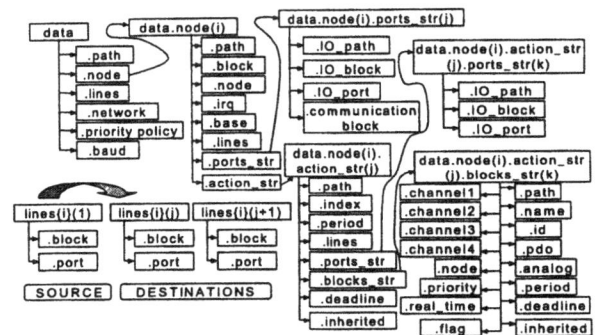

Figure 4. Data structure of Requirements Model

Finally, the system temporal model is derived from the Requirements Model as defined in Marcos et al (1999a, 1999b). Figure 6 shows the elements identified in terms of processors, tasks and events, i.e., accesses to internal resources (shared variables) and accesses to bus (messages).

5. EXAMPLE OF USE

Let us assume a particular implementation of a supervisor/controller for the heat exchanger shown in Figure 7. The system consists of two feedback loops in order to control the temperature at heat exchanger output and the level of the tank in the secondary circuit. A supervisor has been designed in order to handle alarms as well as to decide every sampling period what controllers are active. Its design follows the methodology stated in Marcos and Artaza (1996). The supervisor implements the heuristic knowledge on the process in terms of operation states and the rules that conclude that the system has evolved from one operation state to other. Thus, it is implemented as a state transition diagram that monitors process evolution detecting transitions between operation zones and switching on the control algorithm specifically designed for such operation zone.

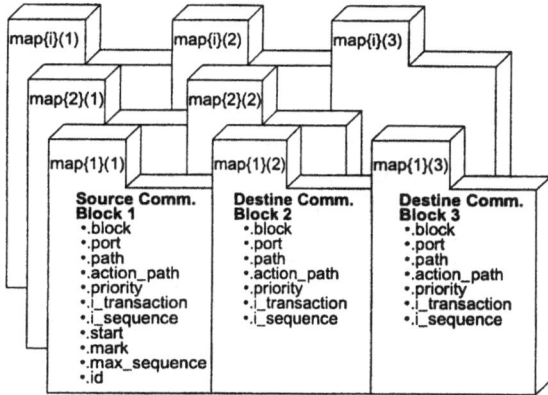

Figure 5. Transactions Table

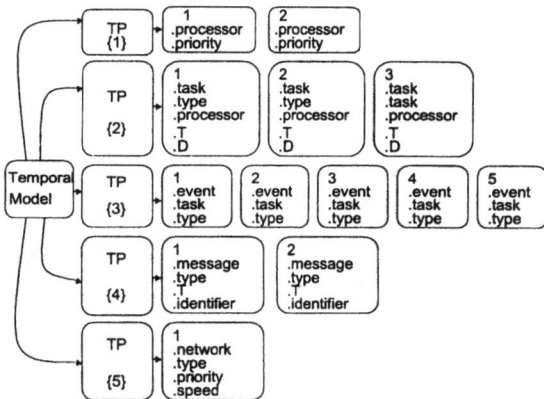

Figure 6. Components of the temporal model

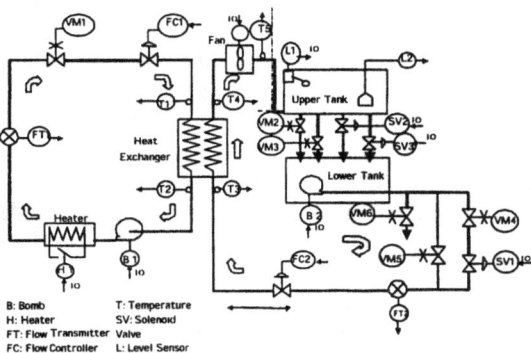

Figure 7. Heat Exchanger

Figure 8 shows the state transition diagram for the supervisor. Two concurrent state diagrams describe the supervision of the level and temperature control loops, respectively. The specification of the

distributed system is presented in Figure 9. There are two intelligent nodes: one for executing the supervisor and the other that contains the set of possible control algorithms. Inside the intelligent nodes, there can be simulink, stateflow (for implementing event driven behaviour blocks) or user S-funcions (code written in c or matlab). Figure 10 shows the internals of controllers node, constituted by send/receive blocks (meaning accessing to CAN bus) and by simulink or user S-functions (temperature and level controllers). RTW is used to translate these diagrams into executable files for both of intelligent nodes.

6. CONCLUSIONS

RTF is the first step towards a general Framework designed to assist the user during all phase involved in the design of real-time distributed applications. A first prototype that is matlab-based has been implemented, supporting CAN bus as communication network. A new library (RTF_lib) has been defined in order to allow the user to express the distributed architecture of the system. The Requirements Model is defined from the simulink model file. It contains all the relevant information of the system in terms of high level architecture, components and messages. This information is complemented via graphical user interfaces in order to define hard and soft real-time transactions within the distributed system. Finally, code for each intelligent node is generated (by means of sequential executions of Real Time Workshop toolbox) containing communication libraries for the selected network.

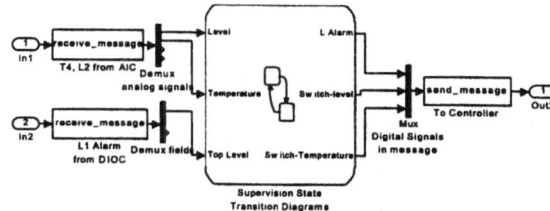

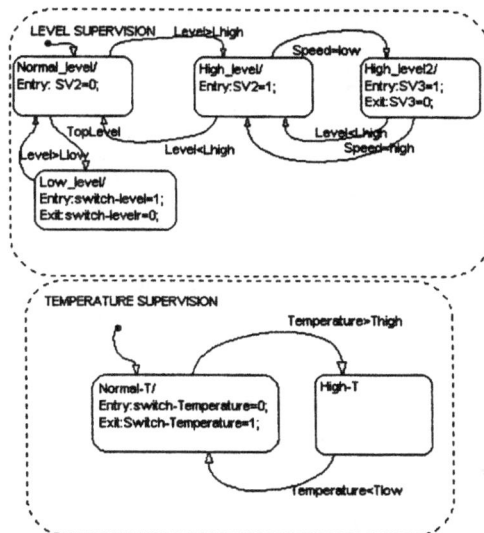

Figure 8. State Transition Diagrams for Supervision

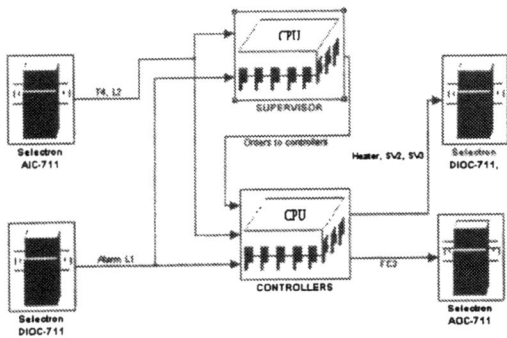

Figure 9. Definition of the distributed system

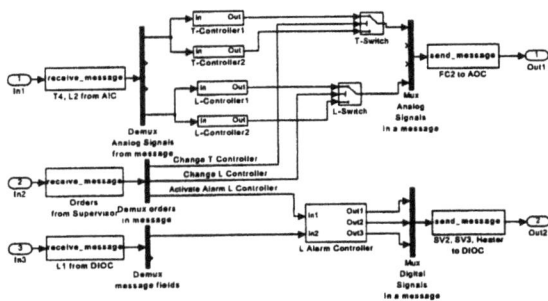

Figure 10. Specification of controller intelligent node

ACKNOWLEDGEMENTS

This work has been supported by CICYT project TAP 98-0585 C03-02, 42 by UPV/EHU UPV 146-345-TB196/98 and by GV/EJ project GV PI-1998/42.

REFERENCES

AEEC (Airlines Electronic Engineering Comittee) 1991. *Multi-Transmitter Data Bus. ARINC Specification 629-2: Part 1, Technical Description*, Aeronautical Radio Inc. Edition, October 1991

Audsley, N., A. Burns, M. Richardson, K. Tindell, A.J. Wellings (1993). *Applying New Scheduling Theory to Static Priority Pre-emptive Scheduling*. Internal Report ICC TB sept 94. Univ. Of York

Bass, J. M., A. R. Browne, M. S. Hajji, D. G. Marriott, P. R. Croll and P. J. Fleming (1994), *Automating the Development of Distributed Control Software*, IEEE Parallel and Distributed Technology, Vol. 2, No. 4, Winter 1994, pp. 9-19.

Buhr, R.J.A. (1984). *System Design with Ada*, Prentice Hall International.

Burns, A. A. Wellings (1995*). HRT-HOOD: A Structured Design Method for Hard Real-Time Ada Systems*. Elsevier.

Cadre Technologies (1990). *Teamwork*

CAN (1991). *Bosch CAN specification, V 2.0 PartA*. R. Bosch Gmbh, Germany

Decotignie J. D., P. Raja, G. Noubir, L. Ruiz and J. Hernandez (1993). *Analysis of polling protocols for fieldbus networks*. Computer communication review, ACM Sigcomm, vol 23, N° 3, pp. 69-90

Douglas, B. P. (1998). Real-Time UML.Developing Eficient Objects for Embedded Systems. Addison Wesley.

European Space Agency (1991*). HOOD Reference Manual Issue 3.1.* HRM/91-07/V3.1.

Hajji, M.S., J. M. Bass, A. R. Browne and P. J. Fleming (1997). *Design Tools for Hybrid Control Systems*, Int. Workshop on Hybrid and Real-Time Systems, HART '97, Oded Maler (Ed.), Grenoble, France, LNCS 1201, Springer-Verlag, March 1997, pp. 87-92.

Harel, D., H. Lachover, A. Naamad, A. Pnueli, M. Politi, R. Sherman, A. Shtull-Trauring and M. Trakhtenbrot, (1990) *STATEMATE: A Working Environment for the Development of Complex Reactive Systems*, IEEE Transactions on Software Engineering, Vol. 16, pp. 403-414, 1990.

Lauber, R.J. (1989). EPOS Overview: Short Account of the Main Features. GPP, Munchen.

Marcos and Artaza (1996). A Methodology for the Design of Rule-Based Supervisors. *European Journal of Control*, Vol 2, No 4, Springer 1996, pp. 305-312.

Marcos, M. M., Bass, J.M., Portillo, J. (1999a). An Integrated Framework for Development of Real-Time Control Software based on CAN bus *Procceedings of the 14th IFAC World Congress*, , pp. 447-452.

Marcos, M. M., Portillo, J. (1999b). BERTA: Basic Environment for Real-Time Analysis using CAN bus. Submitted to the *25th Workshop on Real Time Programming*.

Mathworks (1996). Using Matlab version 5.

Mathworks (1997a). Simulink: Dynamic System Simulation for Matlab V2.1

Mathworks (1997b). Stateflow: For use with Simulink. User's Guide

Saba G., Thomesse J.P. and Song Y.Q. (1993). *Space and tima consistency qualification in a distributed communication system*. Proc. Of IMACS/IFAC Int Symposium on Mathematical and Intelligent Models in System Simulation, Brussels, Belgium, April 12-16 1993, Vol 1, pp. 383-391

Spang, H. A., et al, (1993) The BEACON Block-Diagram Environment, Preprints of the 12th IFAC World Congress, Vol 6, pp105-110, Sydney, Australia, July, 1993.

Tindell, K., A. Burns (1994). Calculating controller area network (CAN) message response times, 12th IFAC workshop on Distributed Computer Control Systems

AN OBJECT-ORIENTED INFRASTRUCTURE FOR DISTRIBUTED PROCESS CONTROL

D. N. Ramos-Hernandez, P.J. Fleming and S. Bennett

Department of Automatic Control and Systems Engineering, The University of Sheffield, Mappin Street, Sheffield S1 3JD, UK.

Abstract: The objectives and design of an object-based infrastructure, namely the Integrated Design Notation (IDN), are presented. IDN supports the design, development and implementation of decentralised distributed control systems. An extrusion process is targeted as a demonstrator application, where object-orientation technology is expected to facilitate the improvement of extruder control in a distributed environment. The use of the Unified Modelling Language (UML) and CASE tools supporting UML is discussed and an initial design for the IDN is presented. *Copyright © 2000 IFAC*

Keywords: Distributed control, object modelling techniques, process control, real-time systems, simulation.

1. INTRODUCTION

"Process Control System Integration" (PiCSI) is an EPSRC-funded programme, involving three universities and eight industrial collaborators. The research objectives are divided into four work packages:

1) Integrated Design Notation (University of Sheffield).
2a) Multiple View Co-Simulation (University of Sheffield).
2b) Co-Simulation with Non-uniform Elaboration (University of Bangor).
3) Hard Real-time Virtual Machine (University of Bangor).
4) Integrated-constraint Controller Selection and Design (UMIST).

This paper describes the Sheffield contribution for this programme. The Sheffield work packages represent a radically new development of a control software design environment, previously reported at IFAC meetings (e.g. Browne *et al.*, 1997, Hajji *et al.*, 1996).

At present, the development process to achieve decentralised distributed control systems for the process and manufacturing industries is very complex. This is hampered by a lack of software tools to support all development lifecycle phases: simulation/modelling, development and implementation. Software packages that exist generally do not conform to standards and this makes integration difficult. This paper proposes an object-oriented infrastructure that supports the integration of multi-disciplinary software tools to facilitate the development process for fully decentralised process control.

Based on simple concepts, object-orientation (OO) facilitates flexibility, adaptability and reusability of the developed software. The purpose of OO is categorisation of objects in the problem domain into object classes combining attributes and behaviour into a single entity (Larsen et al., 2000). In the last couple of decades, object modelling has been replacing strict functional decomposition because it decomposes systems on the basis of encapsulatable characteristics which may be either structural or behavioural (I-logix, 1999).

The organization of the paper is as follows. The objectives of the object-oriented infrastructure are presented in Section 2. A design of the infrastructure is then proposed in Section 3. The UML methodology and CASE tools supporting UML are also presented in this section. Section 4 describes the target application, an extrusion process, where the control can be improved and developed with the object-oriented environment. Section 5 concludes the paper.

2. OBJECTIVES OF THE OBJECT-ORIENTED INFRASTRUCTURE

The planned Object-Oriented Infrastructure is named the Integrated Design Notation (IDN). IDN is an object-based environment that supports the integration of control engineering expertise with design imposed by the underlying hardware architecture. It will establish and redefine open infrastructures that enable integration and co-simulation of continuous and state-event system models that will comprise both abstract (plant architecture independent) and fully elaborated subsystems. It will have well-defined interface standards and an underlying formalism to ensure consistency, enable analysis of the system under development prior to implementation and support co-simulation (Bass, 1998a; Bass, 1998b; Turnbull, 1998). The IDN consists of three models: the requirements model, the software task model and the architectural model.

- ❑ The *requirements model* will provide policies and mechanisms for the hierarchical integration of the selected domain-specific views: transfer-function block diagrams (Simulink), state charts (Stateflow) and IEC 1131-3 standard process control notations. Translation rules for converting each view into the requirements model will also be defined.
- ❑ The *software task model* will integrate information obtained from the requirements and architectural models in a form, which facilitates automatic generation of Java source-code. This model will enable temporal analysis of the system under development, prior to implementation.
- ❑ The *architectural model* will allow that systems specified in the requirements model to be manipulated into a form that may then be implemented.

3. THE INTEGRATED DESIGN NOTATION (IDN)

There are many object modelling methods such as those of Booch, Rumbaugh, Shlaer and Mellor, Jacobson, Ward, Coad and Yourdon. Of these methodologies, Object Modelling Technique (OMT) (Rumbaugh et al., 1991), Object-Oriented Design (OOD) (Booch, 1991) and Object-Oriented Software Engineering (OOSE) (Jacobson et al. 1993) have been combined to create the *Unified Modelling Language* (UML). In 1997, the OMG accepted the UML as its standard for modelling object-oriented systems (Douglass, 1998).

UML supports what information is captured and how it is interrelated, it defines standard notations for representing the information from a graphical perspective, and it supports the use of an interchange format. This modelling language has a rich set of notations and semantics, which can be applicable to a wide set of modelling applications and domains (Douglass, 1998). UML is primarily used to model discrete systems rather than continuous ones, hence its emphasis on finite state machines and activity modelling. However, continuous models can be represented in the UML as well (I-Logix, 1999).

In real-time systems, timeliness and schedulability are two important issues. The modelling techniques required to achieve these requirements involve performance modelling, which relies upon an ability to model the system's architecture and the system's concurrency (Beckwith and Moore, 1999). Although UML is a well-defined and flexible modelling language, it is deficient in certain modelling requirements for real-time systems and the support of a CASE tool is sought.

Currently, there are several Computer-Aided Software Engineering (CASE) tools supporting UML. In general, a CASE tool might cover strategic planning, through domain analysis, system analysis, design, implementation (code generation), and testing, from an object-oriented perspective. Also, a CASE tool should simulate the behavioural dynamics of the system and provides the designer with early warnings of errors, inconsistencies, and unnecessary complexity. In addition, other features can be considered, such as tool connectivity, team/groupware support, heterogeneous hardware/software platform support, reusability support and price (Coad and Yourdon, 1991).

In order to select a CASE tool with the UML notation for the IDN; the following characteristics were deemed important.

- *Round-trip engineering.* This is the ability to both forward and reverse engineer source code.
- *Repository support.* This is the use of files or repository databases to store model information. A repository is necessary for sharing of component designs between developers and is generally built on top of a database, which provides data sharing and concurrency control features. Another way to build the repository is on top of the source code for a project, using a source-code control system to provide concurrency control. The benefit of this approach is a higher degree of synchronisation between the code and the model.
- *HTML documentation.* The tool should provide generation of HTML documentation for an object and its components. HTML documentation provides a static view of the object model that any developer using the model can refer to quickly in a browser.
- *Robustness.* This may be found in applications implemented in Java (JVM run-time protection) or as open-source projects (web-wide, parallel debugging).

Table 1 shows several CASE tools supporting UML. Since, the IDN environment must facilitate the incorporation of new software packages in the future, as well as accessing legacy software, Java on its own or combined with a distributed object technology, such as CORBA (Common Object Request Broker Architecture), allows ready integration. Thus the features considered are the ability to reverse-

Table 1 CASE Tools supporting UML

	Round-Trip Engineering	Windows NT	RTOS	CORBA/IDL Generation	Internet Web Publisher HTML	Repository Integration	Others
Rational Rose Real-Time (Rational Software Corp.)	C++	✓	Tornado (Wind River)	✗	✗	✗	
Rational Rose Java Features (Rational Software Corp.)	Java	✓	✗	✓	✓	✓	
StP/UML (Aonix)	C++ Java IDL	✓	✗	IDL	✓	✓	
Systems Architectures 2001 (Popkin Software)	Java	✓	✗	✓	✓	✓	Support Older Methodologies
Together/J Version 2.2 (Object International)	Java	Multiple OS platforms	✗	✗	✗	✗	Rational Rose (import/export)
Rhapsody (I-Logix)	C++	✓	Tornado VxWorks	✓		✓	
Real-Time Studio Professional (Artisan)	C C++ Java	✓	✓ VxWor	✓	✗	✓	Constraints Diagrams

engineer in Java with Windows NT as the software/hardware platform, a real-time operating system, CORBA/IDL (Interface Definition Language) support, repository support and finally HTML documentation. Regarding the need for Java and real-time operating system support to be important for the development of this environment, Table 1 reveals that Rational Rose (Java and Real-Time) and Real-Time Studio Professional are two potential CASE tools to consider in the design of the IDN.

Figure 1 shows the design for the IDN environment. In this, the three models of the IDN (requirements model, software task model and architecture model) are presented. Through the IDN the different tools can access directly or indirectly the different information of the system. Also, the IDN allows the replacement or integration of tools. The UML CASE tool can be accessed via the IDN as well as legacy software and a repository tool. Finally, a Java code generation tool is illustrated which is connected to the template library and to a Java compiler producing executable code for the target distributed architecture allowing reverse engineering of the system.

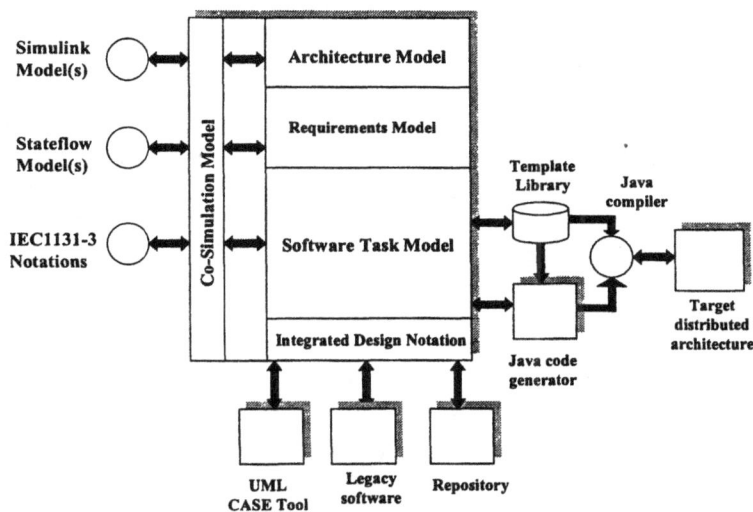

Fig. 1. Integrated Design Notation.

4. THE CABLE EXTRUSION PROCESS

The target application for testing the environment's capabilities is a cable extrusion process. This process is simply the forcing of thermoplastic melted from pellets through a restricted opening to form a continuous-length shape. The extrusion is usually cut to length on-line and then notched and drilled for additional openings. Structural details and surface treatments are limited to the forming direction. This process can be used to replace metal extrusions and roll-formed metal parts.

The plastic material is basically fed through the transport section onto the infinitely rotating screw via the hopper or feeder. The extruder which converts the plastic granules into the homogenous melt is quite similar to the one used in injection moulding. The plastic is heated slowly as it is moved and pressed forward towards the die. The melt is forced and compressed through the die. Once cooled with either water or other coolants, the extrusion is sized and cut to desired length. (see Figure 2).[1]

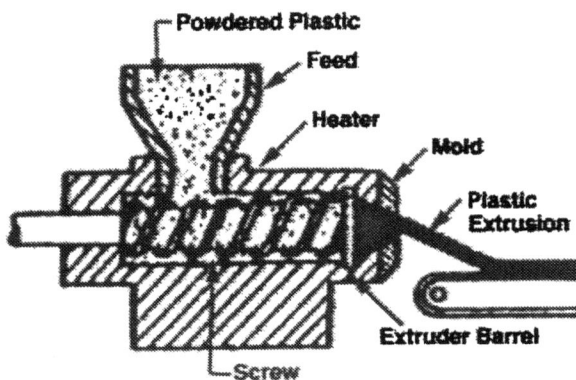

Fig. 2. Extrusion Process[1].

The main aims of extruder control improvement are to minimise start-up scrap and maximise production line speed while maintaining quality. Important issues to observe in extruder control are the effect of interaction between extruder zones, controlling the speed of the screw and the heat generated by the screw. Since, the distribution of the extruder zones is very complicated, OO technology may be applied where system hardware components map naturally onto the concept of objects.

The extruder control will be modelled using the different views (Simulink, Stateflow and IEC1131-3) in a complementary form. Additional models of the application will be obtained with the IDN, such as the software task model, which enables temporal

analysis of the controller and the architectural model, which contains the target architecture of the application. A combined simulation of the different views and models will be obtained via the Co-Simulation model. Once the extruder is modelled and simulated, it is planned to produce Java executable code for the specific distributed architecture.

5. CONCLUSIONS AND FUTURE WORK

An object-oriented infrastructure has been presented in this paper. This is targeted at improvement of the development and performance of decentralised distributed control systems for the process and manufacturing industries. An application of extruder control has been described where an object-oriented approach is needed to deal with the interaction between extruder zones. In order to design the infrastructure, the UML methodology has been considered and a review of the different CASE tools supporting this methodology has been presented. Also, a design of the IDN has been discussed. Further work will include the integration of the different tools and the implementation of the IDN's models, as well as a Java code generator.

ACKNOWLEDGEMENTS

The authors gratefully acknowledge the support of UK EPSRC (Grant GR/M55299) and collaborators in industry, University of Wales, Bangor and UMIST.

REFERENCES

Bass, J.M. (1998a). Proposals Toward an Integrated Design Environment for Complex Embedded Systems, *Euromicro 6th Workshop on Parallel and Distributed Processing*, Madrid, Spain, January 1998, pp. 273-8.

Bass, J.M. (1998b). An Open Environment for the Specification, Design and Code Generation of Control Algorithms, *IEE Colloquium on Open Control in Process and Manufacturing Industries*, London, May 1998, pp. 7/1 - 7/4.

Benkwith, A. and A. Moore (1999). UML for real-time? Yes, but..., *Embedded System Engineering*, April/May 1999.

Booch, G. (1991). *Object-Oriented Design with Applications*. Redwood City: Benjamin/Cummings.

Browne, A.R., J.M. Bass and P.J. Fleming (1997). A building-block approach to the temporal modelling of control software, *Proc 4th IFAC Workshop on Algorithms and*

---

[1]

http://www.core77.com/resource/plastique/extrusion.html

Architectures for Real-Time Control AARTC 97, Portugal, *pp* 433-438.

Coad, P. and E. Yourdon (1991). *Object-Oriented Design.* Prentice Hall, New Jersey, USA. Chapter 9.

Douglass, B.P. (1998). *Real-Time UML Developing Efficient Objects for Embedded Systems.* Addison-Wesley. Reading, Massachusetts, USA.

Hajji, M.S., A.R. Browne, J.M. Bass, P. Schroder, P.R. Croll and P.J. Fleming (1996). A prototype development framework for hybrid control system design, *Proc 13th World Congress of IFAC*, Vol. O, pp 459-464.

Jacobson, I., M. Christerson, P. Jonsson and G. Overgaard (1993). *Object-Oriented Software Engineering, A Use Case Driven Approach.* Revised printing. ACM Press and Addison-Wesley, USA. Chapter 16.

I-Logix, (1999). *Applying UML To Real-Time Embedded Systems.* Based on Developing Efficient Objects For Embedded Systems. Bruce Powel Douglass.

Larsen, M.H., G. Langer and P. Kirkby (2000). A Modelling Approach for improved Implementation for Information Technology in Manufacturing Systems. *International Journal of Intelligent Automation and Soft Computing.* 7(1).

Rumbaugh, J., M. Blaha, W. Premerlani, F. Eddy and W. Lorensen (1991). *Object-Oriented Modeling and Design.* Englewood Cliffs, NJ: Prentice-Hall.

Turnbull, G. (1998). *Open Environment for Distributed Control Systems.* Open Architecture Control Group. Journal article, October 1998.

DLL's Simplify Process Tuning

Andy Doonan, Chris Cox[1],
School of Computing and Engineering Technology, University of Sunderland, Sunderland, UK.

Roger Morris
Automation Partnership Limited, Peterlee, Co. Durham, SR8 2RJ, UK.

Abstract: This paper describes the implementation of a robust tuning algorithm based on the Astrom-Hagglund autotuning algorithm. The development of the software from DOS through Windows95 to DLL is chronicled with screen shots of the software in action. *Copyright © 2000 IFAC*

Keywords: PID Control, Autotuners, Computer-aided design, Implementation.

1   INTRODUCTION

The Control Systems Centre at the University of Sunderland has a long and distinguished history of collaborating with industry on seeking solutions to 'real problems'. Central to much of this research has been the search for a software-based CAD system to aid the choice of the parameters of the ubiquitous PID controller.

As early as 1986, the Centre were commissioned by the multi-national company Bristol Babcock to develop a self-tuning (STP1) module for use with their new 3300 modular series of controllers. In the event three algorithms were developed. These were based on (1) closed-loop cycling, (2) pattern recognition (similar to the EXACT system) and (3) recursive model-based identification. The STP1 module was implemented using ACCOL (Advanced Control and Communication Operating Language) for use with the company's standard PID3TERM module. The STP1 module was an 'invisible attribute' in that it only became visible when it was enabled. It recommended settings but only the individual operator had the power to download them into the target PID module. The software was evaluated by the Oak Ridge National Laboratory (Tennessee) and a report was later published by Tapp (1992).

Around about this time, improvements in microelectronics together with greatly increased computing power were allowing a wide range of controller manufacturers to offer (with their new systems) additional features such as pre-tune, gain scheduling and auto tune. However, the number of controllers these represented was minute compared to the vast number already in operation that did not possess these advanced features. A project was initiated to help re-dress the imbalance by developing some software, initially under DOS, that could provide some of these modern attributes to an otherwise standard controller. The only requirement was that the controller must have a communications port (e.g. RS232, RS422). The result was MasterTune CAD software environment.

The current version of MasterTune operates under Windows 95. While the basic kernel of the software remains the same, advantage has been taken of the multi-tasking nature of the operating system. To extend and simplify the future applicability of MasterTune, advantage is being taken of the Windows DLL concept. This should allow existing SCADA packages (with little code change) and future SCADA developers to easily incorporate the algorithm.

This paper is structured as follows: Section 2 briefly describes the tuning algorithm implemented. Section 3 and 4 describe how this algorithm has been incorporated into user friendly software both in DOS

---
[1] Corresponding author : voice +(44) 191 515 2824, fax +(44) 191 515 2703, email chris.cox@sunderland.ac.uk

and Windows95 respectively. Section 5 describes the latest development in the application of this algorithm by exploiting the Windows DLL mechanism. Finally, Section 6 provides some concluding remarks.

2 RELAY AUTO TUNING

In 1984, Astrom and Hagglund (1984) described an automatic tuning (auto tuning) method which utilised a relay controller plus an integrator to force the system to oscillate with a constant amplitude, fixed frequency oscillation known as a limit cycle. An important feature of the technique is that changing the magnitude of the relay characteristic can alter the amplitude of the oscillation. Also, the addition of the integrator has the benefit of forcing the limit cycle to be sustained about the set point value and helps to ensure 'bumpless' transfer between tuning and control modes. Much has been published on the relay auto tuning technique so it is sufficient here to stress that the design condition requires that K_c and T_i be chosen such that

$$T_i = \frac{P_U \, tan \, \phi_m}{2\pi} \tag{1}$$

and
$$K_c = \frac{2V_m P_U \, sin \, \phi_m}{\pi^2 A} \tag{2}$$

where

P_u is the period of the limit cycle oscillation
A is the peak value of the limit cycle appearing at the input to the non-linearity, and,
ϕ_m is the phase margin.

A block diagram of the system during tuning is presented as Fig. 1.

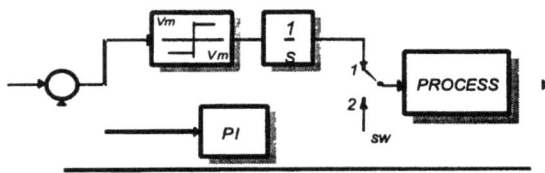

Fig. 1 A block diagram of the system during the tuning phase.

A more in depth discussion of the technique and its advantages can be obtained from the references (Doonan, 1997).

3 CAD SOFTWARE - DOS BASED

CAD software called MasterTune has been developed (Doonan, 1997) to help in the automatic tuning of PI

controllers using the technique described above. As MasterTune is an on-line tuning procedure (responsible in part for its successful results) real-time connection to the process to be tuned must be attained. This interface to the process is achieved via the RS232 serial port that virtually all modern industrial controllers have and is shown as Fig. 2. The rest of this section briefly describes the operation of the software

3.1 Set test parameters

The values of various test parameters used during the tuning phase are automatically set to default values. These values are set by the software to provide 'sensible' tuning parameters however, if the user so chooses these values can be modified to suit individual requirements. The parameters that can be modified include

- The percentage overshoot: the amount of overshoot the closed loop system will exhibit when compensated by the tuned PI controller. The percentage overshoot is used to determine the phase margin required by the design equations (see Eqn 1 and 2).

- The relay characteristics: the amplitude (to control the size of the limit cycle oscillations during tuning) and the hysteresis (used to prevent false switching caused by noisy signals).

- Constraints: control over the allowable level variations of both the process and manipulated variables.

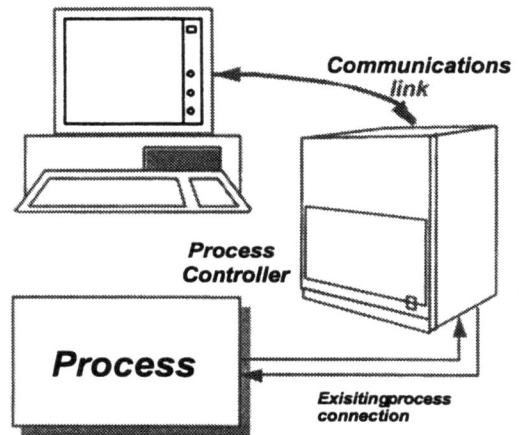

Fig. 2 Hardware configuration.

212

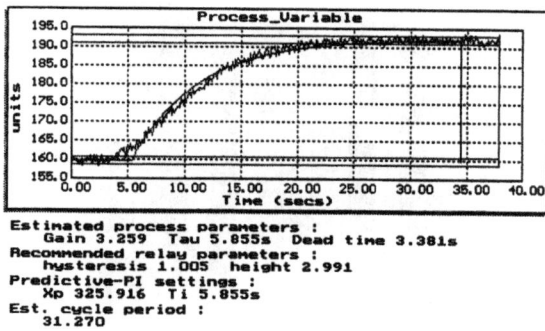

Fig. 3 Typical process analysis result from a laboratory scale flow rig.

3.2 Process analysis phase

The process analysis phase is the preliminary step of the tuning procedure. Here an open loop step test is conducted and a first order plus dead time model automatically calculated using a characteristic area approach (Nishikawa et al, 1984). The process inputs and outputs are visually displayed to the process operator throughout and on completion of the test the model is overlaid onto the process reaction curve allowing an informed judgement to be made about the quality of the model (Fig. 3). If, as recommended, the fast tune procedure is by passed in favour of the relay feedback autotuning, then the relay characteristics must be set. The process analysis phase calculates a value for the relay height that will produce a limit cycle with amplitude approximately equal to that set in the 'Set test parameters' phase above. Additionally, the software will analyse the level of noise imposed on the process variable and use this value to suggest a level of hysteresis.

This stage relieves the process operator of the rather complex task of specifying the relay characteristic. The operator only needs to specify the tolerable process variable swing. Further, it should be noted that this stage might only be required when tuning a process for the first time. In subsequent sessions, the relay parameters can be entered directly based on previously recorded values. If the time delay is not appreciable in relation to the time constant the software recommends that a PI controller be used. This fast tune facility uses the transfer function obtained in the process analysis phase to calculate the controller settings using some empirical formula, typically those recommended by Cox et al (1997). This feature can be used to get a control loop working satisfactorily in a very short period of time.

3.3 AutoTune

When the autotune in invoked, the relay and integrator force the process variable to oscillate with the specified amplitude. After a stable limit cycle is exhibited by the process, a PI controller is calculated which will result with the required level of overshoot being observed during compensated evaluation tests (see Fig. 4).

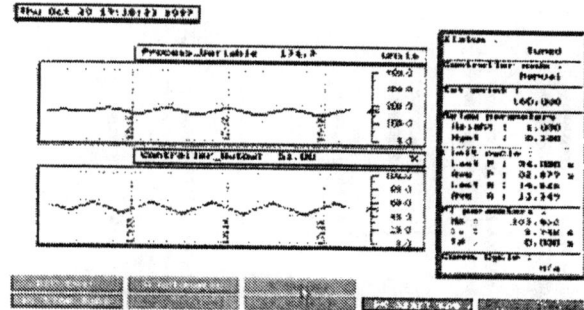

Fig. 4 CAD software screen dump showing Autotuning in process.

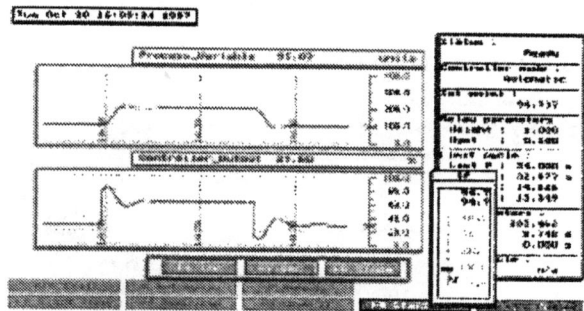

Fig. 5 CAD software screen dump showing evaluation of PI controller.

3.4 Evaluation

Once the appropriate controller has been selected and tuned, the evaluation stage can be implemented. Here the controller is put into automatic mode and, if desired, a step change can be induced in the set-point (see Fig. 5). This, as well as each other stage in the tuning cycle, can be logged and saved in a data file to provide a full record of the work completed.

4 MASTERTUNE – WINDOWS BASED

The current implementation of MasterTune required development of the software to run under the Windows95 operating system. While the basic kernel of the software remains the same (and with stages as described above) advantage is taken of the multi-tasking nature of the operating system allowing, for example, multiple channels to be recorded while tuning one or more loops, or, on the fly scaling of the 'trend' plots. Fig. 6 shows the main screen. Visible in the background of this Fig. is the array of currently logged channels while in the foreground is a trend plot of a selected channel being tuned. Note that many of these trend windows may be opened simultaneously and will each display over 20 minutes of historical data of a particular channel.

Fig. 6 The main MasterTune screen.

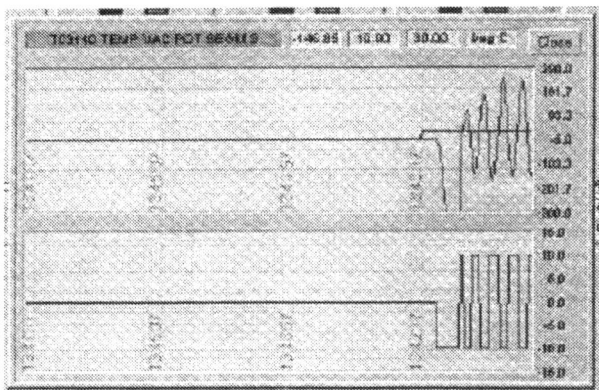

Fig. 7 A trend window showing tuning taking place.

Fig. 7 shows a close up of the trend window. This window shows the history of a particular loop over the last 20 minutes. Note that the trend scale limits are adjusted, then the entire history is also re-scaled. Fig. 8, below, presents a close up of the 'face-plate' display of two control loops. Each loop is displayed on the main screen using such a face-plate at all times.

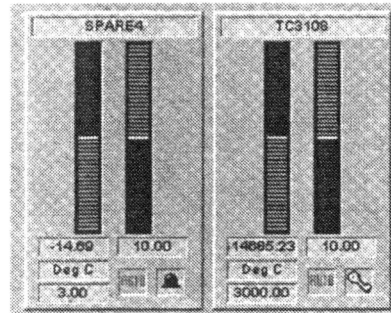

Fig. 8 The 'face-plate' of two control loops.

Critical information regarding alarm status, alarm priority, etc. is clearly visible on the display. Should the user wish to look at the recent activity of the loop, a trend window such as that displayed above (Fig. 7) can be easily opened.

5 THE DYNAMIC LINK LIBRARY

The Dynamic Link Library has always been an important part of the Windows operating system. Although its primary purpose in the days of systems with 1Mbyte of memory or less was to save memory by storing only one copy of commonly accessed code, programmers were quick to exploit the other favorable features of the concept.

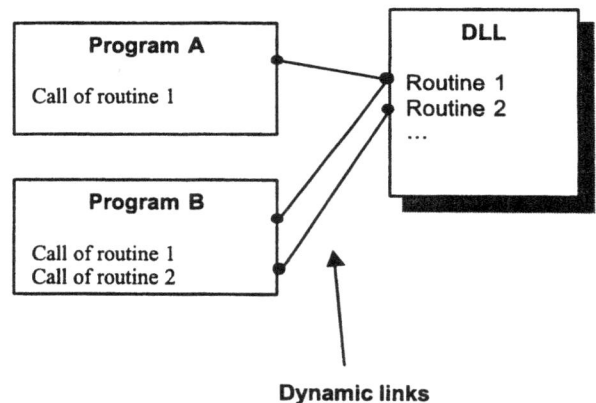

Fig. 9 Block diagram of the DLL mechanism.

In simple terms the DLL can be thought of as being similar to a standard library of routines, these routines could be for drawing, disk accessing, keyboard input, etc. or in the case of this paper, PID tuning algorithms. However, unlike this 'static' library which would be added to a users program at link time (and to every program which need to access the same routines), the dynamic library remains a separate entity and only a series of 'links' into the library are added to the program. It is only when a program using a DLL is executed that the DLL is loaded into memory where it remains until it is no longer required. If a second program requiring the same routines is executed the existing DLL can be used and another copy of the DLL is not required. Hence it is easy to see that for commonly used routines large memory savings (both in disk space and memory) can be made.

The second advantage of the DLL is in the use of links into the library. As long as the order of the routines (and any parameters required by them) in the library do not change, it is possible to change the code in these routines and simply replace the existing DLL with an updated version. The program using the DLL need not be aware of this change and will link into the DLL at runtime as before without any modification. This makes it possible to update sections of code without needing to recompile the entire program and allows independent development of sections of the overall program.

These features of the DLL make it possible for a developer to easily distribute code to other users safe in the knowledge that any future changes can be easily made. Conversely, should a programmer wish to use any routines in a DLL provided the selected programming language (usually the case) supports the mechanism, and the developer has knowledge of the routines and their parameters contained within, then these routines can be utilised.

5.1 The MasterTune DLL

In order to allow easier integration into existing SCADA packages and to allow SCADA developers the opportunity to use the MasterTune algorithm in new programs, a DLL has been developed which contains the functionality of the MasterTune philosophy. Briefly, the procedure required to use the DLL is as follows. Any program requiring to use the DLL must first obtain a current sample of the process variable, this is passed to the tuning algorithms in the DLL which returns the required value of the controller output. The program should send this value to the control element and repeat the above until the tuning algorithm sets a flag to indicate that tuning is complete. The pseudo code for this procedure is given as Fig. 10.

```
Initialise tuning algorithm

begin
        Obtain current sample of process value
        Pass this to DLL tuning routine
        Set controller output to returned value
        Repeat until tuning complete flag set
```

Fig. 10 Tuning procedure algorithm.

5.2 Automation Partnership Limited (APL)

APL is a local SME with around 25 employees specialising in system integration. The development of the DLL has been carried out in conjunction with the company who are looking to integrate it with the Intellution software suite and other SCADA packages.

6 CONCLUSIONS

The DLL development is now complete and an evaluation programme is presently being compiled. The initial testing will use a standard commercial controller, like the Alan Bradley PLC-5 range of controllers to execute the DLL and use a number of laboratory processes in the University's Process Control Laboratory as target systems. Once this phase is complete, further tests will be carried out at sites suggested by APL. Results from these pilot studies will be presented at the Workshop.

REFERENCES

Astrom K.J. and Hagglund T. (1984) Automatic tuning of simple regulators with specifications on phase and amplitude margins. *Automatica*, **20**, pp. 645-651.

Cox C.S., Daniel P.R. and Lowdon A. (1997) QUICKTUNE: A reliable automatic strategy for determining PI and pPI controller parameters using a FOPDT model. *Control Eng. Practice*, **5**, 10, pp 1463-1472.

Doonan A.F. (1997) The Development, Evaluation and Implementation of True Digital Control. *PhD Thesis : University of Sunderland*.

Nishikawa Y., Sannomiya N., Ohat T. and Tanaka H. (1984) A method of autotuning of PID parameters. *Automatica* **20**

Tapp P.A. (1992) A Comparison of three self-tuning control algorithms developed for the Bristol-Babcock controller. *ORNL report*

ON IDENTIFYING AND EVALUATING OBJECT ARCHITECTURES FOR REAL-TIME APPLICATIONS

L. B. Becker [1], C. E. Pereira [2], O. P. Dias [3], I. M. Teixeira [4], J. P. Teixeira [4]

[1] *Institute of Computer Science, UFRGS, Porto Alegre, Brazil*
[2] *Electrical Engineering Department, UFRGS, Porto Alegre, Brazil*
[3] *Escola Superior de Tecnologia, Instituto Politécnico de Setúbal, INESC, Portugal*
[4] *Instituto Superior Técnico, Universidade Técnica de Lisboa, INESC, Portugal*

Abstract: This paper presents a methodology and tool support to the development of distributed real-time object-oriented systems, specially focusing on industrial automation applications. At system level, two different kinds of classes/objects are recognized: (i) application domain objects, which map directly to concepts and components of the problem domain, and (ii) design objects that are related to the functional requirements that the system have to meet. The approach provides a method for automatic identification of possible design objects architectures. A set of quality and testability metrics are proposed to evaluate the generated architectures, allowing the identification of the 'best-fitted' one. *Copyright © 2000 IFAC*

Keywords: object modelling techniques, systems design, software metrics, real-time.

1. INTRODUCTION

The heterogeneity and complexity of today's hardware/software (hw/sw) electronic systems, together with stringent product quality requirements, make system design a difficult process (Lavagno, 96). This becomes even more relevant in real-time systems implemented as complex hw/sw modules, extensively reusing pre-defined Intellectual Property (IP) cores (Gupta, 1997; Dias, 1999a]. All those components come from different suppliers, eventually using different specification languages. When real-time applications are envisioned, time becomes an extra difficulty to be faced. Finally, system testability and reliability needs to be considered as soon as possible in the design flow.

According to traditional object-oriented development methodologies (Rumbaugh, 1991; Booch, 1991; Douglas, 1998), the system architecture (composed by a set of classes/objects connected by means of their logical relationships) is defined according to the development team experience. In this case, an object (class) represents a clear intuitive concept. By doing so, system architecture may be designed in order to comply with structural, behavioral and functional requirements, but lacking of performance considerations. This last aspect could be tackled at system level to constraint architecture topology.

Architecture definition is an important issue in the design flow, mainly because user's requirements usually do not imply a unique solution. As already mentioned, there is a clear trend in structuring modern automation plants as a set of distributed and autonomous computer-based devices (that are usually called 'smart' or 'intelligent'). Unfortunately, it's not a trivial task to distribute the desired functionality among these devices as well as to determine which degree of autonomy (or 'intelligence') will be assigned to each of them. Considering the example of a simple temperature control process, it could be determined at least three different architectures:

1. Sensor-control-actuator (with the control algorithm located in the control object);
2. Smart sensor-actuator (with the control in the sensor);
3. Smart actuator-sensor (with the control in the actuator)

In fact, many system architectures may be chosen, reflecting different design and performance/cost tradeoffs. A tradeoff means that characteristics are compared and some are valued against others. Unfortunately, not all the designers enrolled on a design process value the same characteristics, so that consensus agreement is usually not easily achieved. However, consistent/validated criteria and quality metrics (QMs) are indeed required to compare, select and eventually reconfigure the initial system arch., in order to lead to a high-quality, cost-effective product (Dias, 98).

In this work it's presented a new approach called MOSYS, a methodology to help designers in order to find the 'best-fit' architecture for the problem under analysis. A set of metrics and criteria, usually accepted by hardware and software design teams as characterizing a good architecture, is presented. These metrics and criteria allow the selection of a target architecture based on a small set of architectural properties. The objects/classes that constitute the different architectures are automatically generated, based on a graph representation of the different tasks that the system is expected to carry out.

The proposed methodology is supported by a CASE tool that allows the systems structure, behavior and functional description. It also provide the capacity to make automatic object partition from the functional specification, and to assist designers in order to archive the most effective architecture from reliability and testability points of view.

The paper is organized as follows: section 2 describes the overall methodology; its tool support and a related case study are respectively described in sections 3 and 4. Finally, in section 5 some conclusions are outlined.

2. DESIGN METHODOLOGY

This section presents a new methodology, MOSYS – *Methodology for Object identification from SYstem Specification,* intended to help the design of heterogeneous object-oriented distributed real-time systems (DRTS). The present methodology is innovative because it evaluates the quality of the architecture previously from design phase. It also incorporates the feature of automatically generating a set of possible architectures. The development cycle of the proposed methodology is depicted in fig. 1.

According to the methodology proposed, as a first step for the system requirements understanding, actors and their interaction with the system are identified using UML *use-cases*, as showed in fig. 2. Each use-case identified encapsulates one major functionality of the system. Moreover, use-cases can be detailed, either by using activity diagrams or extended data flow diagrams (E-DFD). E-DFD conveys special information on process complexity and timing constraints.

After that, in the next methodology step actors identified in previous phase are mapped to a class diagram. The class diagram constitutes the basis of an object-oriented approach, and can also contain additional

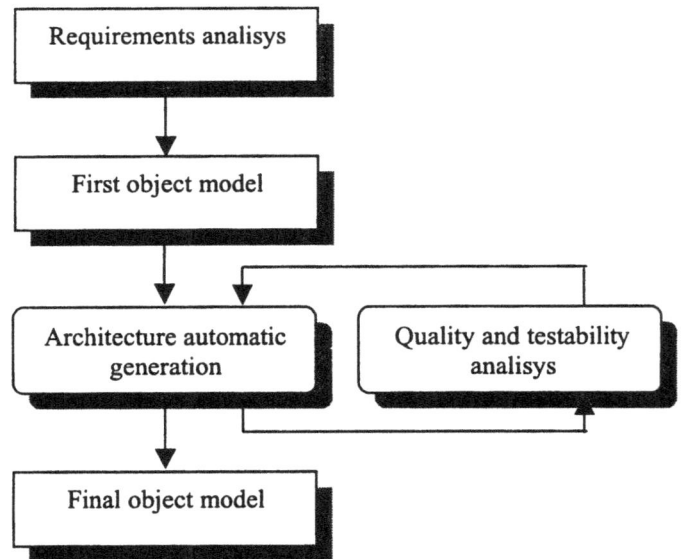

Fig. 1. MOSYS development cycle.

classes, representing other domain elements of interest, according to the approach proposed by Pereira *et al.* in (Pereira,1994a) (see fig. 3). At this time, it's possible to identify the dynamic behavior of each element, through the specification of a state-machine for each object. Timing constraints like cycles, deadlines and timeouts can also be defined during these phase, being associated to the classes operations.

In the third step, the tool support provided by the methodology assists the designer to re-arrange the original architecture, or class diagram. This is done through the mapping of the functional model (union of the E-DFDs) to a graph description, on which nodes are tasks and edges are task interconnections. Graph partition, according to different strategies and algorithms, will lead to different possible task clustering. Each cluster of tasks and their attributes is defined as an object. At this point, design and test oriented metrics and criteria are used to ascertain the quality of the architectures obtained and also to conduct the selection of the most adequate architecture.

With the identification of the target architecture, the development cycle proposed by MOSYS is completed. By the way, the software development cycle isn't finished yet. In order to tackle this problem, there are other related works being conducted, as exposed in section 4. Next subsection details the metrics and criteria's adopted with MOSYS approach.

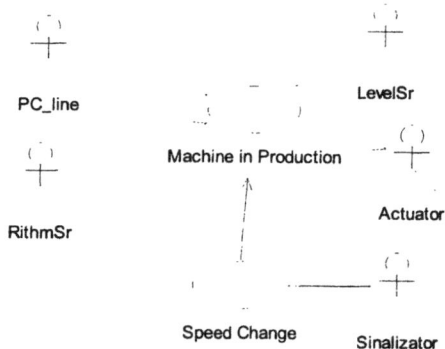

Fig. 2. System context identification.

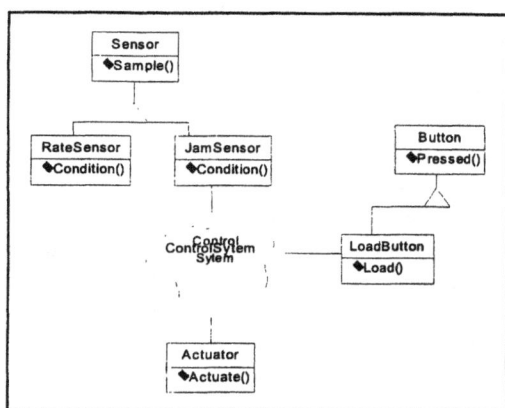

Fig. 3. Mapping of external objects in the object model.

2.1. Quality Metrics and Decision Criteria

Considering the integrated development of *hw/sw*, there is a set of constraints that must be taken into account for the definition of the adequate quality metrics and decision criteria (Dias, 1999b; Dias, 1998). According to these, there are two main points of interest to be take into consideration: system performance and testability.

By definition, each metric adopted evaluates or the system performance or its testability. The former evaluates/guides the choice of the 'best fit' architecture in relation to performance point of view. The last quantifies the testability, measuring the cost and overhead of additional process into the system, that are necessary to improve testability.

The resulting metrics for performance and testability measurement are grouped according to the following division:

- **Performance metrics**: object coesion, system coupling, object autonomy, number of loops;
- **Performance and testability metrics**: systems neighborhood controllability/observability and systems controllability/observability
- **Performance metrics**: increment of resources tasks, resources interconnections, resources primary entrances, resources primary exits and degradation of critical path.

Once the target architecture is defined through the application of performance metrics, it's evaluated the impact of the introduction of testing mechanisms. For effect, the other metrics stated previously are used. As a result it's presented the technique that most complies to testability with less resource increase and performance degradation.

2.2. Related Work

In a comparison conducted by the authors, it was realized that there isn't any other work under development that makes automatic generation of architectures and that includes testability support under design phase. The work that is closer to the one reported in this paper is showed in (Hayek, 1996; Hayek, 1997).

Next section presents the tool support used with the methodology proposed. After that, a case study is presented to highlight the major features of the current approach.

3. TOOL SUPPORT

The tool support for the MOSYS methodology, presented in the last section, consists in the integration of SIMOO-RT environment with SysObj partition tool. The resulting environment is called SIM2SYS. The main components of each individual tool and the environment resulted from their integration are presented in fig. 4.

SIMOO-RT (Becker, 1999a; Becker, 1999b) is a framework for the development of discrete object-oriented simulation models. It offers to the users a *Model Editing Tool* (MET), that allow the development of UML-based object-oriented models. One difference to UML notation concerns to the hierarchical modeling

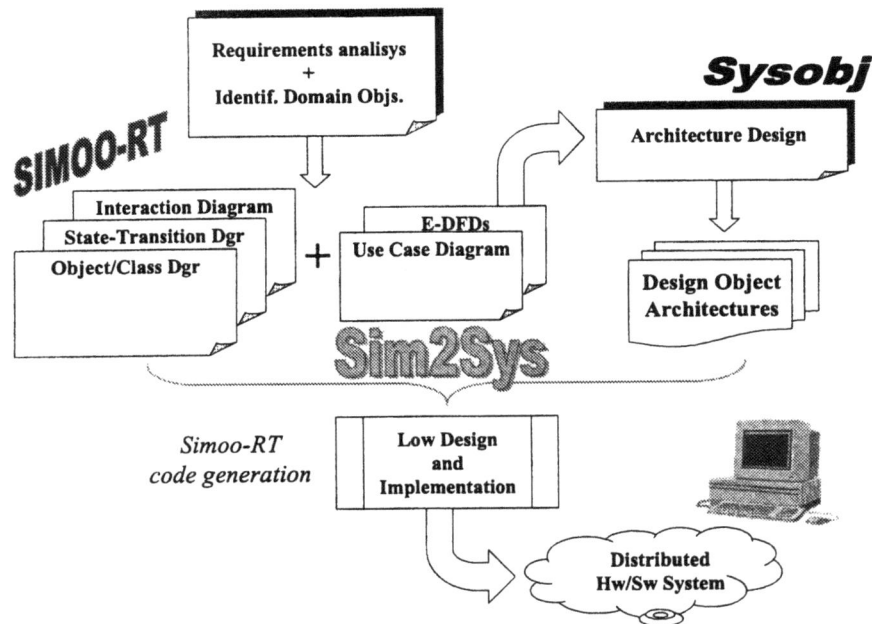

Fig. 4. Sim2Sys integrated environment.

approach, dividing the model in several levels of detail. Besides that, the user must specify an instance diagram, which will guide in the generation of the executable simulation/application model. This tool also includes SIMOO-RT also offers support to make automatic code generation for a real-time operating system, where the adopted target language is AO/C++ (Pereira, 1994b).

The other tool used is SysObj that implements the aspects of the methodology related with the automatic object generation. It accepts as input a description of the DFDs containing the systems requirements (as provided by SIMOO-RT for example). It can also accept a modified DFD including testability features. As output, it provides a set of architectural solutions, as well as the values of the different quality metrics that will help the design team in the selection of the final architecture.

As a result from the integration of the tools presented previously it was conceived SIM2SYS. It keeps all SIMOO-RT functionality, adding extra support to the specification of UML use-cases and E-DFDs. Each use-case modeled by the designer can be detailed in an E-DFD. After definition of the whole system functionality, the designer can ask for automatic object partition. At this time, SysObj, which is running in background, is invoked and receives as input parameter the use-case/E-DFD information. After processing the data, the tool returns the object partition information that is converted to a SIMOO-RT model. The resulting object structure is

specific features that allow definition of temporal constraints like deadlines, timeouts and periodic operations. To describe the model dynamic behavior, the environment encourages use of state machines. encapsulated in a higher-level class that makes part of the original class diagram defined by the designer.

4. CASE STUDY: FILL SYSTEM

In this section, the MOSYS approach for deriving system architecture is applied to a real world case study. The system, in which the case study is embedded, is a wine bottling production line. This case study focus in the design of a real-time control system, referred as Fill, responsible for the automation of the production line (Ribeiro, 1998). Fig. 5 depicts the pre-defined blocks for the electronic system coupled with each line machine. The dark boxes denote already existing devices before line automation. From here, it will only be addressed the aspects that are relevant for methodology assessment of the machine electronic system.

The actual architectural solution for Fill has been derived using the methodology steps presented in section two. First, the application domain objects were defined and organized in a class diagram. Fig. 6a depicts the class diagram obtained for the present case study.

Latter, the system functionalities were characterized through the definition of some use cases. In the present case study, 9 different use cases were defined. The detailing of the use cases in E-DFDs, resulted in a total of 25 different high-level process, that interact among them and with the system actors. In the next step, design objects were generated using the automatic partitioning tool. Here, both design and testability oriented metrics were considered for evaluating the impact of different testability strategies on the system cost overhead. The metrics evaluated for assessing the quality of the architecture are the following: object autonomy, direct coupling, system coupling, number of loops and the number of independent graphs inside each module. According to

the partitioning process, architecture constituted by 4 objects is the one that best fits within the desired characteristics.

Taking in consideration the SIMOO-RT modeling rules, the design object architecture is encapsulated within a higher-level class. In this case, the class chosen for aggregating the design architecture is the *ControlUnit*, as can be observed in fig. 6a. The design objects automatically generated by system are showed in fig. 6b.

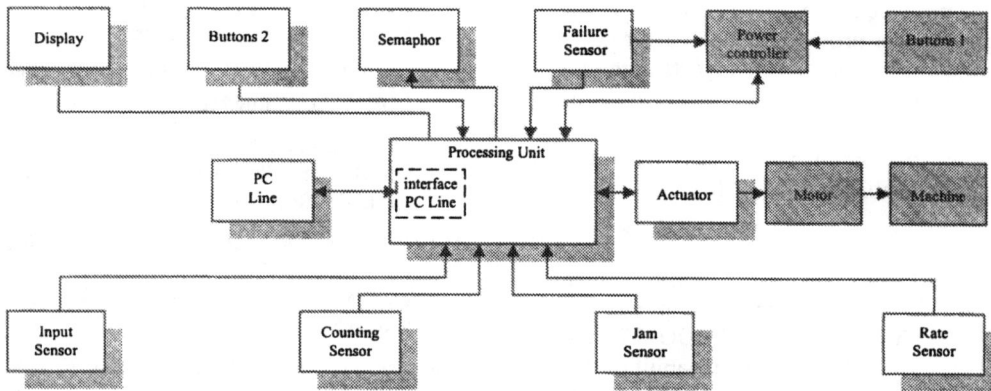

Fig. 5. Block diagram of Fill system.

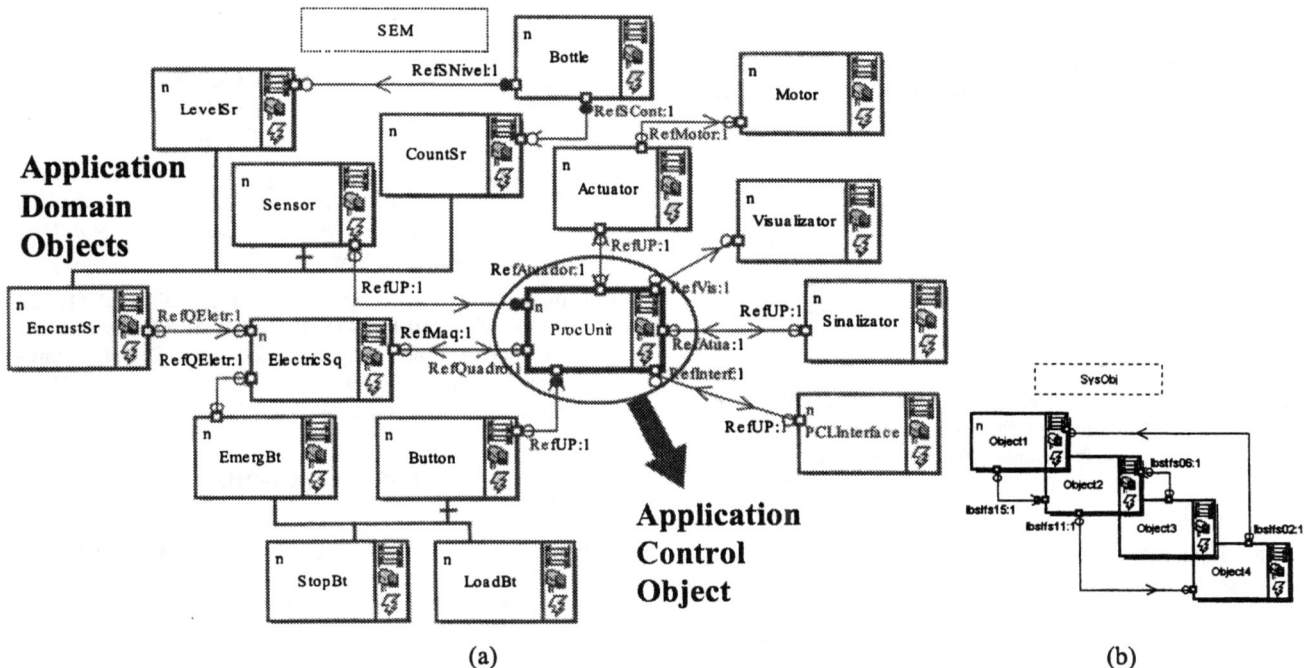

(a)

(b)

Fig. 6. (a) Fill class diagram; (b) design objects automatically generated.

The final model distributes system basic tasks among the objects, that is, distribute minor intelligence through the *active objects* (e.g. sensor periodic data acquisition and conditioning). With this strategy, we have expected to achieve greater flexibility, to improve reusability and to ease system reconfiguration.

5. CONCLUSIONS

The paper presented MOSYS, a methodology for developing distributed real-time applications based on distributed autonomous objects. Besides of providing support for generating an executable object-oriented specification, the approach also provides a method for automatic generation of alternative design objects architectures. A set of quality and testability metrics are proposed to evaluate the generated architectures, allowing the identification of the architecture that constitutes, from the design and test points of view, the 'best-fitted' one.

MOSYS is supported by the SIM2SYS tool, that is an integrated version of two software tools previously developed by the authors, namely SIMOO-RT and SysObj. It combines the SIMOO-RT capabilities for modeling, simulation and code generation of object-oriented applications, with the automatic object identification and testability enhancement features of the SysObj tool.

MOSYS has been successfully applied to some case studies, including 'real' industrial applications and the results obtained so far are very encouraging.

6. ACKNOWLEDGMENTS

This project has been financially supported by CAPES, CNPq and FAPERGS (Brazilian Research agencies).

7. REFERENCES

Becker, L, Gergeleit, M., Nett, E. and Pereira, C. (1999) An Integrated Environment for the Complete Development Cycle of an Object-Oriented Distributed Real-Time System. In: *Proc. of IEEE International Symposium on OO Real-Time Comp.*, Saint-Malo, FR.. pp. 165-171.

Becker, L. and Pereira, C. (1999) SIMOO-RT: An Integrated Object-Oriented Environment for the Development of Distributed Real-Time Systems. *In Proc. of 1". International Symposium on CAD/CAN Robotics and Factories of the Future*, Águas de Lindóia, São Paulo, Brasil. pp. mw13-18.

Booch, G. (1991) **OO Design with Applications**, The Benjamin/Cummings Pub. Co., Inc.

Dias, O.P., Teixeira, I.C., Teixeira, J.P. and Pereira, C.E. (1998), Performance and Test-Oriented Metrics and Criteria for Hw/Sw Design at System Level, Proc. of *Design and Diagnostics of Electronics and Systems Workshop (DIPES)*, pp. 219-225, Polland.

Dias, O.P., Teixeira, I.C., Teixeira, J.P. and Semião, J. (1999), Introducing Test Requirements in Real Time Systems at Architectural Level, In Prof. of *Design of Integrated Systems (DCIS)*, Palma de Mallorca.

Dias, O.P., Teixeira, I.C. and Teixeira, J.P. (1999), Metrics and Criteria for Quality Assessment of Testable Hw/Sw System Architectures, *Journal of Electronic Testing, Theory and Application (JETTA)*, vol. 11, nº. 1/2, pp. 149-158, Kluwer Academic Publishers.

Douglas, B.P. (1998) **Real-Time UML: Design Efficient Objects for Embedded Systems**. Addison Wesley.

Gupta, R.K., Zorian, Y. (1997), Introducing Core-Based System Design, *IEEE Design & Test of Computers*, vol. 12, pp. 15-25, Sept./Oct.

Al-Hayek, G. and Robach, C. (1996), One Adequacy of Deriving Hardware Test Dated from the Behavioral Specification, in Proc. of the *Euromicro Symposium*, pp. 337-342, September.

Al-Hayek, G., Traon, Y. and Robach, C. (1997), Impact of System Partitioning on Test Cost, *IEEE Design and Test of Computers*, pp. 64-74.

Lavagno, L. Sangiovanni-Vincentelli, A. And Hsieh, H. (1996), Embedded System Codesign, Nato ASI-Series, Kluwer Academic Publishers.

Pereira, C. and Darscht, P. (1994), Using Object-Orientation in Real-Time Applications: an Experience Report. Proc. Of *TOOLS Europe 94*. Versailles, France.

Pereira, C. (1994), Real Time Active Objects in C++/Real-Time UNIX. In: Proc. of *ACM SIGPLAN Workshop on Languages, Compiler, and Tool Support for Real-Time Systems*. Orlando, EUA.

Ribeiro, R., Dias, O.P., Teixeira, I.C. and Teixeira, J.P. (1998), Automation and Real-Time Control of a Wine Bottling Production Line: an Object Oriented Based Hardware/Software Co-Design Case Study, Proc. *8". IFAC Conf. On Large Scale Systems (LSS)*, vol. 2, pp. 1211-1216.

Rumbaugh, J. et al. (1991), **Object-Oriented Modeling and Design**, Prentice Hall.

Vranken, H.P.E. Wittman, M.F. and van Wuijtswinkel, R.C. (1996), Design for Testability in Hardware-Software Systems, *IEEE Design & Test*, vol. 13, Nº. 3, pp. 79-87.

SUPERVISED REAL-TIME CONTROL WITH PLCS AND SCADA IN CERAMIC PLANT

E. Jiménez, J.M. Miruri, F.J. Martínez de Pisón, M. Gil

Department of Electric Engineering. System Engineering and Automation Group
University of La Rioja. Spain

Abstract: This paper describes the design of an automatic control system in an industrial ceramics production plant of the latest generation. The automatic control is implemented on PLCs and is monitored through a powerful package of SCADA software. A PC supervisory real-time system is designed and installed to optimise the adaptive control devices to their intended purposes, or to take complete control of the automation if it detects system failures. In addition, a supervisory system captures the most significant control parameters, for later off-line analysis. Furthermore, a telecontrol system allows taking total control over the real-time automation via modem. *Copyright© 2000 IFAC.*

Keywords: Adaptive Control. Supervisory Control and Data Adquisition (SCADA). Industrial Control. Manufacturing Systems. Process Simulators. Real-Time System.

1. INTRODUCTION

The wide range of control methods available today sometimes makes it difficult to choose the most suitable model. The outpouring of relatively recent concepts such as Fuzzy, Genetic Algorithms, Neural Nets, etc., has opened up a wide range of new control possibilities, to add to classical controls still commonly used, like PID control, which has been adapted to technological innovation and has evolved with concepts such as autotuning or setting up in the programming of PLCs.

And the real-time control of all these automation systems can be developed with a wide range of software algorithms and hardware architectures (distributed control, parallel processing, multi-layer control, multi-processor systems, etc.), to obtain the specific requirements of the application.

For example, in a PID control system, its setting up offers many possibilities, since it can have specific PID controllers, a distributed PID control system -as part of an intelligence distributed system-, also processor outlying card for PLCs -with autonomous functioning from the CPU to the PLC-, functional blocks included in the programming languages for medium and high range PLCs, software applications on PCs which carry out the PID control as just another part of the programming, etc.

All these options have pros and cons, and the choice of the most appropriate depends on the type of control and the characteristics of the process to be automated and regulated.

2. PROCESS FEATURES

System automation has been applied to a modern plant for the production of ceramic elements for building: brick, ladri-plaster, blocks of thermal clay, etc. This is a totally automated plant, so it is evident that it require various types and models of controls throughout the plant (oven, drier, removers, loaders, robots, etc.). These are controls that, in general, do not have large dynamic requirements, as it will be seen in a moment, but are very reliable as regards their security.

Broadly speaking, the plant's production system consists of the breaking of the clay, the shaping of ceramic elements in a shaper (by inlayed), the drying of these elements in the drier and the later baking in an oven. All this, together with the processes of heaping up and removing materials, loading and discharging onto trucks, packing, etc.

The newly shaped clay elements are placed in the dryer to remove part of the water they contain. Throughout their journey in the drier, they have to pass through areas with certain parameters of temperature, humidity and air speed; these are variable along the journey and follow a predefined curve. The heat and humidity required are obtained by forced air heating, provided by the residual heat produced in the escaping gases and in the refrigeration water of the co-generation system's gas, used in this case to produce electric power.

The pre-dried products are baked in the oven to get the brick or ceramic elements. The heat needed is obtained by burning gas inside the refractory oven and circulating it around the bricks. As above, the process must follow a temperature curve along the journey; this is divided into three zones: pre-heating, baking and rapid cooling.

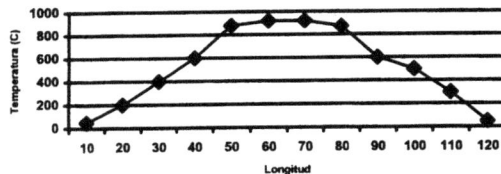

Fig . 1. Typical temperature curve for oven baking.

These two automations are the main ones in the plant (Ezio, 1993), and have the special characteristic of having some parameters (temperature, humidity and air speed) which vary along the journey but which must remain constant (at least in each cycle) at each point, following the indications of the pre-set functioning curves. These functioning curves are of great relevance, since a deviation at some of the points could bring about a breaking of the material. In addition, an excessive speed is not necessary in the control of this (it is not important to optimise the response time); however, it is important to restrict the over-oscillations at certain points and also very important to follow the curve faithfully (that is to say, that the error is low). An over-oscillation in the temperature of the breaking zone could lead to fusion of the material. In the cooling area, special care is needed to carry out a gentle transition to the temperature of 573 °C, since this is the transition between the α crystallization and the β crystallization; in this area, it is therefore essential to totally eliminate error. In the drying, incorrect following of the indications during the first 5% of the process, which is when the contraction is produced, would cause the crazing of the material or its breaking.

Given the above, it is clear that regarding the controls for this part of the plant, the main condition is to follow the curve attempting to eliminate the error. This means that special attention must be paid to the determination of K_i when calculating the parameters of the digital PID controllers (Koivo, 1991).

Natural gas is burnt in the oven through injectors located sideways; at all the injection points, and in all axes, there are sensors for all the required parameters. Therefore, speaking of a curve of temperature or of humidity is not speaking of only one PID control, but of one control for each injection point.

Although the curves are constant for each production cycle, they have to be depending on the load (the mass that passes per unit of time), the product (the type of ceramic element being manufactured at that moment) and the material (the quality and kind of clay used). The reference curves for each should therefore be recorded according to these characteristics, as well as their behaviour. The need for adaptive controls is evident, as is that of providing storage systems for the great quantity of parameters that compose the curves.

Apart from the dryer and the oven, are many points with different characteristics (minor rise time, over-oscillations, etc.) where other controls can be specified. The control system should therefore be capable of adopting diverse models and configurations (Aström, et al., 1992).

With all the requirements described for the regulation process, and considering that PLCs are required for the control of discrete events in this automation, as well as a SCADA system for the monitoring, the kind of control system was decided upon. It was felt that among various forms of real-time control systems available, the most appropriate (in this plant, of course) would be based on PLCs, and the SCADA was to be used too in order to enlarge and complete the functions the control.

3. REAL-TIME CONTROL WITH PLC

A real-time system could be defined as a system which must respond to externally requested stimulus, within a specific and finite time period. To ensure the real time operation of the controllers, the time of exploration must be much smaller than the dominant constant of time of the process, usually $T_{exp} < \tau/10$ (Balcells and Romeral, 1997). As exploration time is limited by the filtering of the analogue signals in the interfaces and the cycle time, usually PLCs do not have very good exploration times. In spite of this, to control in real-time oven and drier devices, which are the most frequent in the plant, they are adequate.

But some of the processes need lower response times. As the response time depends on the delay-in time, the delay-out time, and the cycle time, in these happenings appropriate input and output interfaces (usually with less filter) have been used to reduce the delay times, and fast routines (activated by periodic signals) to reduce their partial cycle time.

However, the overwhelming majority of real-time controls are in the oven and in the drier, and its precision is most important than its response time. To get this reliability it is possible to use a supervised or a redundant system. It is not difficult to develop, because PLCs with multi-processor do it automatically. But more processors are needed, and the price grows up. The best solution in this case is to control the system with PID functional blocks

224

included in the programming language of the PLC, and to use the SCADA, that has PID functions too, like a supervisory system.

As previously discussed, the PLC could carry out the PID control in two main ways: from instructions included in the code language (software) or from peripheral processor cards (hardware).

The functional blocks are in fact a macro-instruction that represents all the corresponding calculation algorithms. They do not normally follow the program execution cycle, but rather follow a separated process activated for interruption according to the exploration time; this should be significantly lower than the system processing time (Silva, 1985). They have the advantages of low price, simplicity, and they do not needing a card for each regulator. On the other hand, they increase the cycle time of the PLC, and their refreshment time on the controller outputs is high.

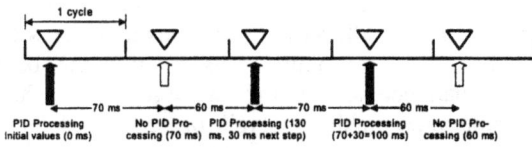

Fig. 2. PLC cycle of the PID control.

The regulation processors are much more rapid and the user can adjust their scanned interval. They can receive the operation parameters on-line from the programming unit, by transference from the program, or directly from the outside. Their operation is independent of the CPU of the PLC, except in order to receive parameters in the adaptive controls (Morriss, 1999; Malpica, 1998), and the CPU is therefore relieved of this calculation process.

The process and exploration times have been analysed for the plant under study, and the most appropriate solution in this plant is the software-control, which has been therefore adopted. This is logical because these regulations do not have many problems with dynamic properties. However, if there had been any critical -in speed- processes, it would certainly have been included some peripheral processor cards.

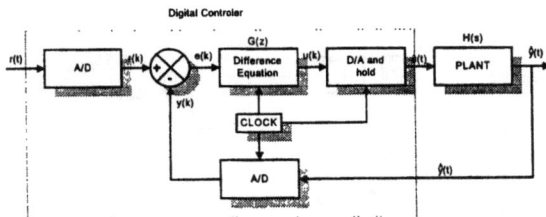

Fig. 3: Closed loop discrete control

The use of PLC has also allowed other functions to optimise operation: isolated, generation of ramps, calculation of averages, hysteresis, etc., which were very useful in the installation.

4. REAL-TIME CONTROL WITH SCADA

The SCADA (Supervisory Control And Data Acquisition) was included principally for monitoring, in real time, the plant. But the used SCADA has too blocks and instructions in its programming language.

These PID blocks maintain balance in a closed loop by changing the controlled variable (an analogue output) in response to deviations from a user-defined set point. The difference between the actual value (an analogue input) and the set point value is the error, or deviation.

In response to errors, the PID block calculates an appropriate control output signal, which attempts to reduce the error to zero. The adjustment that the PID block carries out is a function of the difference between the set point and the measurement, in addition to the values of the proportional band, the reset, and the rate. The steady state PID block algorithm is defined below:

$$\Delta y_n = K_P \cdot \beta \cdot (E_n - E_{n-1}) + \frac{T}{T_I} \cdot (F_n - y_{n-1}) + \tag{1}$$
$$+ \frac{T \cdot K_P \cdot E_n}{T_I} + \frac{K_P \cdot T_D \cdot \gamma}{6 \cdot T \cdot (T_D \cdot \alpha + 1)} \cdot (E_n + 3E_{n-1} - 3E_{n-2} - E_{n-3})$$

Δy_n	=	Current output.
Δy_{n-1}	=	Previous output.
K_P	=	100/PB.
E_n	=	Current error.
E_{n-1}	=	Error at previous scan time.
F_n	=	Feedback tag value.
T	=	Sample time.
T_I	=	Reset time constant.
T_D	=	Derivative time constant.
α	=	Derivative mode filter.
β	=	Proportional action constant.
γ	=	Derivative action constant.

The new PID block output will be:

$$y_{n+1} = y_n + \Delta y_n \tag{2}$$

The PID block can prevent reset-windup through the use of the Feedback Tag. Reset-windup typically occurs when there is a hardware failure in the measurement device. This causes the PID error, or deviation, to be larger than the parameters that the block would normally produce in a full output swing.

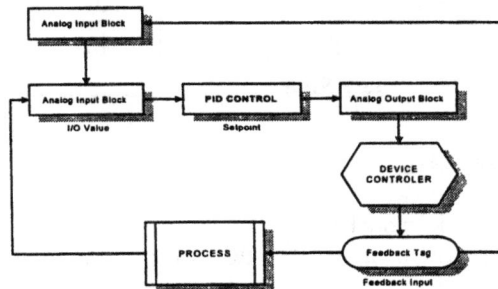

Fig. 4: Using the PID Block's Feedback Tag

Figure 4 illustrates how the Feedback Tag ensures that the controller does not get saturated under these conditions. By indicating that the output of the PID block is not maintaining the set point, the Feedback Tag can help to avoid serious control problems (Ogata, 1998).

If the processing scheme requires a cascade loop, the PID block can include master and slave PID blocks, as shown in Figure 5. The slave PID block can fetch the master PID block's adjusted output value for its set point.

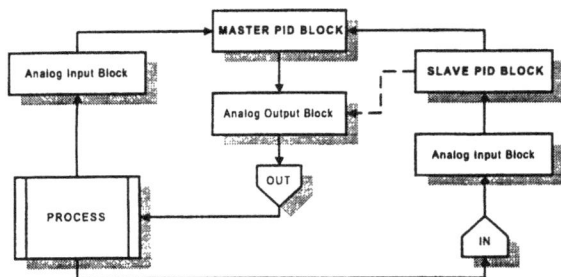

Fig. 5. Cascade control with master-slave PID blocks.

The cascade loop, shown in the figure, illustrates control in which the output of one PID block (master) provides the set point value for another PID block (slave). The Master PID block drives the set point of the slave PID block, which in its turn (Aström and Wittenmark, 1997), controls the input value based on the reading from a process input transmitter.

5. SUPERVISION AND TELECONTROL WITH SCADA

The possibilities of real time control both with PLC and with the PC using SCADA software have already been discussed. However, the automation has been built maintaining knowledge levels in classic automation, where the PLC is in contact with the plant, and the PCI is on a higher level, assuming tasks of management, control, administration, etc.

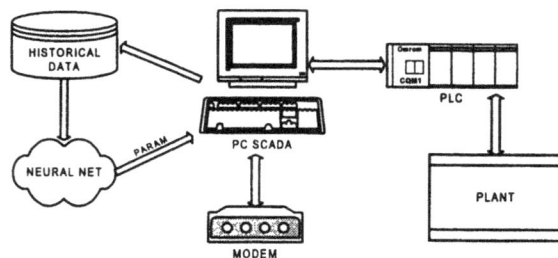

Fig. 6. Overall system.

To do this, the model installed is the one shown in Figure 6, where it is the PC which files the parameters of the many operation curves of the variables (because there is a great quantity of information). In function of the three parameters (load, model of fabrication and raw material) the

corresponding follow-up value is sent to the PLC for its use throughout the productive cycle.

The PC also simulates the theoretical outputs that appear in the real-time control. In case of irregularities, it will activate the corresponding alarm or emergency and act sending new control parameters to the PLC, or even assuming the control of the automation until normality is recovered. Furthermore, the SCADA has a very high capacity, and great facility, for handling emergencies (Burns and Wellings, 1996; Laplante, 1993; Levi, 1990). This means that different types of alarms and emergencies can be depending on the required parameters (response times, etc.) the outputs (accidents, etc.) or the errors in PLC operation (overflow in the watchdog, etc.).

Another characteristic developed in this work, also based on the SCADA, was off-line control analysis, capturing the main operation parameters in a historical data that allows their analysis and use with a training system of neural nets to give the best operation parameters. This off-line analysis also has great importance in the analysis of product defects (fissures, breaks, etc.), since the product of these elements is characterised by never be the same (new shapes of production are designed, the veins of material from the quarry change frequently, etc.). The system is not complicated but it is very useful: it indicates when each item was produced in order to be able to determine its causes.

Finally, an additional function of supervised control is the possibility of remoting execution of all the above actions, via modem, and with almost no difficulty; the SCADA is prepared for simple use of this type of connection.

If any processes had been very important for people security, it would also have been included a direct redundant connection between the PC and the plant; it would then have been totally secure in the case of a total breakdown of the PLC. This connection requires A/D and D/A conversion cards, as shown in the figure 7; it was not felt to be necessary in this case.

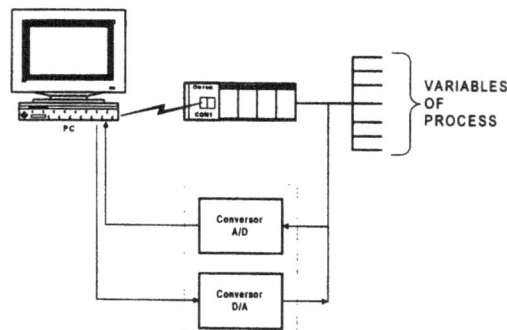

Fig. 7. Redundant system PLC-PC for security in case of total breakdown

6. RESULTS

The results reached with this control type have been fully satisfactory. The test has been made by stopping the operation of the PLCs, simulating a stop in its operation, and the SCADA has carried out its mission of detecting a malfunction, taking the control and sending all the necessary alarms (included phone calls, to responsible people, using the Dial Manager Program). The control operation driven by the SCADA has proved a worse dynamic behaviour, as could be expected since the time of exploration was longer, but as there was a great margin, it has continued within the established limits. This operation has been tried in the most important systems, that fortunately are the ones which have the simplest dynamic requirements: the oven, the drier and the deposits and mixers. The result is shown in the following figures:

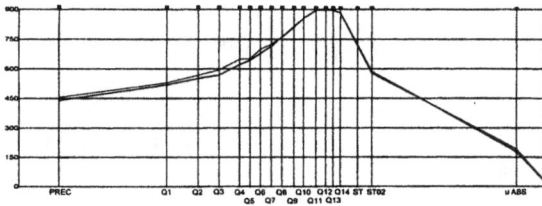

Fig. 8. Temperature and temperature reference along the oven.

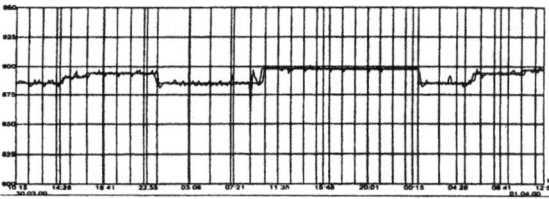

Fig. 9. Temperature and temperature reference depending on the time in a specific point in the oven (the 5th burner).

The figure 9 shows simultaneously the following of the reference value working under PLC or under SCADA. The period of working under PLC is comprised between the 11h and the 24:30h, and clearly be appreciated visually a better working, although the behaviour with control under SCADA is enough. Quantitatively, in the latter case the over-oscillations rise a 21% and the average quadratic error a 18%.

There are many types of products, and each kind of product has several models. Each model is piled up in a particular way, and the circulation of the drying air depends much of it.

In figure 10 is appreciated the humidity in drier, in figure 11 the evolution of humidity in a specific point. The humidity is produced by the water that the material gives up, something very important, but it is not controlled directly, but indirectly by means of temperature and the speed. The humidity fall appearing in the figure, in a change of material,

produced a batch of wrong product. After the data analysis the load model (wagon, material and piling up form) was modified and it was achieved that in the next material change, realized eight hours later in the figure of the test, the behaviour were right and the material became alright.

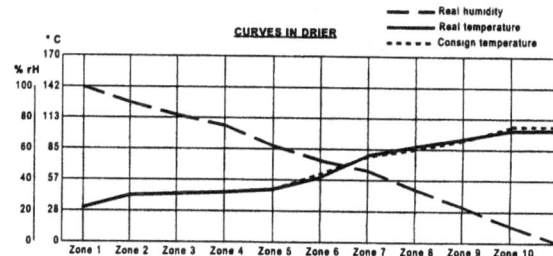

Fig. 10. Temperature, humidity and temperature reference along the drier.

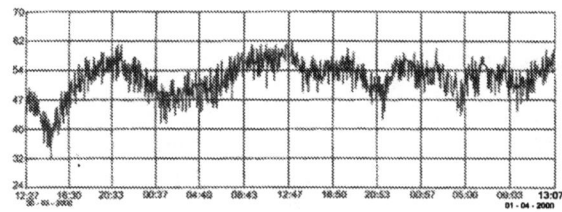

Fig. 11. Humidity and humidity reference depending on the time in a specific point in the drier.

In the other hand, has been demonstrated that besides using the comparison between the variables obtained for the plant and those simulated with the SCADA, to regain control or to take the system to a state of security in the case of failure, it is very useful to have the information on how is this difference under normal operation. This has allowed us to check whether the model has a good or a bad behaviour, and how and when it is necessary to modify it. This has been a good tool to improve the adaptive control. For example, it has been demonstrated that the humidity of the atmosphere influences less than predicted with our model, and however, the external temperature (and therefore the seasonally influence) have much more influence.

In fact, the possibility to modify in real time the parameters of the regulators has been very interesting. In many cases the behaviour of the system changes considerably depending on the pint of work, and the use of a regulator for all of them, independently of the operation type of the plant, means missing the abilities of the system.

These variations in the operation mode are realized in a double way. From PLC, parameters are changed when the activity changes (nominal work, load descompensation, emergency, etc) to achieve an autonomous operation. From SCADA, parameters are modified (operating curves and controllers parameters) depending on the material type, although in case of emergency SCADA may realize both tasks.

In figure 12, an example of this way of working can be seen. The pattern corresponds to the PLC program. The subroutines in parallel correspond to the working ways in plant. SCADA modifies the parameters of those subroutines to adapt them to the type of material, production, weather, etc. Besides SCADA provides the reference values to achieve the corresponding operation curve, from the hundred of curves that supports the hard disk.

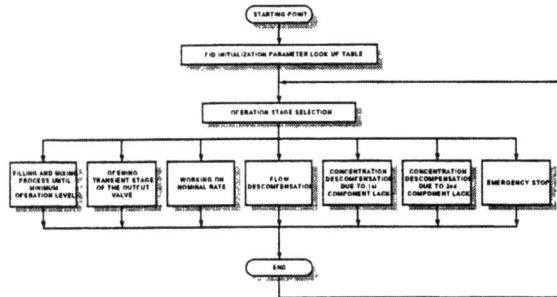

Fig. 12. Flowchart for PLC programme.

This form of adaptive control, in which the own SCADA can replace the control parameters for the more appropriate ones, has supposed an improvement in all the behaviour parameters. It has also been tried to modify these control and operation parameters from outside, via modem, from the laboratory where the operation parameters are received for their analysis. This analysis off-line has not been fully developed yet, and at present the results of the analysis of failures is being studied by means of neuronal nets.

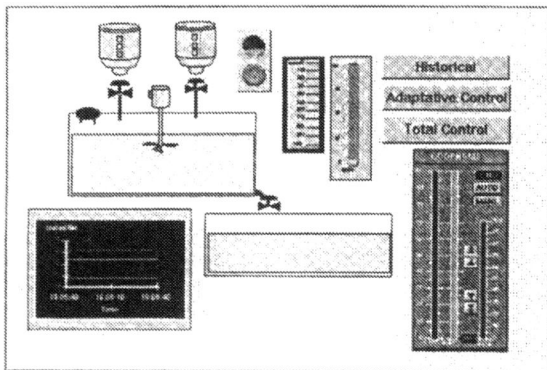

Fig 13. Local SCADA screen.

These results correspond to the specific development, supervised and adaptive which has been realized, but it should not be forget the advantages which any SCADA offers: process simulation facilities, real-time data observation, access from any work node to the production global graphics or to any process in particular, trend graphics, and all the advantages which make easier plant work or production management.

7. CONCLUSIONS

This paper reports the selection, development and installation of control systems within the automation chain of a modern industrial production plant. The type of control required in each part of the whole process was analysed and then most suitable model of control architecture was selected.

The possibilities offered by both PLC and SCADA systems for real-time controls were studied, and an appropriate system for the plant control was chosen, linking software algorithms and hardware architectures in a multi-layer control. This was developed using a supervised, remote driven and adaptive control that fulfilled the plant control requirements and provided a high degree of operation security and reliability, without needing excessive resources.

8. REFERENCES

Aström, K.J., C.C. Hang, P. Persson and W.K. HO (1992). Towards intelligent PID control. *Automatica*, **28**, 1-9.

Aström, K.J., and B. Wittenmark (1997). *Computer controlled systems*. Prentice-Hall, Englewood Cliffs, New Jersey.

Balcells, J. and J.L. Romeral (1997). *Automatas programables*. Marcombo Boixareu Editores, Barcelona.

Burns, A. and A. Wellings (1996). *Real-Time Systems and Programming Languages*. Addison-Wesley, California.

Ezio, F. (1993). *Ceramic technology*. Iberian Faenza Editrice, Milan.

Koivo, H.N. and J.T. Tanttu (1991). Tuning of PID controllers: survey of SISO and MIMO techniques. *IFAC Intelligent tuning and adaptive control Proceedings, Singapore*, pp. 75-80.

Levi, A. (1990). *Real-Time System Design*. McGraw-Hill, New York.

Laplante, P.A. (1993). *Real-Time Systems Design and Analysis: An Engineer Handbook*. IEEE Press.

Malpica, J. A. (1998). *Introducción a la teoría de autómatas*. Ed. Universidad de Alcalá, Madrid.

Morriss, B. (1999). *Programmable Logic Controllers*. Prentice Hall, New Jersey.

Ogata, K.(1998). *Engineering of Modern Control*. Spanish American Prentice Hall, Mexico.

Silva, M. (1985). *Las Redes de Petri : en la automática y la informática*. AC, D.L., Madrid.

REAL-TIME VIDEO FOR DISTRIBUTED CONTROL SYSTEMS

J.A. Clavijo * M.J. Segarra * R. Sanz ** A. Jiménez **
C. Baeza ** C. Moreno ** R. Vázquez *** F.J. Díaz ***
A. Díez ***

* SCILabs Ingenieros
** Universidad Politécnica de Madrid
*** Unión Fenosa Ingeniería

Abstract: Automation based on distributed control systems (DCS) technology increases enormously the capability of plant supervision of a single human. DCSs over broadband networking makes possible the integration and operation of a country-wide electricity generation and distribution system. Out-of-plant supervision is, however, more risky due to the impossibility of direct plant observation and exclusive reliance on digital data systems for decision making. This paper describes a system that the authors have developed for providing operators of such a large supervision system with real-time visual feedback of their actions in the plant. *Copyright © 2000 IFAC*

Keywords: Distributed control, supervision, safety, intelligence, real-time video, broadband networking.

1. INTRODUCTION

The objective of reduction of human personnel in all types of industrial plants is pervasive. In the case of power plants, and specifically under the pressure of electric market de-regulation, this objective is critical to reduce operational costs. The present trend in electric utilities and generation systems is to gather separate DCSs into single, integrated ones, thus reducing the need of operation personnel to a minimum. This paper presents a real-time video system used by plant operators to *see* in real-time the effects of their actions in relevant plant subsystems.

2. INTEGRATION OF DCSS

Conventional DCS, like those of Honeywell, Siemens or Wonderware are applications based typically on a fixed hierarchical structure where there are some data sources (sensors), data sinks (actuators, user interfaces), processing elements (controllers, optimizers) and storage (historical databases). These systems run on mostly heterogeneous computing equipment over a communications infrastructure that can be proprietary or standardized, being a global trend the use of TCP/IP networks at higher levels and specialized networks in lower ones.

The global trend to total integration and human personnel reduction makes necessary the fusion of separately operated systems into global solutions. This can be done by integration of heterogeneous manufacturer systems, or more commonly by replacement using unified systems (this is due to the difficulties found in integrating system from heterogeneous vendors).

An example of this trend is the reduction of personnel done in electricity generation and distribution systems (in particular in hydroelectric or wind-powered plants due to their simplicity). New DCSs are built to operate country-wide collections of plants by small human teams (See Figure 1).

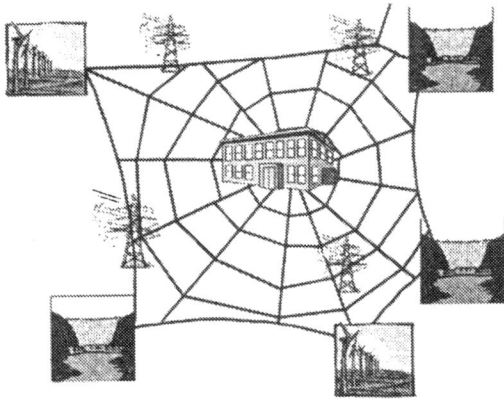

Fig. 1. Individual DCSs for electric power plants and distribution systems are being integrated into one-big DCS that can span whole countries.

This can reduce the cost of operation and enhance the application of system-wide policies.

This paper describes Hydra-Visión, a systems for real-time video distribution to support the operation of all the hydroelectric power plants owned by Unión Fenosa. The system has been developed by SCILabs Ingenieros (a software house), Universidad Politécnica de Madrid and Unión Fenosa Ingeniería.

3. AGENTS FOR PROCESS CONTROL

Component-oriented and agent-oriented development are a best-fit for heterogeneous and complex environments (Sanz *et al.*, 2000). Each component can be developed using different methodologies and can be used in a distributed way from different platforms and languages, using flexible and reliable communications systems (Pfister, 1998). Common environments for component development are Java RMI, COM+ or CORBA.

The approach taken for the design and construction of the Hydra-Visión application is based in the **ICa** agency model (de Antonio, 1999). This is a development model for complex control systems based in the construction of active objects over a CORBA compliant middleware. CORBA is a well established specification for distributed object-based systems construction (OMG, 1998). As COM+ or Java RMI it is a technology that simplifies the development and deployment of complex distributed applications. It is however different in the sense that CORBA tries to provide a broader spectrum of functionality, getting even into the business of real-time systems. This is the reason why we have selected this specification as a base component of ICa technology.

The objective of the **ICa** research line is the development of control systems technology for im-

proving construction processes of complex integrated controllers. From our perspective, *improved* means that the process is:

- **Repeatable**: While repeatability is a good measure of the development process quality, this has been traditionally difficult to achieve in the intelligent/complex control area due to the type of requirements for these applications.
- **Low-defect level**: Reliability is of extreme importance when the control system in charge of safety-critical system operation (like nuclear plants).
- **Reduced development effort**: The effort needed for building a heterogeneous intelligent controller in enormous; because it requires personnel knowledgeable in quite separate domains, and the development is somewhat exploratory, with an spiralized lifecycle.
- **Openness**: The control modules in a complex controller are not isolated entities performing their dedicated tasks. They form part of a supersystem into which they must interact with masters, slaves or peers. The usually overlooked integration problem is one of the biggest nightmares of the system deployment and final verification.
- **Portability**: The open system nature of today's computing equipment makes a need of component portability. Adhering to open standards, specially ISO and POSIX standards (accepted or emerging (POSIX.21, 1995)), are good ways to achieve broad portability of the code.

A basis for reuse is the use of an extensible architectural model in what we call a *meta-architecture* (Sanz *et al.*, 1999*b*) that can be accommodated to new tasks and deeply reused. **ICa** agents don't possess an unique and specific internal architecture. The basic framework is designed to support a wide variety of agent models, and they are provided in the form of agent libraries, where specific agent architectures are implemented. The **ICa** model for agent construction is based in the incorporation of layers of behavior through inheritance for specific purposes.

ICa stands for the Integrated Control Architecture. These three keywords describe the main characteristics of the ICa approach to systems development:

- "Architecture" is a software design at a particular abstraction level, focused on patterns of system organization, describing how system functionality is partitioned and how these different elements are interconnected. The architecture is a partition model and, at the same time, the architecture is a coordination model of these components. ICa promotes an architecture-based development model with a component oriented methodology. At this level, a component is a software module that interacts with others through well-known interfaces.

- When we say "**Control**" we are speaking about the world ICa was initially developed for. ICa has extensions to cope with industrial process control, with real time restrictions, fault tolerance extensions and with hard high speed requirements. Obviously the framework, covering all of these restrictions, is useful for environments with only some of the requirements.

- Finally, the "**Integrated**" term refers to one of the more important ICa characteristics: it is able to integrate heterogeneous technologies, platforms and languages. We use four different integration mechanisms: in-thread, in-process, in-host and in-net integration. ICa pretends to be a more flexible integration tool, allowing to modify this integration mechanism on any of the development steps, with the minimum effort.

What ICa really is depends on the point of view and the way you are using it. It is a development methodology, assisting the developer in system architecture implementation task (Sanz *et al.*, 1999*a*). But ICa is a software development environment too. It provides libraries for developing components using different languages, compilers that help to build them and tools for managing component communities. It is a framework for the development of new tools that will help to build new systems with different restrictions.

4. THE INTEGRATED CONTROL ARCHITECTURE BROKERING SERVICE

We have selected CORBA as the middleware to develop distributed industrial environments because it fits best the type of applications needed in industry. We have made some modifications to the basic specifications of the OMG to adapt them to our specific niche. We have made modifications to the specification for real-time and fault tolerance. At the same time we removed others mainly related to dynamic invocation and interface elements. The result of this work is ICa (de Antonio, 1999). The ICa Broker basically offers a superset of the minimum CORBA specifications [1]

ICa Broker is an implementation of the CORBA specifications (see Figure 2) and it is CORBA 2.0 compliant at the interface level. But it adds

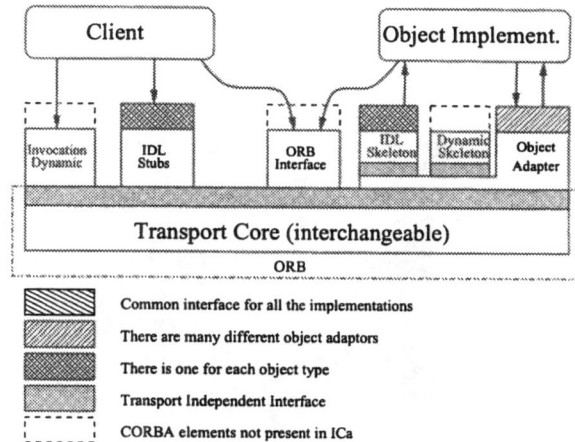

Fig. 2. ICa broker architecture is specifically tailored to applications in automation and has been developed for the distributed process control industry.

new important industrial environment capabilities such as real time, fault tolerance, data streaming, etc. and it simplifies CORBA specifications to improve predictability and functionality for industrial environments.

As in all CORBA implementations, the core of the system is a broker (See Figure 2), a common element that acts as a bus to transfer data among all the components. ICa has an additional abstraction layer, named transport, allowing the developer to change the communication mechanism with the minimum effort, even in runtime. These feature allows to move from an application using shared memory to another using a network transport such us TCP/IP.

5. HYDRA-VISION

The Hydra-Vision system has been developed using **ICa** technology to provide operators with real-time video from remotely operated plants.

The system is composed by a collection of nodes that can generate video streams (usually plants) and another collection of nodes that are clients of those streams. These last nodes con be operation stations, supervision stations, security offices, etc.

Video and audio streams are generated using remotely-controlled video cameras and microphones and compressed using real-time MPEG compressors based on hardware. These systems are organized into *zones* that are controlled by a single system. The core equipment (controller, MPEG compressor, multiplexers, etc) of a zone receive the name of Local Control Center (CCL).

---

[1] Minimum CORBA is the proposal of the OMG for reliable and embedded systems.

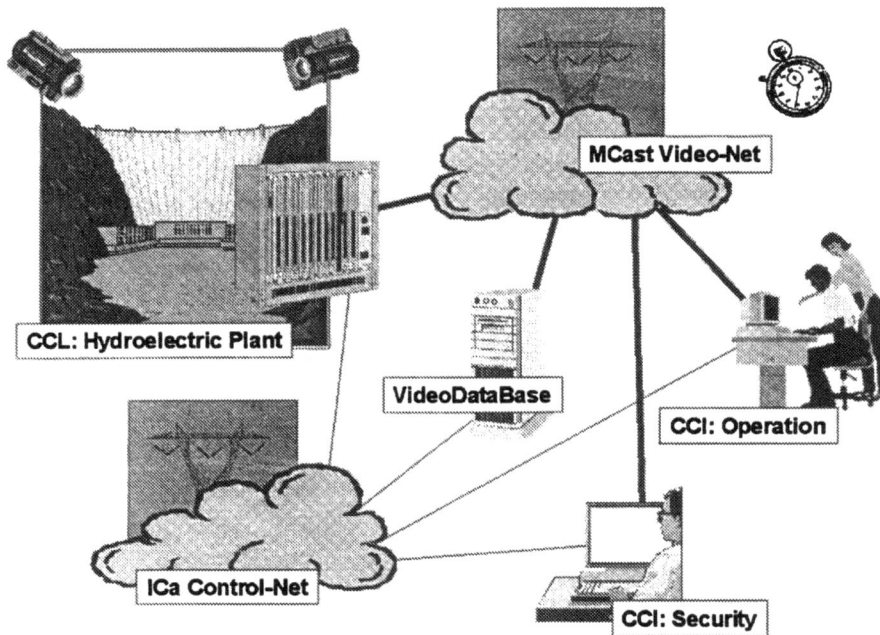

Fig. 3. Two separate networks are used for the application. The Video Net employs multicasting for video feeding to remote sites. The Control Net employs ICa to handle the collection of video generating and video consuming agents.

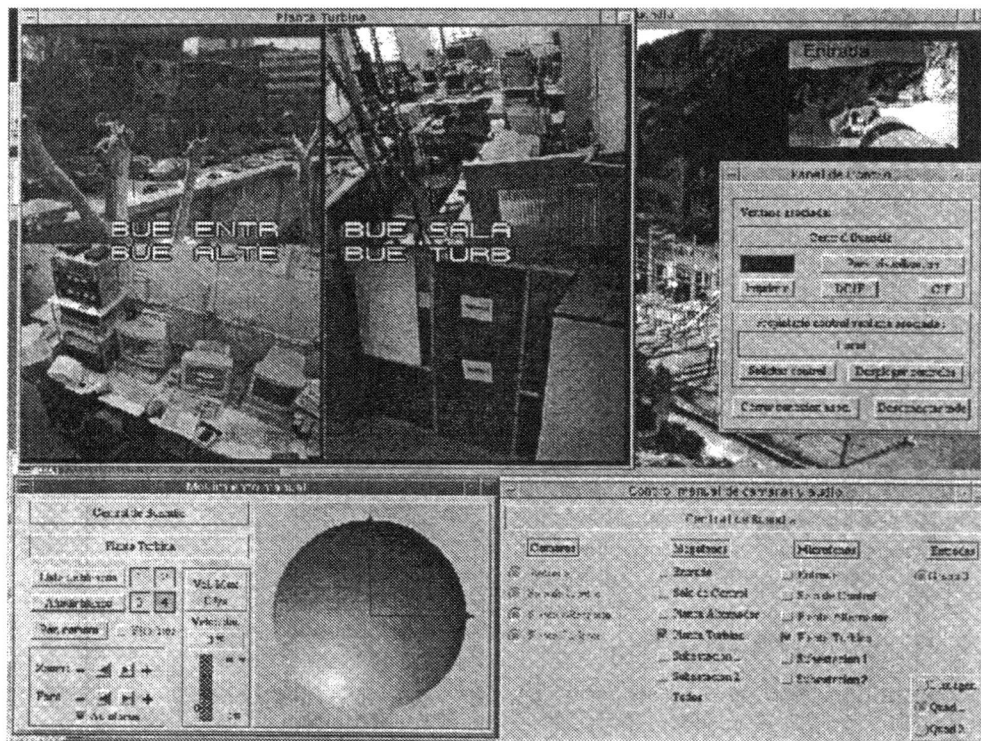

Fig. 4. Client user interface for the Hydra-Vision system.

All these elements are built as ICa agents, offering CORBA services to other agents in an open architecture.

Hydra-Visión employs ATM broadband networking as the base communication infrastructure. The system builds two virtual networks on top of it:

- The *control network* handles all service negotiation between Hydra-Visión agents using **ICa** technology. This is the network that all Hydra-Visión agents use to interchange information and service requests with all other agents.
- The *video network* handles all video and audio traffic. This network employs multicasting to minimize bandwidth use, making possible the concurrent visualization of the same video stream by different clients.

Remote clients are usually placed in Integrated Control Centers (CCI) and receive MPEG video feeds, decompress them and visualize in operator stations.

Video feeds can also be visualized in any other station in the network (even if they are placed in CCLs).

The main graphical user interface is shown in Figure 4 and it operates in three different modes:

Local: Used when a user wants to visualize a video feed in a station in the CCL where that camera is placed. The user selects the camera, the target and visualization parameters and the system interacts with all the equipment to serve the video feed to the client. The user can operate the camera and visualize the video like in a conventional CCTV system.

Remote Manual: Used when a user wants to visualize video form a camera in a different CCL. The operation is exactly the same with the exception of the distance between the camera and the visualization screen that can be hundreds of kilometers.

Remote Automatic: This mode is used for normal operation. Hydra-Visión is integrated with the DCS used to operate the plant to be able to respond automatically to plant events. Used when the DCS detects an special situation in the plant, it requests the video system to generate automatically a video stream from a specific plant target to a specific visualization station. For example, the triggering of an alarm in a transformer will produce the targeting of a camera to that transformer, the generation of the compressed video stream, the reception in the operation station responsible of that alarm and the visualization of the video. All this with a delay of seconds (mainly due to camera positioning).

As the number of concurrent users is not determined in advance, it is necessary to employ policies that control the response of systems to remote and/or local requests, to handle priorities between users. This is done to minimize the impact of non-critical video feeds to critical ones. Criticality is determined by the originating cause and priority level of the video feed.

The system runs over a network of UltraSparc workstations running Sun Solaris and Pentium based computers running Microsoft Windows NT (to support MPEG compression boards).

6. BEYOND SAFETY

The potential uses of this type of technology are not restricted to safety applications. This system is going to be employed also in:

- **Security**: Plants without personnel can be target objectives for thieves or vandalism. This system can be used by security personnel in remote sites as a CCTV using bi-directional audio to make-up a fake presence in plant.
- **Maintenance**: Experienced maintenance personnel is scarce and using this technology it is possible to send less experienced personnel to plants that can keep contact with the expert using bi-directional audio and video.
- **Videoconference**: The homogeneous nature of the system (all nodes share the same functionality) makes also possible the use of the systems for videoconferencing.

7. CONCLUSIONS

Broadband integrated DCSs for electric utilities are one of the main applications of distributed object technology. At the time of this writing, the system is being being tested and installed in some hydroelectric plants in Spain.

The system demonstrates the capability of CORBA technology for industrial application. New work of the authors is focused in the embedding of CORBA technology in intelligent electronic devices (IEDs) as a base infrastructure for automatic control and teleoperation systems.

8. ACKNOWLEDGEMENTS

The development of the **ICa** framework was partially funded by the Comission of the European Union and the Comisión Interministerial de Ciencia y Tecnología through ESPRIT project DIXIT.

Fig. 5. In the figure, two of the authors in front of the operator stations of the system developed for Hydra-Vision project.

The construction of the Hydra-Visión application has been funded by Unión Eléctrica Fenosa, who is planning to deploy it over all his hydro plants in Spain using to operation centers.

9. REFERENCES

de Antonio, Angel (1999). Arquitectura de Control Inteligente: Un Enfoque basado en Componentes para Sistemas de Control. PhD thesis. Departamento de Automática, Universidad Politécnica de Madrid. Madrid.

OMG (1998). Common object request broker architecture and specification. Technical Report 2.2. Object Management Group.

Pfister, Cuno (1998). Machine tool control and component software. Technical report. Oberon Microsystems AG. Zurich. Project KTI (Kommission für Technologie und Innovation).

POSIX.21 (1995). Interface requirements for real-time distributed systems communication. Technical report. IEEE.

Samad, Tariq (1998). Complexity management: Multidisciplinary perspectives on automation and control. Technical Report CON-R98-001. Honeywell Technology Center. Minneapolis, MI.

Sanz, Ricardo, Fernando Matía and Eugenio A. Puente (1999a). The ICa approach to intelligent autonomous systems. In: *Advances in Autonomous Intelligent Systems* (Spyros Tzafestas, Ed.). Chap. 4, pp. 71–92. Microprocessor-Based and Intelligent Systems Engineering. Kluwer Academic Publishers. Dordretch, NL.

Sanz, Ricardo, Miguel J. Segarra, Angel de Antonio and José A. Clavijo (1999b). ICa: Middleware for intelligent process control. In: *International Symposium on Intelligent Control*. Cambridge, MA.

Sanz, Ricardo, Walter Schaufelberger, Cuno Pfister and Angel de Antonio (2000). Software for complex control systems. In: *Complex Control Systems* (Karl Åström, Alberto Isidori, Pedro Albertos, Mogens Blanke, Walter Schaufelberger and Ricardo Sanz, Eds.). Springer. To appear.

THRUST VECTORING SYSTEM CONTROL CONCEPT *

Abel Jiménez
Daniel Icaza

Industria de Turbo Propulsores S.A. (ITP)
Parque Empresarial San Fernando
Edificio Japón
Avenida de Castilla, 2
San Fernando de Henares
28830 Madrid, Spain

Abstract: Thrust Vectoring can provide fighter aircraft with advantages regarding performance and survivability. ITP's research programme on this technology has recently met an important milestone as is the ground testing of a prototype, where ITP's partner company MTU developed the electronic Control System. The ITP concept consists of a design featuring the so-called "Three-Ring-System", that allows all nozzle functions with a minimum number of actuators in a single system. The overall system of nozzle, controller and actuators has to meet stringent performance, reliability and safety requirements derived from the use of Thrust Vectoring as an additional control surface and from the interaction between nozzle and engine operation. *Copyright © 2000 IFAC*

Keywords: Actuators, Aircraft Control, Engine Control and Systems, Safety Critical

1. INTRODUCTION TO THRUST VECTORING AND VECTORING NOZZLES

Engine thrust maximisation in fighter aircraft requires variable nozzle geometry in order to optimise jet expansion with engine setting and flight conditions. Maximum thrust is achieved when jet exit pressure and ambient pressure are equal (adapted nozzle). The most common solution in modern fighter aircraft equipped with reheated engines is convergent divergent mono-parametric nozzles. In those nozzles, throat area constitutes the parameter to be controlled according to the following criteria: maximum thrust is envisaged, surge margin should be kept within reasonable limits and nozzle actuator loads can not be exceeded. Logic to cater for these criteria is implemented within the engine controller. However, ideal maximum thrust can not be achieved with this configuration for every flight condition and engine setting as the exit area follows a mechanically pre-defined relationship to the throat area. The problem can be overcome adding a second degree of freedom leading to a bi-parametric nozzle, i.e. throat and exit area are controllable. For every particular case an assessment is needed to estimate if the advantage of the extra thrust / reduced specific fuel consumption compensates for the additional systems for exit area actuation plus the increased overall system complexity and weight.

Thrust Vectoring constitutes the next step in nozzle optimisation and increased functionality. The nozzle is intended to direct the jet in directions other than the engine axis in order to generate lateral forces and moments around the aircraft centre of gravity that can be used for aircraft manoeuvring. In 2-D Pitch only nozzles the jet can be deflected within the vertical plane, so the nozzle complements horizontal control surfaces. Pitch vectoring can be achieved by both rectangular (better stealth characteristics) and round nozzles. In 3-D Pitch and Yaw nozzles, the jet can be deflected in any direction so complementing both horizontal and vertical control surfaces.

Regarding the mechanical design, jet deflection can be achieved by either deflecting the whole nozzle or deflecting the divergent section. The second, although more complex from a kinematic point of view, is preferred as it does not have any effect in engine controllability and performance (throat area is normally choked so changes in downstream geometry do not impact upstream flow) and requires a lighter actuation system. A third configuration can be considered for technology demonstration vehicles that consists of external flaps or paddles deflecting the jet, like in the X-31 vehicle. This solution is not deemed adequate for a production standard due to the low efficiency of the system. In figure 1 a schematic of the three systems is shown.

* Presented at the joint IFAC AARTC'2000 and WRTP'2000 Workshops.

DEFLECT WHOLE NOZZLE | EXTERNAL FLAPS | DIVERGENT SECTION

Fig. 1. Existing types of 3-D TVNs.

Thrust Vectoring capability is envisaged as an optional feature in future fighter aircraft as well as a retrofit for existing ones. Therefore changes required in engine and aircraft should be kept to the strictly necessary. Several studies for different applications have shown that aircraft/engine structure requires little redesign and strengthening. However, major changes are required in the flight control system and engine control system, plus the new design of the nozzle controller. An enhanced Air Data System capable of operation at high angles of attack and low speeds is also necessary.

Thrust Vectoring technology has been developed in different programmes. The X-31 Enhanced Fighter Manoeuvrability experimental programme demonstrated feasibility and operational benefits of Thrust Vectoring using a paddles system. Proper vectoring nozzles were developed and flight tested by General Electric in the F-16 MATV programme and by Pratt & Whitney in the F-15 ACTIVE programme (Wood, 1999). In Europe, ITP have designed and developed a 3-D Pitch and Yaw vector nozzle that has already been bed tested at Sea Level Static in an EJ200 (EF2000 Typhoon engine manufactured by the EUROJET consortium) vehicle. EUROJET partner company MTU have developed a vector nozzle control system aimed at future integration in the EJ200 control system and implementation in the EF2000.

2. ADVANTAGES OF THRUST VECTORING

2.1. Extended flight envelope.

Conventional flight control surfaces require high dynamic pressure and therefore high speed (higher as altitude increases and air density decreases) to be operative. Traditional fighters flight envelopes do not cover low speed areas. In thrust vectored aircraft, lack of control surfaces action at low speeds is traded by the nozzle lateral forces, allowing for stationary flight in low speed regions (Rauh and Jost, 1999). Figure 2 shows qualitatively the new envelope areas reachable by having Thrust Vectoring capability.

THUST VECTORED AIRCRAFT EXTENDED FLIGHT ENVELOPE

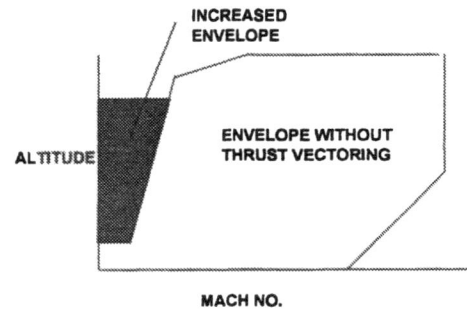

Fig. 2. Extended flight envelope.

This also means that higher angles of attack can be flown, even in post stall conditions. Normal maximum angle of attack in a non thrust vectored fighter is around 30°, while up to 70° angle of attack was demonstrated for the X-31. This capability represents a very clear advantage for air superiority, specially for close-in combat.

2.2. Use of Thrust Vectoring for aircraft trimming.

Having a somehow new aircraft control surface allows for further optimisation of Thrust Vectoring and surfaces deflection hence improving the lift/drag characteristics of the aircraft in all phases of flight.

In cruise. Thrust Vectoring can be traded with aerodynamic surfaces deflection reducing drag. However, depending on the aircraft configuration, there might not be a great potential for drag reduction if aerodynamic surfaces are already operated close to the minimum drag position (i.e. aircraft with canard).

In take off and landing. Runs can be reduced as, for landing, Thrust Vectoring allows for lower approach speeds and quicker de-rotation and, for take-off, the aircraft can be rotated earlier and lift off at a lower speed.

In sustained and instantaneous manoeuvres. Thrust Vectoring also allows for a lower deflection of the aerodynamic control surfaces so reducing drag and increasing manoeuvrability potential.

2.3. Potential for safety improvement.

Although Thrust Vectoring constitutes a complex system susceptible of suffering failures leading to critical situations, it can also constitute not just a complementary control surface but a redundant one if adequate control modes are implemented within the flight control system. These modes would come into place once a failure in an aerodynamic control surface is detected and would comprise logic for fault detection and isolation as well as reversionary laws.

Once the Thrust Vectoring system is validated as an alternative to aerodynamic control surfaces it might be possible to gradually reduce them and replace their functionality with Thrust Vectoring. This approach will lead to two additional advantages: mass reduction and improved stealth characteristics of the aircraft.

THRUST VECTORING:

POTENTIAL MASS REDUCTION
IMPROVED STEALTH FEATURES

Fig. 3. Reduced Aerodynamic Control Surfaces.

2.4. Nozzle exit area control

Although this is not a pure vectoring capability, most 3-D vector nozzles (GE, P&W and ITP) incorporate it. As discussed before, in conventional military mono-parametric nozzles the ratio exit area to throat area is given by mechanical design and optimised only for certain engine settings and flight conditions. Having this capability represents a potential for enhancing engine and aircraft performance due to two reasons: exit area to throat area ratio can be optimised and hence thrust maximised; and afterbody drag can be reduced by increasing the exit area in supersonic flight. Figure 4 represents nozzle exit area vs throat area, comparing a traditional fixed schedule with the flexibility of variable exit area.

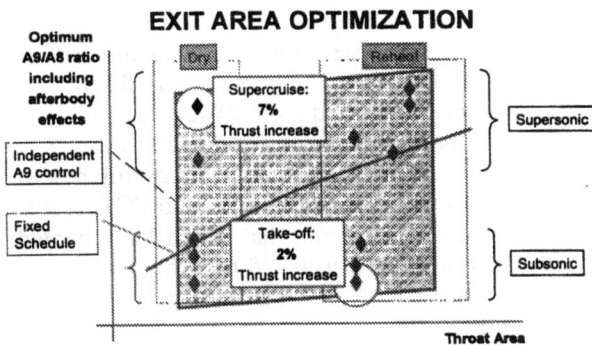

Fig. 4. Nozzle Area Ratio, conventional vs variable.

3. THE ITP VECTOR NOZZLE

ITP's R&D programme on Thrust Vectoring technology started back in 1991. In 1995, a "Technology Demonstration Phase" was launched aiming to design, build and test a vector nozzle prototype for an EJ200 engine. That effort led to Sea Level Static testing of engine and nozzle back in winter 1999. Altitude testing is planned for May/June 2000.

The ITP Thrust Vector Nozzle is convergent-divergent round nozzle with 3-D vectoring capability achieved by mechanical deflection of the divergent segment (Icaza, 2000). Therefore, impact in upstream flow is negligible and requires less power when compared to a whole nozzle deflection system.

The baseline design consists of three concentric rings linked one to each other as shown in figure 5.

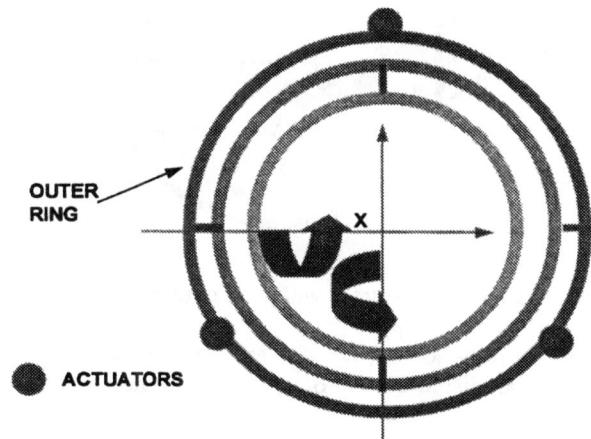

Fig. 5. Three Ring System.

The inner ring slides inside along the engine rear jet pipe, and is linked to the convergent section of the nozzle. A system consisting on rollers and cam profiles on the nozzle master convergent petals causes the throat area to close as the inner ring moves forward (towards the exit area) and to open when moving backwards (towards the engine hot sections).

The intermediate ring is linked to the inner ring by two vertical pins so it can be tilted over the vertical axis allowing for yaw thrust deflections.

The outer ring is linked to the intermediate ring by two horizontal pins so it can be tilted over a horizontal axis. Combining the horizontal deflection of the intermediate ring with respect to the inner ring and the vertical deflection of the outer ring with respect to the intermediate ring, deflections up to the nozzle mechanical limits can be achieved around the whole circumference (cardan joint). The divergent section, whose petals are linked to the outer ring by reaction bars, is able to generate a jet beam capable of covering a cone with a semiangle equal to the nozzle maximum deflection.

Such a ring system is actuated by three actuators independently controlled. Synchronised actuators movements lead to changes in throat area (synchronised displacement of the whole rings system). Uneven actuator position demands provoke tilting of the outer ring plane and therefore divergent section deflection and vectoring.

An additional degree of freedom can be achieved by splitting the outer ring into two halves (top and bottom) and adding a fourth actuator. The two halves use the outer to intermediate link pins as hinges and the actuators are located as shown in figure 6.

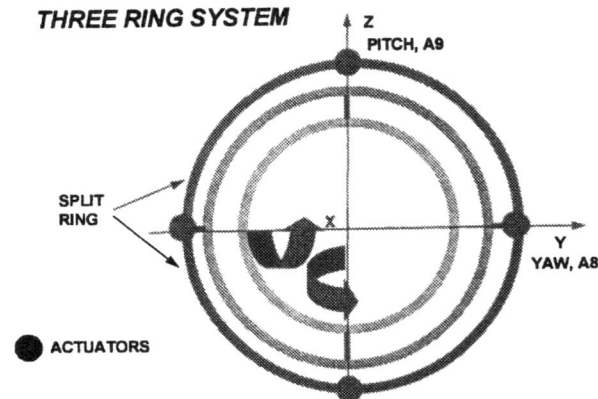

Fig. 6. Three Ring System with split outer ring.

This configuration allows for exit area control by differential displacement of vertical and horizontal actuators. With horizontal actuators remaining in the same position, extension of the vertical ones causes the exit area to diminish while a retraction causes an increase in exit area. Advantages of this feature have been already discussed.

The main characteristic of ITP concepts when compared to other vector nozzles is that the same system governs both the nozzle throat area (of extreme importance for engine control and safety) and the vectoring function (of extreme importance for aircraft controllability). This presents the advantage of requiring a lower number of actuators and therefore a less complex system.

Failures in the system could lead to problems derived from the inability to control engine throat area (risk of surge and inadequate thrust levels) and to loss of the vectoring function, that might be critical under certain flight conditions. Although the ITP nozzle self-centres in case of loss of control power and this is considered to be a fail safe mechanism, a non demanded absence of vectoring capability could lead to a lack of aircraft controllability.

Depending on the aircraft/programme application, adequate redundancies should be provided and/or flight restrictions applied in order not to run into hazardous failures as a consequence of thrust vector system failures.

4. ITP NOZZLE ACTUATION SYSTEM BASELINE

As already explained, the ITP nozzle can be actuated by either 3 or 4 actuators depending on the configuration. Each actuator is controlled by a two stage servovalve and a position transducer provides position feedback to the nozzle controller. All development carried out to date has pointed to implementation of the nozzle in EJ200. Use of the engine hydraulic pump is envisaged for nozzle actuation (as in standard EJ200 configuration). Depending on aircraft manufacturers requirements the current pump could be deemed sufficient or, in the other hand, a higher power one needed. An alternate pressure source coming from the aircraft should also not be discarded. Figure 7 shows an schematic of the control and actuation systems.

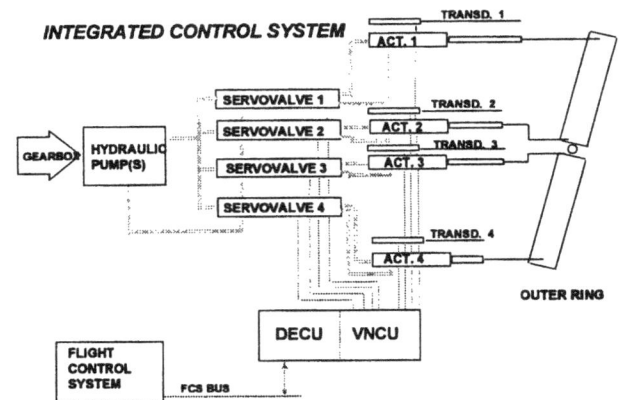

Fig. 7. ITP Thrust Vector Nozzle Basic Actuation System.

5. SAFETY REQUIREMENTS, REDUNDANCIES

As already mentioned the risks caused by incorrect control of the nozzle are loss of aircraft controllability plus potential inability to modulate thrust. When defining a configuration, enough redundancies should be provided in the system to avoid entering into failure modes leading to those consequences.

For the ITP Vector Nozzle plus actuation and control systems, the following assumptions are used:

- in case of loss of hydraulic power, the nozzle is centred by the internal aerodynamic forces. The failure is detected and alternate control modes are invoked in engine controller and FCS. ITP is also studying a system that would be activated by pressure loss that would take the nozzle to a predetermined safe centred position.

- risk of nozzle seizure or actuator jamming (i.e. permanent vector) is deemed negligible. A detailed analysis of the design plus the experience accumulated in the test bed confirm this statement.

- the nozzle control system is capable of detecting self and actuation system malfunctions and invoking corrective reversionary actions. The only reversionary action considered so far on the nozzle side consist of commanding bypass (setting to low pressure) of all nozzle actuators. The nozzle would be centred by aerodynamic forces. For configurations where a redundant actuating system is available, the corrective action would consist of a system change.

Studies have been carried out to assess the potential implementation of ITP thrust vector nozzle in many different aircraft applications, all of them using the EJ200 engine as the propulsion vehicle. Therefore, minimising the impact on engine redesign is vital in order to be able to offer an interesting retrofit/option package.

5.1. Twin engined aircraft / production (example: Eurofighter Typhoon).

Realisation of this configuration has been the driver for the thrust vector programme. There exist a working group formed by DASA, MTU and ITP developing a feasibility study where implementation of thrust vector capability into Eurofighter Typhoon is being analysed.

Availability of two engines and two vectoring systems contributes to some relaxation in systems reliability as the vectoring systems can be designed such as the aircraft can recover from failures in one of them just using the other. Loss of control in one system can also be covered by simply bringing the thrust on that engine down or cancelling it. Considering all this aspects the resulting system would consist of a dual electronic control system and a simplex hydraulic system using the engine pump as a pressure source

5.2. Twin engined aircraft / demo.

The objective of a demo thrust vectored aircraft would be demonstration of technology: nozzle control capability, enhanced flight mechanics ... Some other aspects of extreme importance in a production aircraft can be sacrificed for this purpose, namely maintainability and mission abort rate. Some relaxation can also be admitted in system safety taking into account that fleet life is limited to a few hours.

Feasibility of a demo concept greatly depends on costs and, therefore in minimising the changes from the original aircraft/engine baseline.

For this configuration the proposed baseline is as follows:
- nozzle concept similar to test bed demonstrator.

- simplex hydraulic system driven by the engine pump.
- simplex electronic control system. In order to be able to implement nozzle control within the current EJ200 DECU a suitable solution could consist of a single lane control for the afterburner functionality. That would release enough interfaces for thrust vector control including all functionality in one box.
- using non redundant control for afterburner and thrust vector functions would imply certain restrictions for the use of those functions within some areas of the flight envelope:
  - afterburner takeoffs would be forbidden
  - Thrust Vectoring would not be allowed in takeoff, approach and landing and under a safety height to allow aircraft recovery in case of malfunctions.

5.3. Single engined aircraft / production.

In this configuration the nozzle system is double critical as there is not a second engine to cover loss of thrust in the failed nozzle. A "full redundant" system might be needed:
- dual electronic control system.
- dual hydraulics comprising dual pressure source, duplicated servovalves and changeover system. Tandem or duplicated actuators might also be considered.

The level of change required with respect to the original engine/nozzle configuration is major, specially if the second pressure source has to be engine driven as well.

5.4. Single engined aircraft / demo.

The configuration would be as for twin engined aircraft / demo: single electronics and single hydraulics. Same flight restrictions would apply. A higher level of risk would have to be assumed in the demo programme as the additional "redundancy" of a second engine is not available.

6. THE THRUST VECTOR NOZZLE CONTROL SYSTEM

Since the beginning of the "Technology Demonstration Phase", Eurojet partner company MTU have been working on the development of a control system suitable for test bed and Airshow applications, and, at the same time, preparing the way forward for integration of vector nozzle control within the current EJ200 FADEC system (Rausch and Lietzau, 2000). The current EJ200 FADEC system consists of a Digital Engine Control Unit (DECU), the Ignition System, the Main Fuel Control, the Afterburner Fuel Control, the Air Flow Control System (AFCS) composed of Variable Inlet Guide Vanes (VIGVs) and Variable Nozzle and all the related sensors. Seven actuator control loops are used for normal engine control: two for the AFCS (VIGVs and Variable Nozzle), two for the dry engine fuel flow (Main Metering Valve and Emergency Spill Valve) and three for the afterburner fuel flow (primary, core and bypass valves); and there exists some extra capability for growth. The DECU (developed by MTU) houses all logic and functionality of the system. It consists of two identical lanes with independent access to all sensors, actuators (position feedbacks and torque motor drives) and data links. Communication between both lanes is realised via a high speed serial link. Each lane has full control capability. A lane is in control when connected to the output drives, while the other is actively shadowed. Automatic or operator commanded (under DECU supervision) lane change switches the lane in control. Each lane has two Mil–Bus 1553 interfaces. One per lane is used for communication with the Flight Control System (FCS). The second Mil-Bus interface is used for communication with the Engine Monitoring Unit (EMU) in lane 1 and for communication with test facilities when required in lane 2.

6.1. The Thrust Vector Demonstrator Control System.

ITP are really thankful to MTU for developing a control system that made possible the Thrust Vector Nozzle Sea Level Static test campaign with such a tight time schedule. Requirements imposed over the control system were certainly tough. Safety concerns affected not only the nozzle, but also integrity of the engine and test facility, all in a context full of uncertainties as it corresponds to an unexplored technology.
Basically the nozzle control system had to be able to:

Calculate nozzle direct and inverse kinematics. From demanded values of throat area, exit to throat area ratio and vector deflections the controller comes up with the actuators demanded position. The other way around also works: from the read actuators positions, the control system derives vector and areas values.

Account for nozzle envelope limitations. Although the nozzle has four degrees of freedom, not all possible actuators position are compatible. As the actuators are dimensioned to cater for aerodynamic forces that exceed the nozzle mechanical limits, the nozzle could be damaged if actuators are commanded in not compatible positions.

Account for vectoring and area rate limitations in line with aircraft and engine requirements.

Perform adequate nozzle actuators control.

Perform nozzle actuators supervision. Any deviation from standard actuator performance could lead to excursions into the forbidden nozzle envelope and cause damage to the nozzle or risk of engine surge.

Command recovery actions for engine and nozzle in case of failure and emergency deactivation of nozzle actuators when required.

Provide new test features. Aimed to perform an extensive post-test analysis of nozzle and control system performance.

Given the facts that the system was going to be operated in a benign lab environment and that the current EJ200 DECU does not have enough control loops for the four additional actuators, it was decided to implement nozzle control in a separated box, designated as Thrust Vector Control Unit (TVCU). This strategy also allowed for development of digital actuator control loops. Communication between DECU and TVCU was realised by a Mil–Bus 1553. This bus is shared by a third unit: the Thrust Vector Data Acquisition Unit (TVDAU). The TVDAU diverts DECU and TVCU parameters to the Data Acquisition System (DAS). A schematic of the engine and nozzle control configuration can be seen in figure 8.

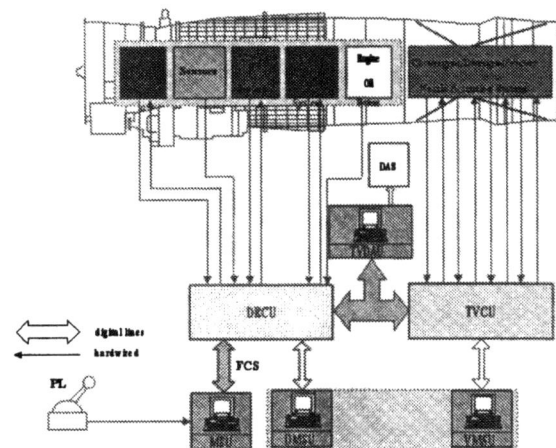

Fig. 8. Thrust Vectored Engine Control System.

The baseline is the standard EJ200 testing configuration with additional units for the Thrust Vectoring functionality and software modifications in the standard EJ200 units. Namely:

MSU (Mil-Bus Simulation Unit). It is a PC simulation of the Flight Control System Mil-Bus 1553 interface. It has been modified so vectoring demands are introduced through it.

DMSU (DECU and EMU Monitoring and Set Up Unit). It is the interface PC with the DECU (RS232) during bed testing or maintenance operations, allowing alteration of engine control parameters. Software modifications for the thrust vector application allow for changing ovalization (nozzle exit to throat area ratio) demands.

VMSU (Vector Monitoring and Setting Up Unit). It is the interface PC with the TVCU. It allows for nozzle control parameters change and visualisation.

Although nozzle actuators were controlled by a single lane, the TVCU incorporates two identical synchronised channels each of them consisting of a Mil-Bus 1553 interface (described above), two microcontrollers Motorola MC68332 (Control and Output Computers, also synchronised) and position sensing and torque motor drives electronics. Communication between Control and Output computer is realised by a Dual Port Ram, and communication between homologue computers in different lanes by a serial data link. Both computers can be accessed from the VMSU. The Output Computer houses the actuator control loops while the Control Computer holds the rest of logic, calculations and Mil-Bus 1553 interface with DECU.

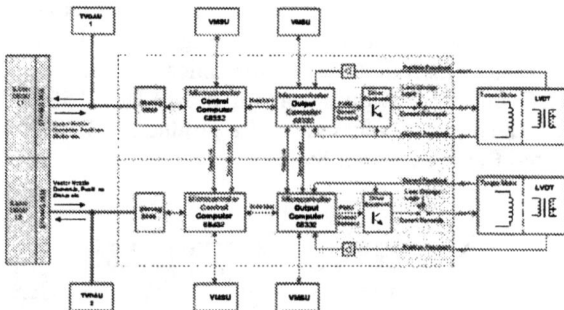

Fig. 9. TVCU HW Architecture.

The Output computer cycle time is 2.5 ms. This allows for implementation of digital actuator control loops where the torque motors of the actuators servovalves receive a pulse-width modulated signal from the controller with a basic frequency of 9.6 kHz. Figure 10 shows a schematic of the digital actuator control. The position controller calculates the current demand as a function of the position error. Either this current or a user pre-defined current (for test purposes) is input into the current controller, where the current demand is compared with the current feedback to calculate a pulse width modulation ratio which is converted into a PWM signal for the torquemotor.

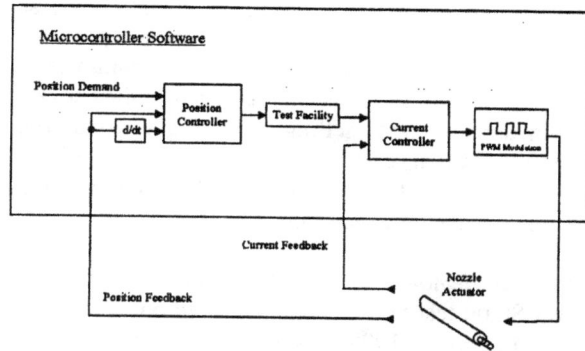

Fig. 10. Digital Actuator Control Schematic.

Digital control loops constitute an innovative design with respect to the rest of EJ200 actuators, where position control loops are realised in analogue circuitry. This feature enhances system versatility and flexibility. Demonstration of this technology paves the way forward for its implementation in future engine controllers.

Also, actuators supervisory (of extreme importance in actuators involved in aircraft controllability) are enhanced with respect to standard EJ200 due to the following facts:
- the reduced Output computer cycle time of 2.5 ms allows for accurate modelling of torque motors current and, therefore, for implementation of reliable current model checks
- MTU have developed enhanced actuator model checks capable of detecting actuator failures to follow demanded positions in very reduced times, allowing for the controller to take corrective actions before the failure can reach catastrophic effects. The Control Computer cycle time of 10 ms has been demonstrated to be sufficient to achieve an accurate modelling. Therefore, the improved model checks are susceptible to be applied to standard EJ200 actuators without any hardware change.

Validation of the system was performed in three steps:

Actuators hydraulic rig. A dedicated set up was built to perform tests on real actuators under different load conditions. This testing allowed for verification of the actuators, validation of the actuator control in closed loop and further identification of the actuator system.

Real Time test rig. One of the existing DECU and EMU Acceptance Rigs (DEAR) used for DECU validation in MTU was modified in order to test the vector nozzle control system. Basically the DEAR consists of a computer simulation of engine, sensors and actuators that works against a real DECU in real time. For validation of the thrust vector control system new items were needed in the simulation, namely an actuators model, a nozzle inverse kinematic model (given the actuators position, provides the nozzle vectoring and area parameters) and an actuators loads model (given the nozzle parameters, engine setting and flight condition, provides loads in each actuator). This facility allows for extensive testing of the control system without taking any risk on development hardware.

Test bed demonstration. The purpose of the Sea Level Static test bed campaign carried out since July 1998 to February 1999 at Ajalvir (Madrid) was triple: on one side validation of nozzle features, on the other validation of the integrated control system and, finally, validation of the whole engine-thrust vector nozzle control system integration. Some achievements to be mentioned from the test campaign are:

- 80 running hours, 15 with afterburner operation.
- Vectoring in all 360° at different dry and reheat settings.
- Maximum vector of 23.5° (at partial reheat).
- Maximum slew rate of 110°/s.
- Maximum lateral force of 20 kN.
- Sustained 20° vector at reheat for 5 minutes for thermal validation.
- More than 100 performance points run.
- Engine slams (and therefore quick throat area gradients) with sustained vector.
- Demonstration of thrust improvement by exit area modulation (up to 2% increase).
- Endurance tests: more than 6700 vectoring cycles at constant throat area and more than 600 throttle cycles with a sustained vector.

Fig. 11. Thrust Vector Nozzle ground test at ITP.

Next step towards validation/integration will be the altitude testing to be carried out in the Altitude Test Facility (ATF) in Stuttgart where other than Sea Level Static conditions will be simulated. This new test campaign will also allow for acquiring enough data to elaborate a lateral thrust model of the nozzle (together with Ajalvir data and CFD calculations).

This item is of extreme importance for building thrust vectored aircraft control laws.
Final validation will necessarily require a flight test programme, still to be defined.

6.2. Future developments in thrust vector control.

The key issue of Thrust Vectoring resides in the integration of thrust vector control with the Flight Control System. At a first glance, it could be thought that nozzle control should be a Flight Control Computer function. However, this might not be the best approach, specially for the ITP nozzle where vector deflection and throat area (a parameter intrinsically linked to engine performance and safety) can not be independently controlled. Thrust vector control should therefore be integrated within the FADEC system. The FADEC would receive lateral forces demands from the flight control system, that would be translated into geometrical deflection demands (also dependant on engine setting and flight conditions) obtained from a lateral thrust model. The FADEC would return to the flight control system the actual lateral forces (also extracted from the thrust model) as well as the resultant application point for aircraft momentum calculation. The FADEC would also inform the Flight Control System of the maximum available lateral forces for that particular flight condition and engine setting as well as indications of malfunctions. Communication between FADEC and FCS would remain a Mil-Bus 1553. Figure 12 shows an schematic of a production Thrust Vectored Engine Control System.

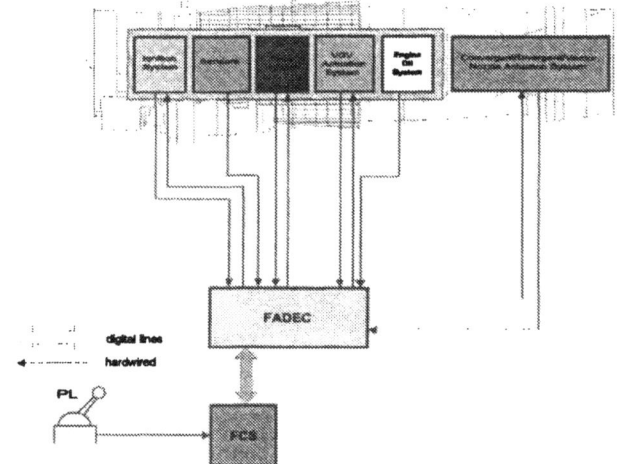

Fig. 12. Thrust Vectored Engine Control System.

Not only a new architecture, but also new requirements with respect to a test bed demonstrator apply to a Thrust Vectoring flight application:

Response characteristics. Thrust Vectoring system should have similar response characteristics to those of other flight control surfaces. This has been analysed and found to be particularly dependent on actuators servovalves and synchronisation between FCS and FADEC.

Accuracy of the lateral thrust model. As already pointed out, elaboration of a lateral thrust model is a complex task that requires extensive testing and CFD calculations. However, the accuracy required by the FCS is not too stringent as lateral forces are closed loop controlled. Aircraft manufacturers have identified that accuracy to be in the order of 5% maximum lateral forces).

Operational and mechanical limitations. The thrust vector controller will prevent from commanding deflections that could compromise aircraft, engine and nozzle integrity.

Recovery actions for failure cases. It might be the case that little has to be changed from the current thrust vector controller demonstrator recovery actions, at least on the nozzle side. Aerodynamic analysis and test bed experience show that the nozzle self-centres in case of hydraulic pressure loss, either unintended or commanded by the controller upon failure detection. Elimination of lateral forces in case of system malfunction is a desired effect from an aircraft point of view. However, the engine should remain in a surge free condition and with the maximum modulation capability in spite of having an uncontrollable nozzle. Therefore, some kind of fail safe mechanism might be needed in order to lock the centred nozzle in a safe position for engine control.

Potential increased hydraulic power. Depending on system performance requirements, higher that currently available in EJ200 hydraulic power might be needed. That would also lead to changes in the cooling pattern.

7. FUTURE OF THE PROGRAMME

As already mentioned, the next important milestone will be the ATF testing at Stuttgart in May-June 2000, with MTU and ITP participation and the two other EJ200 PCs (Rolls Royce and Fiat) supporting as required.
A dynamic display of the ITP vector nozzle controlled by the MTU demonstrator controller will be exhibited in the Berlin's ILA and Farnborough air shows (June and July 2000 respectively).
On the mean time ITP keep running optimisation studies looking into improving nozzle features: mass reduction, maintainability, stealth characteristics, improved mechanisms and mechanical design ...
MTU work hard towards integration of the thrust vector control within the FADEC as well as looking into simplified solutions for a potential demonstrator.

The feasibility study for thrust vector application in Eurofighter with both companies and DASA will continue, aiming to validate the technology with a flight programme.

8. CONCLUSIONS

Thrust Vectoring offers numerous advantages for fighter aircraft. Modern fighters incorporate some way of Thrust Vectoring. Future fighters will have to incorporate full pitch and yaw 3-D Thrust Vectoring capability in order to be competitive, as the technology is attaining a high maturity level.
ITP pitch and yaw nozzle is a mechanism capable of providing maximum vectoring and exit area modulation capabilities and performance with the most simple actuation system, required safety and minimum impact into the engine.
MTU has successfully solved the problem of thrust vector and engine controls integration for lab/test bed applications and is actively working on integration with the flight control system.
Moreover, ITP/MTU demonstrator has set Western Europe in a privileged position to afford future thrust vectored flight demonstrators. The existing collaboration atmosphere within the aerospace industry and the support given by the institutions, invites to think in a flight programme in a short to medium term.

9. ACKNOWLEDGEMENTS

The success of ITP's programme has been made possible only with the help and contribution of the following institutions:

MTU, of Munich, Germany, developed the control system for the demonstrator and works towards integration of the system in a flight programme. Without their help and contribution, ITP presence in this forum would have not been possible.

DASA, of Munich, Germany, as partner in the Eurofighter feasibility study, giving the guidance only an aircraft manufacturer can provide.

Eurojet and PCs (Rolls Royce and Fiat), who supported the EJ200 – TV nozzle test campaign.

CESA, of Getafe, Spain, provided the TV nozzle actuators.

SENER, of Las Arenas, Spain, greatly contributed to the engineering work.

Spanish Ministries of Industry and Defence for their support and funding to the R&D programme.

REFERENCES

Icaza, D. (2000). Thrust Vectoring Nozzle for Modern Military A/C. *Applied Vehicle Technology Symposium on "Active Control Technology for Enhanced Performance Operation Capabilities of Aircraft". Braunschweigh, Germany, 8th –11th May 2000.*

Rausch, C. and Lietzau, K. (2000). Integrated Thrust Vectored Engine Control. *Applied Vehicle Technology Symposium on "Active Control Technology for Enhanced Performance Operation Capabilities of Aircraft". Braunschweigh, Germany, 8th –11th May 2000.*

Rauh, G. and Jost, M. (1999). Thrust Vectoring Growth Potential for Advanced Aircraft. *7th European Propulsion Forum Aspects of Engine/Airframe Integration. Pau, France,10th-12th March 1999.*

Wood, C.B.(1999). P&W Multi Axis Thrust Vectoring Engine Exhaust Nozzle Design, Development, Flight Test Techniques and Results. *14th Synposium on Air Breathing Engines. Florence, Italy, 5th – 10th September 1999.*

AUTHOR INDEX

Continued from outside back cover

Title/Year of publication	Editor(s)	ISBN
1999		
Nonsmooth and Discontinuous Problems of Control Optimization and Applications (W)	Batukhtin & Kirillova	0 08 043237 9
Linear Time Delay Systems (W)	Dion, Dugard & Fliess	0 08 043047 3
Large Scale Systems: Theory and Applications (S)	Koussoulas & Groumpos	0 08 043034 1
Automatic Control in Aerospace (S)	Jang Gyu Lee	0 08 043041 4
Automation in Mining, Mineral and Metal Processing (S)	Heidepriem	0 08 043031 7
Low Cost Automation (S)	Chen & Chai	0 08 043027 9
Distributed Computer Control Systems (W)	DePaoli & MacLeod	0 08 043242 5
Analysis, Design and Evaluation of Man-Machine Systems (S)	Nishida & Inoue	0 08 043032 5
Control in Natural Disasters (W)	Sano & Ishii	0 08 043240 9
Motion Control (W)	Georges *et al*	0 08 043044 9
Artificial Intelligence in Real Time Control (S)	Pao & LeClair	0 08 043227 1
14th Triennial World Congress (TWC) - CD-ROM	Chen, Cheng & Zhang	0 08 043248 4
14th Triennial World Congress (TWC) - Paperback (18 vols)	Chen, Cheng & Zhang	0 08 043247 6
2000		
Space Robotics (W)	Rondeau	0 08 043050 3
Control Applications in Marine Systems (W)	Kijima & Fossen	0 08 043033 3
Intelligent Manufacturing Systems (W)	Kopacek & Pereira	0 08 043239 5
Real Time Programming (W)	Frigeri, Halang & Son	0 08 043548 3
Adaptive Systems in Control and Signal Processing (W)	Bitmead, Grimble & Johnson	008 043238 7
Computation in Economics, Finance and Engineering: Economic Systems (S)	Holly	0 08 043048 1
Multi-Agent-Systems in Production (W)	Kopacek	0 08 043657 9
Programmable Devices and Systems (W)	Srovnal & Vlcek	0 08 043620 X
Lagrangian and Hamiltonian Methods for Nonlinear Control (W)	Leonard & Ortega	0 08 043658 7
Modelling and Control in Biomedical Systems (S)	Carson & Salzsieder	0 08 043549 1
Digital Control: Past, Present and Future of PID Control (W)	Quevedo & Escobet	0 08 043624 2
Power Plants and Power Systems Control (S)	Waha	0 08 043252 2
Algorithms and Architectures for Real-Time Control (W)	Hernandez & Irwin	0 08 043685 4
Real –Time Programming (W)	Crespo & Vila	0 08 043686 2
Instability Resolution in Regions of Long Confronted Nations (W)	Dimirovski	0 08 043690 0
Control in Transportation Systems (S)	Schnieder & Becker	0 08 043552 1
Fault Detection, Supervision and Safety for Technical Processes (S)	Edelmayer	0 08 043250 6
Advanced Control of Chemical Processes (S)	Biegler	0 08 043558 0
Automated Systems Based on Human Skill (S)	Brandt & Cernetic	0 08 043254 9
Control Systems Design (C)	Kozak & Huba	0 08 043546 7
Robust Control Design (S)	Kucera & Sebek	0 08 043249 2
System Identification (S)	Smith	0 08 043545 9
Control Application of Optimization (S)	Zakharov	0 08 043550 5
Technology Transfer in Developing Countries: Automation in Infrastructure Creation (C)	Craig & Camisani-Calzolari	0 08 043553 X
Management and Control of Production and Logistic (C)	Binder	0 08 043621 8
Modelling and Control in Agriculture, Horticulture and Post-Harvest Processing (C)	van Straten, Bontsema & Keesman	0 08 043251 4
Manufacturing Modelling, Management and Control (S)	Groumpos & Tzes	0 08 043554 8
Future Trends in Automation of the Mineral and Metal Processing (W)	Vapaavuori & Jamsa-Jounela	0 08 043622 6
Manoeuvring and Control of Marine Craft (C)	Blanke, Vukic & Pourzanjani	0 08 043659 5

Customers wishing to obtain details of all available IFAC volumes, should contact their nearest Elsevier Science office or check the IFAC Publications website (www.elsevier.com/locate/ifac).

www.ingramcontent.com/pod-product-compliance
Lightning Source LLC
Chambersburg PA
CBHW082306210326
41598CB00028B/4450